1994

Mammalian Neuroendocrinology

Paul V. Malven, Ph.D.
Department of Animal Sciences
Purdue University
West Lafayette, Indiana

CRC Press
Boca Raton   Ann Arbor   London   Tokyo

Library of Congress Cataloging-in-Publication Data

Catalog record is available from the Library of Congress.

This book represents information obtained from authentic and highly regarded sources. Reprinted material is quoted with permission, and sources are indicated. A wide variety of references are listed. Every reasonable effort has been made to give reliable data and information, but the author and the publisher cannot assume responsibility for the validity of all materials or for the consequences of their use.

Direct all inquiries to CRC Press, Inc., 2000 Corporate Blvd., N.W., Boca Raton, Florida, 33431.

International Standard Book Number 0-8493-8757-4

Printed in the United States of America 1 2 3 4 5 6 7 8 9 0

Printed on acid-free paper

PREFACE

Neuroendocrinology is the scientific discipline devoted to the interaction between the two major integrative organ systems of the body. It began as a branch of endocrinology and continues today as an important element of that field. As a result of increased knowledge about the nervous system in the last few decades, neuroendocrinology has also become one of the essential components of neuroscience.

The present book was designed as text for a graduate course in neuroendocrinology such as has been taught by the author for the past 20 years. Because of the interests of the author and the population of graduate students taking his course, neuroendocrinology in this book was limited to mammalian species. A great deal of neuroendocrine information exists for invertebrate species as well as for non-mammalian vertebrates, and readers interested in the neuroendocrinology of these species will have to find it elsewhere. The author has sought to provide a comparative mammalian approach to many of the topics in this book. Whenever possible, this comparative approach involves (1) rodents reared for laboratory research, (2) domesticated ungulates reared for agricultural production, and (3) primates studied either clinically (humans) or in the laboratory (monkeys). At the conclusion of many chapters, a section is included that discusses applications of the neuroendocrine information to clinical medicine and/or to agricultural production.

Because this book is intended as a graduate text, the author has tried to explain the bases on which many well-established concepts are founded. Therefore, enough details are provided so that students might be better able to integrate future discoveries that might challenge these well-established concepts. The author begins each semester of his graduate course with the admonition to the students that much of what he will tell them may eventually turn out to be incorrect, but that he does not know what topics fall into that category. Certainly, that admonition holds for this book. To enable readers to evaluate the specific conclusions made in this book, many original references were cited, especially in Chapters 5 through 15. References of a more general type were used in Chapters 1 through 4 because the subjects covered seemed to the author to be more stable, and the concepts were less likely to change.

The author expresses his gratitude to Purdue University for granting him a sabbatical leave to write this book and to the University of South Florida for appointing him a Visiting Professor of Physiology and Biophysics. Sincere appreciation is also extended to the following current and former colleagues who made valuable suggestions on one or more chapters: J. Albright, H. Bryant, M. Diekman, and H. Head. Special appreciation is also extended to S. Grabowski and D. Rasmussen who carefully reviewed all the chapters. Of course, none of these people are responsible for the shortcomings of the book which rest solely with the author.

The scientist/philosopher, Sir Isaac Newton, wrote *"If I have seen further it is by standing on the shoulders of Giants."* The present author believes strongly in the sentiment of that opinion. Researchers in neuroendocrinology, as in all other fields of scientific endeavor, must begin their quest for new knowledge at the point at which their predecessors brought the field. Although this author appreciates all of his neuroendocrine predecessors, he is additionally appreciative of two neuroendocrine "*giants*" who were also scientific mentors to him. Professor William Hansel helped kindle an early interest in comparative neuroendocrinology involving the domestic ungulates. Professor Charles H. Sawyer, through his personal counsel and by providing an exciting interdisciplinary neuroscience environment, contributed greatly to development of neuroendocrine skills and confidence. The author is most grateful for the graduate and postdoctoral training provided by these two outstanding scientists and for their continuing friendship. Therefore, this book is dedicated to these two scientific mentors and also to the author's wife, without whose love and support neither his research career nor this book would have been possible.

Paul V. Malven
Purdue University
July 1992

TABLE OF CONTENTS

Mammalian Neuroendocrinology

Chapter 1

PRINCIPLES OF NEUROENDOCRINOLOGY

The discipline of Neuroendocrinology examines the interactions between the nervous system and the endocrine system. During the development of Endocrine Physiology and Neurophysiology as scientific disciplines, the distinction between neural and endocrine systems was very clear, but in the last 30 years the clearcut differences have become less apparent. The scientific discipline of Neuroendocrinology has developed in the interface between strictly endocrine and strictly neural mechanisms. The origin of the field was the discovery over 30 years ago that certain neurons secreted chemical messengers into blood (Scharrer and Scharrer, 1963). This characteristic was previously reserved for hormones secreted into blood by endocrine glands. The term ***neurohormone*** was coined to describe a hormone produced by a neuron. Restrictions on the use of that term have become less stringent in recent years. In current usage, the proof that a chemical messenger produced by neurons acts as a true hormone (i.e., is secreted into blood) has not always been rigorously enforced. In the opinion of this author, to qualify as a hormone or neurohormone a chemical messenger ***must*** have an endocrine mode of action. With such a strict definition, the chemical messengers produced by neurons can be described as having one or more of the following types of action on other cells:

1. **Endocrine action:** Enter the blood stream to reach and alter activity of distant target cells.

2. **Paracrine action:** Diffuse locally through interstitial spaces to reach and influence neighboring cells.

3. **Neurocrine action:** Cross a synaptic junction to either activate or inhibit the postsynaptic cell.

Chemical messengers that act in an endocrine manner are generally known as hormones or neurohormones. Chemical messengers that act in a neurocrine manner are generally known as neurotransmitters. Descriptive terms for chemical messengers produced by neurons and that act in a paracrine manner include (1) neuromodulator and (2) localized hormone, but there is no generally accepted term. Moreover, it is not always known whether a particular chemical messenger acts in a paracrine or neurocrine manner, and some compounds can act in more than one manner.

The chemical structures of neuronal products known as chemical messengers can vary considerably. Peptides constitute a large class of chemical messengers produced by neurons. The size of these neuropeptides is almost always much smaller than peptides produced by the endocrine system. Some of the greatest advances in Neuroendocrinology have involved these neuropeptides, including their discovery as secretory products of neurons and emerging knowledge about their physiological functions within the nervous system and elsewhere in the body. The other major type of chemical messenger produced by neurons consists of modified amino acids and includes many of the aminergic neurotransmitters that were discovered during the early days of Neurophysiology. Examples of modified amino acids produced by neural elements are catecholamines (norepinephrine, dopamine), indolamines (serotonin), acetylcholine, and others. Some neuronal amino acids do not seem to require modification to function as chemical messengers (e.g., gamma aminobutyric acid, glycine, glutamate, and aspartate).

It is not possible to make generalizations about the type of action that a chemical messenger may exert based on its chemical structure. For example, both the neuropeptide somatostatin and the catecholamine dopamine appear to exert all three types of actions depending on the target tissue. Somatostatin acts in an endocrine manner to inhibit secretion of somatotropin by the pituitary gland and in a paracrine manner to inhibit secretion of insulin and glucagon by cells of the pancreatic islets. Somatostatin also acts in the central nervous system (CNS) on adjacent neurons (either paracrine or neurocrine action). Dopamine acts in an endocrine way to inhibit secretion of prolactin by the pituitary gland. It also acts within the CNS in a neurocrine manner, and within the mediobasal hypothalamus, dopamine may act in paracrine manner on adjacent neural elements.

Neuroendocrine Transduction

One critical process in the discipline of Neuroendocrinology is called ***neuroendocrine transduction***. This process transforms neural information (i.e., action potentials) into chemical messengers secreted into blood (i.e., hormones) where they exert endocrine effects. The small number of identified neuroendocrine transducers have been studied in detail. The diagrams in Figure 1-1 illustrate the two general categories and four specific types of neuroendocrine transducers. Neuron A in Figure 1-1 typifies the simplest form of ***secretomotor innervation*** in which a single CNS neuron innervates a secretory cell. One example is found in the adrenal medulla where neurally derived chromaffin cells are innervated by axons of the sympathetic nervous system. In response to synaptic release of the neurotransmitter acetylcholine, the chromaffin cells discharge epinephrine and norepinephrine into blood. Another less well-known example represented by neuron A in Figure 1-1 involves hypothalamic neurons sending axons to innervate non-neural cells of the pars intermedia of the hypophysis. The chemical messenger released by these axons (probably dopamine) inhibits the release of pars intermedia products. Neuron B of Figure 1-1 represents a modified secretomotor innervation involving a two-neuron chain in which the axon of the second neuron innervates the secretory cell. Innervation of the pineal gland by the sympathetic neurons typifies this situation. Postganglionic neurons originating in the superior cervical ganglion release norepinephrine at their secretomotor terminals adjacent to the pinealocyte and this activates the release of melatonin into blood and cerebrospinal fluid (CSF).

Neurosecretory neurons depicted as C and D in Figure 1-1 release their neuronal products into blood. The two types differ only in the type of blood vessel into which they secrete. Hypothalamic neurons, which send axons into the pars nervosa of the hypophysis, release chemical messengers (e.g., vasopressin, oxytocin and others) into the general circulation (neuron C). Other neurons in the hypothalamus and adjacent regions have axons that extend to the median eminence where they discharge their chemical messengers into the capillaries of the hypophysial portal veins. These chemical messengers travel in the portal blood a few millimeters to the capillaries of the hypophysis where most of them probably act in an endocrine manner to stimulate or inhibit the secretion of adenohypophysial hormones into the general circulation. Of course, the hypophysial portal blood then enters the general circulation and any neurohormones that remain also enter that circulation. There is no clear evidence that neurohormones secreted into hypophysial portal blood reach the general circulation in

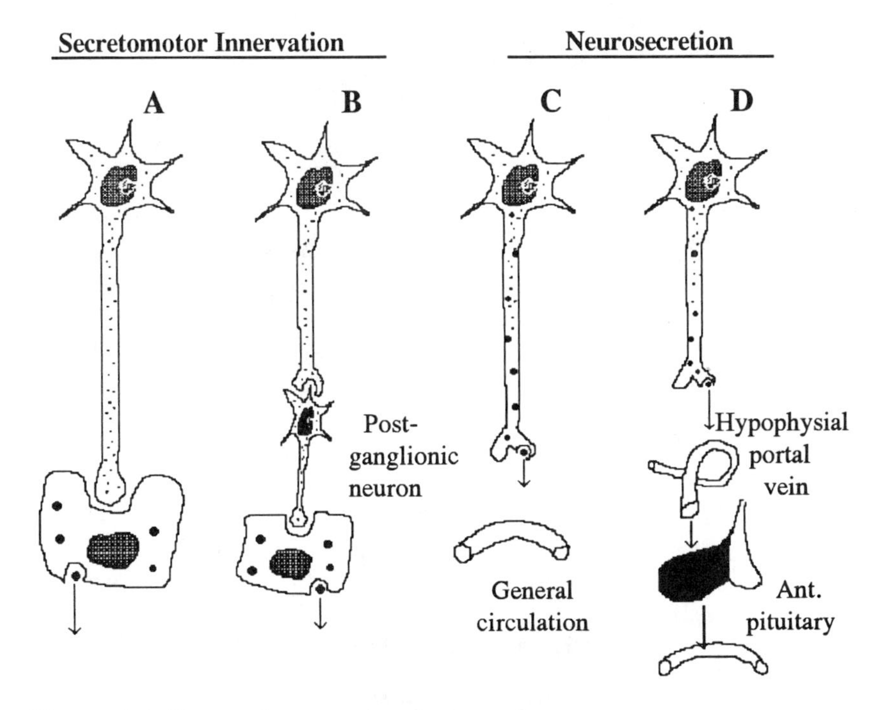

Figure 1-1. Types of neuroendocrine transducers.

Neuroendocrine transduction is depicted in diagrams which illustrate the secretory principles for two types of secretomotor innervation (A and B) and two types of neurosecretion (C and D).

physiologically relevant concentrations, but the possibility remains. In summary, neuroendocrine transduction can involve either secretomotor innervation or neurosecretory neurons, but in both types the transduction from neural to hormonal signal involves substantial amplification of that signal as well as more sustained generalized actions than are possible within the nervous system.

Neuroendocrine Integration

In addition to the transduction of neural signals into endocrine signals as just described, the neuroendocrine system mediates the cooperation between the nervous and endocrine systems to regulate in an optimum manner the physiological functions of the organism. This function can be described as ***neuroendocrine integration*** and is illustrated diagrammatically in Figure 1-2. In addition to a neurosecretory neuron that transduces the information, other ordinary (i.e.,

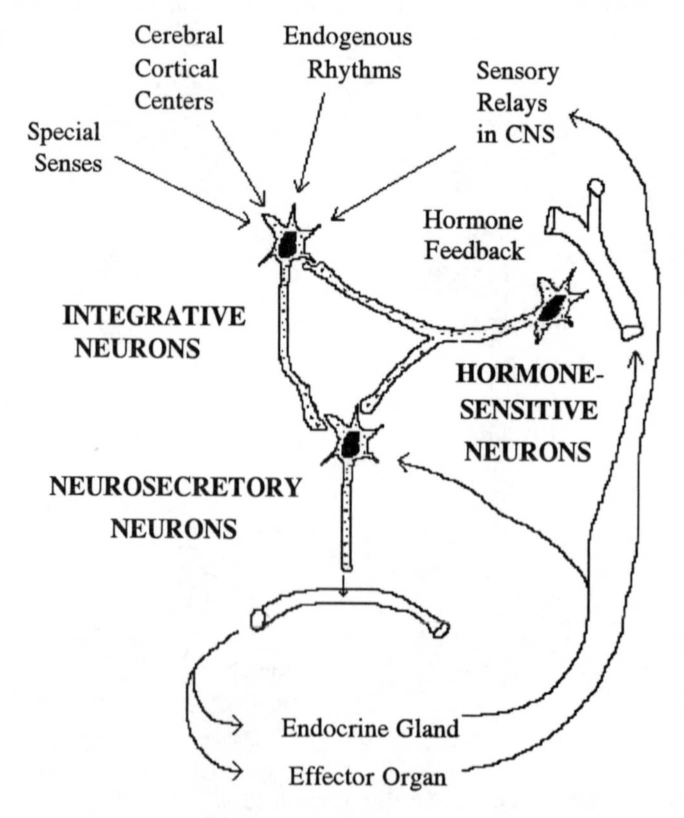

Figure 1-2. Neuroendocrine integration.

Schematic representation of the various elements involved in neuroendocrine integration.

non-neurosecretory) neurons play important roles in the integration of information. Two such neurons represented in Figure 1-2 are (1) integrative neuron and (2) hormone-sensitive neuron that each have direct input to neurosecretory neurons. Figure 1-2 also illustrates that integrative neurons of the neuroendocrine system receive a variety of inputs. These may include (1) information about the ambient environment obtained through the special senses, (2) integration of current inputs with the learned or conditioned information stored in higher cortical centers, (3) endogenous free-running rhythms (e.g., circadian or ultradian), (4) neurally mediated sensory information from internal organs (e.g., reproductive tract) and sensors (e.g., blood osmolarity, pH and pressure), and (5) neural signals from specific hormone-sensitive neurons (e.g., feedback from endocrine glands).

Hormonal Products of Neurosecretory Neurons

The number of chemical messengers of neurosecretory neurons for which endocrine actions are proved is relatively low (Table 1-1). Small peptides

Table 1-1. Specific neurohormones.

Neurohormone	*Species Distribution*	*Site of Synthesis*	*Site of Release Into Blood*
Arginine vasopressin	Most mammals	Hypothalamus	Neurohypophysis
Lysine vasopressin	Pig and hippopotamus	Hypothalamus	Neurohypophysis
Oxytocin	All vertebrates	Hypothalamus	Neurohypophysis
Arginine vasotocin	Most vertebrates except mammals	Hypothalamus	Neurohypophysis
Hypophysiotrophic hormones (i.e., releasing and inhibiting hormones)	Most vertebrates	Hypothalamus and adjacent areas of the diencephalon	Median eminence

secreted by the neurohypophysis following their synthesis in the hypothalamus represent a significant proportion of known neurohormones. The chemical structures of oxytocin and vasopressin were identified many years ago and in recent years details of their biosynthesis and gene structures have been forthcoming. Oxytocin, with some structural modifications, is found in submammalian vertebrates, but vasopressin is not. In these submammalian species, arginine vasotocin appears to functionally replace vasopressin in the control of water balance. More details about these neurohypophysial peptides will be presented in Chapter 3.

Hypophysiotrophic hormones are listed in Table 1-1 as a general class of neurohormones common to most vertebrate species. They are also known as ***releasing hormones (factors)***, but they also include neurohormones which are inhibitory to secretion/release of adenohypophysial hormones. The site of secretion into blood for neurohormones that influence the adenohyphysis is the median eminence where they enter the portal blood vessels (see D in Figure 1-1). Most of the hypophysiotrophic hormones are synthesized within the hypothalamus, but some, such as luteinizing hormone-releasing hormone (LHRH), may be synthesized in regions just rostral to the hypothalamus and still be secreted into portal blood. Hypophysiotrophic hormones will be covered in greater detail in Chapter 4, and readers interested in learning more about the history of their discovery are directed to articles in McCann (1988).

REFERENCES

McCann, S.M. (ed). *Endocrinology: People and Ideas.* (Amer. Physiol. Soc., Bethesda, MD, 1988) pp 471.

Scharrer, E., and B. Scharrer. *Neuroendocrinology.* (Columbia Univ. Press, New York, 1963) pp 289.

Chapter 2

NEUROENDOCRINE MORPHOLOGY

The unique morphological aspects of the ***hypophysis*** (also known as the ***pituitary gland***) provided some of the earliest clues to the discipline that would become known as Neuroendocrinology. Moreover, the following descriptions of hypophysial morphology in different species (Purves, 1961) will provide the basis for later understanding of neuroendocrine mechanisms. The hypophysis is divided into neurohypophysis and adenohypophysis based on embryological development from neural and epithelial substrates, respectively. Adenohypophysial tissue (pars tuberalis, pars intermedia, and pars anterior) develops specifically from an outgrowth of ectodermal epithelium of the primitive oral cavity called Rathke's pouch. The neural tissue denoted in Figure 2-1 represents the entire neurohypophysis consisting of pars eminens and pars nervosa. Although these neural tissues could be categorized as either hypothalamus or hypophysis, most modern authors classify pars nervosa as part of the hypophysis and pars eminens as part of the hypothalamic infundibulum, where it is also known as the ***median eminence***. The demarcation between pars eminens and pars nervosa cannot be precisely defined, but it occurs somewhere in the hypophysial stalk.

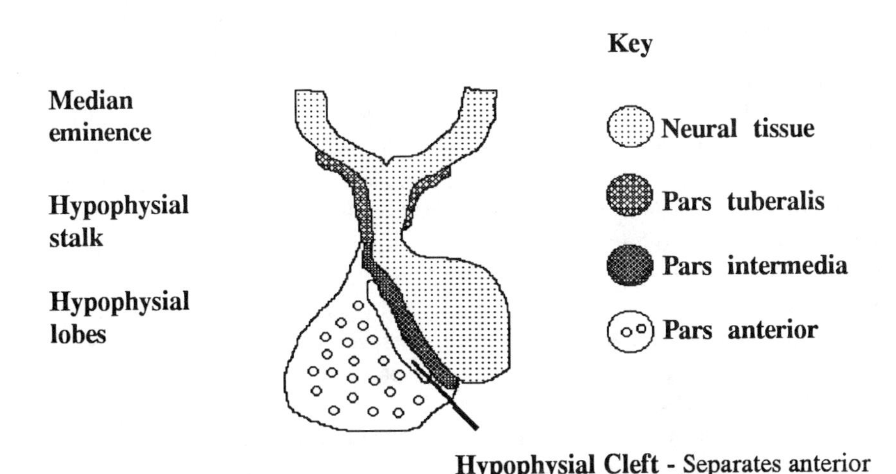

Hypophysial Cleft - Separates anterior lobe from the neurointermediate (posterior) lobe.

Figure 2-1. Divisions of the hypophysis.

This generic diagram illustrates various tissue types (key on right) and regional names (listed on left) for the hypophysis.

Subdivisions of adenohypophysial tissue are the pars tuberalis, the pars intermedia and the pars anterior. A thin rim of pars tuberalis tissue adheres to and encircles the neurohypophysial tissue of the pars eminens and the neurohypophysial stalk. Pars intermedia tissue is located adjacent to the pars nervosa in the lobular part of the hypophysis and is separated from the pars anterior tissue by the hypophysial cleft which derives from the lumen of Rathke's pouch. The commonly used terms, ***anterior pituitary*** and ***posterior pituitary***, refer to tissues anterior (i.e., rostral or ventral) or posterior (i.e., caudal or dorsal) to the hypophysial cleft. The term anterior pituitary always includes pars anterior and may also include the pars tuberalis. The term posterior pituitary includes the pars nervosa and the pars intermedia (when it exists as separate entity). A more appropriate term for the posterior pituitary is neurointermediate lobe (NIL) because this name reflects the combined presence of pars nervosa and pars intermedia tissue.

Hypophysial morphology differs among species and Figure 2-2 illustrates some examples of this diversity. The hypophysis of the rat represents the conventional morphological divisions as shown generically in Figure 2-1, but the axis of the rat hypophysis is rotated so that the neurointermediate lobe lies dorsal to the pars anterior. The hypophysial cleft in the rat is much narrower *in vivo* than in the diagram, but upon postmortem dissection the tissue will readily separate along the hypophysial cleft. The hypophysis of the cow (B in Figure 2-2) is somewhat unique because it usually contains pars anterior tissue known as the Cone of Wulzen located on the pars intermedia side of the hypophysial cleft in the caudal part of the hypophysis. Therefore, bovine tissue of the neurointermediate lobe will contain variable amounts of pars anterior tissue which would complicate the interpretation of some types of *in vitro* studies. Only some sheep hypophyses have a Cone of Wulzen, whereas another ungulate species, the pig, does not possess this unique feature. The diagram of the cow hypophysis in Figure 2-2 also illustrates that the bovine median eminence is a thin-walled tubular structure with a lumen continuous with the 3rd ventricle of the brain.

Figure 2-2 also illustrates the human hypophysis as an example of species that lack a well-defined hypophysial cleft. Although the human hypophysis may sometimes contain colloid-filled remnants of an embryonic cleft, a variety of mammals (blue whale, porpoise, elephant, beaver) completely lack a hypophysial cleft, and in some there may also be physical separation of the neurohypophysial and adenohypophysial lobes. In species that lack a hypophysial cleft, pars intermedia cells mix together with pars anterior cells in what is then appropriately called the ***pars distalis.*** Sometimes the term pars distalis is misused to describe the pars anterior tissue in those species such as the rat which do have a hypophysial cleft. Colloid-filled remnants of the hypophysial cleft are quite variable in the human hypophysis, and in their absence, the cells of the pars distalis and pars nervosa are closely adhered and the border is often less smooth than shown in Figure 2-2.

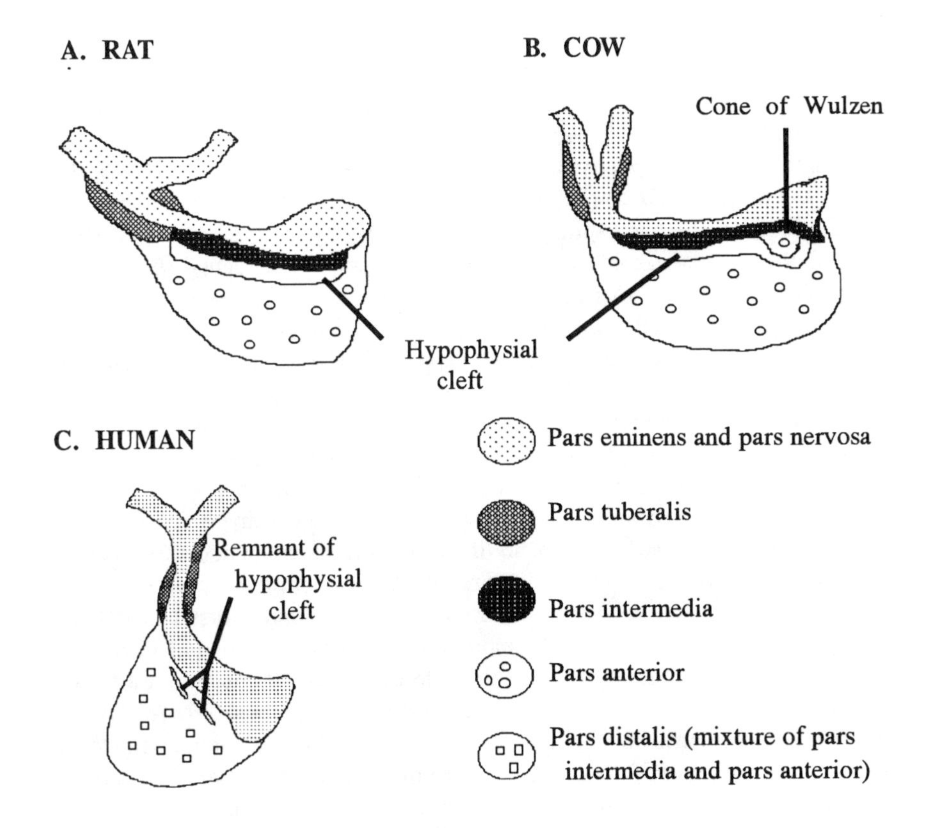

Figure 2-2. Examples of hypophysial morphology.

These diagrams illustrate the location of various hypophysial tissues in three different species (key in lower right). Pars anterior and pars intermedia tissues occur only in rat and cow whereas the pars distalis of humans combines these two tissues.

Vasculature of the Hypophysis

The unique blood supply of the hypophysis provided early indications for hypothalamic control of the adenohypophysis via blood-borne compounds. Blood vessels now known as ***long*** portal veins were described as connecting the pars eminens/pars tuberalis tissue of the median eminence with the pars anterior tissue of the hypophysial lobes in a large number of vertebrate species (Green, 1951). In the absence of a direct arterial supply to pars anterior tissue, the concept arose that all the blood supply of the adenohypophysis passed through two capillary networks, one conventional network in the median eminence and a second sinusoidal network in the adenohypophysis. The primary portal capillaries in the median eminence possess a fenestrated endothelium similar to peripheral capillaries but different from most brain capillaries. The fenestrated endothelium of the capillaries in the median eminence is responsible for this tissue being outside the blood-brain barrier.

Another type of hypophysial portal blood vessel was discovered during the 1950s, and these vessels connected pars nervosa with pars anterior. Because the distance between the two capillary networks of these portal veins was so much less than the one discovered earlier, these blood vessels were called ***short*** portal veins. Their discovery was probably delayed because they are not nearly as visible as the long portal veins. The functional significance of short portal veins was emphasized by the results of surgical transection of the hypophysial stalk in various species. In all species, tissue of the pars anterior or pars distalis that is adjacent to the pars nervosa survives much better than other adenohypophysial tissue, and it is precisely this tissue that would be supplied by short portal veins after transection of the long portal veins which traverse the hypophysial stalk.

Venous outflow from the adenohypophysis and neurohypophysis occurs through conventional veins similar to those draining the brain and other cranial structures, with one possible exception. Several authors have suggested that some portion of hypophysial outflow may ascend the hypophysial stalk and reach the hypothalamus. There is both morphological and hormonal evidence to support delivery of hypophysial hormones to the brain without dilution in the systemic circulation. This hypothesis is sometimes called ***retrograde flow*** of hypophysial hormones to the hypothalamus where they might participate in feedback regulation of factors that influence hypophysial secretion. This postulated feedback is referred to as ***short*** feedback in contrast to ordinary feedback by hormones produced in the periphery and delivered to the hypothalamus in systemic blood.

The morphological evidence consistent with retrograde flow is as follows: (1) confluent capillary network of the pars nervosa and pars eminens which could shift venous outflow, as well as reversed flow in the short portal veins, toward the hypothalamus if vascular pressures were appropriate, and (2) subependymal veins located on the ventricular wall of the median eminence into which injected dye was observed to ascend subsequent to its downward flow in the long portal veins of a anesthetized preparation.

The hormonal evidence in support of retrograde flow was provided by concurrent quantification of hypophysial hormone concentrations in cannulated long portal veins and the systemic vasculature (Oliver et al., 1977). Portal concentrations of several hypophysial hormones were up to 100-fold greater than in the systemic circulation where they are diluted by blood from other sources. If all hypophysial secretions drain directly into the systemic circulation, it is very difficult to explain the concentration gradient between portal and systemic blood without retrograde flow to deliver enriched levels of hypophysial secretions to the hypothalamus for subsequent entry into the long portal veins. Some authors have suggested that the technique of collecting blood from long portal veins may create an unphysiological pressure which would abnormally draw hypophysial effluent back up the cannulated vein. Others have suggested that certain hypothalamic neurons may secrete small quantities of hypophysial hormones, and this could explain the concentration differential between portal and systemic blood. In summary, the hypothesis of retrograde flow to deliver enriched levels of hypophysial hormones to the hypothalamus should be considered unproven but still tenable.

Species Differences in Cranial Vasculature

The arterial supply to the brain and hypophysis differs among species. Some differences involve only the nomenclature of arteries that are homologous between species (e.g., anterior hypophysial in rat versus superior hypophysial in human). The arterial supply to the pars nervosa and par eminens region of the hypophysial stalk in various species is known by different names. It is known as peduncular artery in rat, artery of lower infundibular stalk in sheep, and artery of the trabecula (also middle hypophysial) in human. Despite species differences in nomenclature, arteries that supply blood to the hypophysis in most species originate from the arterial arrangement known as the Circle of Willis that also supplies blood to the brain. The manner in which blood is supplied to the Circle of Willis represents the greatest difference in cranial vasculature among species.

Selected species of ungulates (cow, pig, sheep, camel) and carnivores (cat, dog) have been shown to possess a unique arterial vasculature called a ***carotid rete*** (sometimes also called rete mirabile) that is located at the base of the brain and provides blood for the Circle of Willis (Daniel et al., 1953). The carotid rete consists of a dense network of many small arterioles, but no capillaries, and the network is located within the venous sinus known as the cavernous sinus. The arterioles of the carotid rete subsequently reunite into larger arteries which supply the Circle of Willis. Because the carotid rete lacks capillaries for exchange with tissue fluid or venous blood, early theories regarding its function emphasized blood pressure modulation. However, it is now clear that one very important function of the carotid rete is temperature exchange between the venous blood in the cavernous sinus and arterial blood in the carotid rete (Hayward and Baker, 1969). Because much of the venous blood in the cavernous sinus drains nasal structures and other peripherally located tissues, the temperature of that venous blood is less than core body temperature. The arterial blood entering the carotid rete reflects core body temperature, and during transit through the carotid rete, it is cooled by heat exchange with the cooler venous blood of the cavernous sinus. Through this heat exchange, animals that possess a carotid rete have the ability to maintain the temperature of their brain lower than core body temperature. This ability is used primarily during situations of ambient heat stress or exercise when maintenance of brain temperature lower than core body temperature represents a distinct biologic advantage. During situations of exercise or heat stress, evaporative cooling in nasal structures becomes very important, and the venous drainage from nasal structures into the cavernous sinus may become cooled even more. Laburn et al. (1988) demonstrated in hyperthermic sheep that experimental deflection of airflow away from the upper respiratory tract increased brain temperature abruptly to that of core body temperature.

Cellular Components of the Hypophysis

Neurohypophysis. The cytology of the neurohypophysial tissue consists only of axons, axon terminals, and neuroglial cells because there are no neuronal cell bodies (perikarya). The axons arise from very large (i.e., magnocellular) perikarya located in the adjacent hypothalamus. It has been estmated that about 10,000 perikarya in the rat hypothalamus project their axons into the pars nervosa, and the estimate for the human is much larger. Each axon gives rise to multiple branches, each with a terminal for release of neuronal products into blood (see Figure 1-1.C). There are also numerous capillaries with especially large perivascular spaces, which is consistent with the high levels of blood flow and the secretion of neuronal products into blood.

The neuroglial cells of the neurohypophysis (also known as ***pituicytes***) share many characteristics with neuroglia known as astrocytes elsewhere in the brain, but recent observations suggest that they may serve unique functions in the neurohypophysis. For example, the physical relationship among axon terminals, capillaries, and pituicytes changes during extreme states of altered neurohypophysial secretion (Hatton et al., 1988). Second, receptors for opioid peptides exist in pituicytes rather than in neurohypophysial axons (Bicknell et al., 1989). Third, mRNA transcripts that encode the synthesis of neurohypophysial secretory products continue to exist in the isolated pars nervosa (Murphy et al., 1989), and the pituicyte is the only cell type thought to be present in the isolated pars nervosa that would be capable of gene transcription. In summary, the pituicyte of the mammalian neurohypophysis functions as a neuroglial cell in support of axons, but it may also have unique functions that are not fully understood.

Adenohypophysis. Great progress has been made in understanding the cytology of the adenohypophysis due in large part to the following three techniques: ***immunocytochemistry*** (Polak and Van Noorden, 1986); ***in situ hybridization*** (Morrell, 1989); and ***reverse hemolytic plaque assay*** (Frawley et al., 1985). Space limitations do not permit detailed discussion of these important techniques, but the interested reader is directed to the cited references. However, the type of knowledge derived from each technique applied to adenohypophysial cytology will be summarized. Immunocytochemical analysis, using both light and electron microscopy, permits the definitive identification of intracellular (or intragranular) peptides against which the antibodies used were directed. *In situ* hybridization allows the identification of cells containing mRNA transcripts of specific peptide-producing genes, and specificity depends upon the nucleic acid probe used for hybridization. The intracellular presence of the mRNA capable of encoding a peptide represents strong evidence that the hybridized cell actually synthesizes the peptide. For secretory cells of the adenohypophysis, intracellular content of the peptide as determined by immunocytochemistry and of the mRNA as determined by *in situ* hybridization almost always provide confirmatory results. In other tissues where cells may take up the secretory product of other cells as well as synthesize their own products, application of both techniques provides insight into which intracellular peptides are synthesized and which are taken up.

The reverse hemolytic plaque assay is performed under slightly less physiological conditions than the techniques described above because the secretory cells of the adenohypophysis must be enzymatically dispersed prior to *in vitro* culture. If one assumes that this dispersion and culture do not alter the cellular function, this technique can determine which cell types actually release specific peptides. When combined with immunocytochemistry of the dispersed and cultured cells, it is possible to determine what proportion of peptide-containing cells actually release that peptide (Kineman et al., 1990). When the hemolytic plaque assay is combined with *in situ* hybridization, it is possible to quantify gene transcription in cells that release a specific peptide (Scarbrough et al., 1991).

Pars Intermedia. This tissue is present as a distinct structure in only some species, but its secretory cells are present within the pars distalis of other species. When present as a separate tissue, the pars intermedia is innervated by secretomotor axons from hypothalamic perikarya (see Figure 1-1.A) and perhaps from the pars nervosa. Pars intermedia tissue is very poorly vascularized, and the sources of its blood-borne nutrients are not fully understood. The secretory cells of the pars intermedia appear to all synthesize the precursor protein known as ***pro-opiomelanocortin*** which can be proteolytically cleaved to yield many hormonal peptides (to be discussed in Chapters 5 and 6).

Pars Tuberalis. The vasculature of the pars tuberalis is not well understood, but there may be diffusion of secretory products and nutrients to and from the adjacent neural tissue of the pars eminens of the median eminence and hypophysial stalk. Although there are fewer cell types, pars tuberalis cytology is similar in some ways to that of the pars anterior. Relatively few stainable secretory vesicles are seen with light microscopy, but ultrastructural analysis does reveal the presence of secretory vesicles. Pars tuberalis cells that contain stainable luteinizing hormone (LH) and thyrotropin (TSH) have been observed in several species. There is even some evidence for secretion of these hormones from the pars tuberalis into blood. Many cells of the pars tuberalis do not stain for any known adenohypophysial hormones. Recent evidence that almost all cells of the pars tuberalis cells contain receptors for the pineal gland secretory product, melatonin, has raised the possibility of some unique function for the pars tuberalis in relation to pineal-mediated processes (de Reviers et al., 1989). It has been suggested that melatonin regulation of LH secretion by the pars tuberalis modulates release of LHRH in the adjacent pars eminens by diffusion (short feedback) of LH produced by pars tuberalis (Nadazawa et al., 1991). In this regard, it should be noted that almost all cells of the ovine pars tuberalis contain mRNA encoding the synthesis of LH, whereas only a fraction of the cells contain stainable quantities of LH (Pelletier et al., 1992).

It should also be noted that ***hypophysectomy*** (surgical removal of the hypophysis), as it is commonly performed in rats and other species, does not remove pars tuberalis tissue from the animal. Therefore, small quantities of adenohypophysial hormones that are sometimes detected in the blood of hypophysectomized animals could potentially have come from pars tuberalis tissue. Moreover, hypophysectomy may stimulate hormone synthesis in cells of the pars tuberalis (Ordronneau and Petrusz, 1980; Gross, 1983).

Pars Anterior. Secretory cells of the pars anterior are very diverse because there are at least six different peptide hormones secreted into blood from this tissue. At one time, it was thought that there was one adenohypophysial cell for each secreted hormone. It is now known that a single cell may secrete more than one hormone. The corticotroph cell synthesizes pro-opiomelanocortin which can be cleaved to yield corticotropin (ACTH), β-lipotropin, β-endorphin, and several forms of melanotropin (MSH). Corticotrophs comprise about 4% of the cells in the rat pars anterior, and their secretory vesicles show diverse sizes and shapes.

Thyrotroph cells that secrete TSH comprise about 3% of pars anterior cells in rats, and their secretory vesicles are very small (i.e., diameters of about 50 nm). Thyrotrophs appear to be strictly monohormonal (i.e., secrete a single hormone). In contrast, gonadotrophs may be either monohormonal or bihormonal, secreting LH and/or follicle-stimulating hormone (FSH), and there are species differences as regards the relative proportions of these monohormonal and bihormonal gonadotrophs. Moreover, the proportions of LH only, FSH only, and LH + FSH gonadotrophs may also depend on physiological status. Secretory vesicles present in gonadotrophs have diameters of intermediate size, and it is unclear whether LH and FSH always exist in separate secretory vesicles within individual gonado-troph cells.

Although the coexistence of LH and FSH in one gonadotroph was initially surprising, each hormone contains one subunit that also occurs in the other hormone. The discovery that the evolutionarily related prolactin (PRL) and somatotropin (GH) could sometimes occur in a single cell (Frawley et al., 1985) was more surprising because these hormones are different molecules arising from different genes. Cells that contain only PRL are known as mammotrophs because PRL has strong actions on mammary function. The secretory vesicles of mammotrophs found in the rat pars anterior are unique because of their irregularly shaped (i.e., pleomorphic) secretory vesicles located peripherally within the cytoplasm. However, pleomorphic secretory vesicles are not a general feature of mammotrophs because the secretory vesicles in sheep mammotrophs are spherical rather than pleomorphic. Cells that contain only GH are known as somatotrophs and usually have spherical secretory vesicles whose size varies with species. Cells that contain both PRL and GH are known as mammosomatotrophs, and they have been observed in several species. Rat mammosomatotrophs are quite small cells containing inconspicuous rough endoplasmic reticulum (RER) and Golgi apparatus. Individual secretory vesicles of rat mammosomatotrophs appear to contain both PRL and GH. In contrast to the rat, bovine mammosomatotrophs are not as small and are often binucleated cells (Kineman et al., 1990).

Cytological Features of Adenohypophysial Secretion. Hormones are synthesized, packaged into secretory vesicles, and secreted from adenohypophysial cells in ways that appear to be similar to those of other hormone-secreting cells. Gene transcription and RNA processing occurs in the nucleus followed by transport of the mRNA to the RER where synthesis of the prohormone occurs. The prohormone is cleaved into the products for secretion, and occasionally subunits produced from different gene transcripts are covalently linked. The secretory products are packaged into dense-core secretory vesicles in the Golgi apparatus from which they migrate to a position near the plasma membrane to await release. Release into the extracellular space occurs by exocytosis wherein (1) the membrane of the secretory vesicle fuses with the inner surface of the plasma membrane, (2) the double layer of fused membrane breaks down, and (3) the contents of the secretory vesicle enter the extracellular space. The speed of the entire process of synthesis, packaging, and exocytosis probably depends upon the intensity of stimulation of the adenohypophysial cell. Studies with strongly stimulated mammotrophs from rats suggest that the entire process can occur in as little as 50 min, but this may represent a lower limit.

Brain Morphology Related To Neuroendocrinology

Selected aspects of neuroanatomy necessary to understand neuroendocrinology will be presented here. Most of the important brain regions are part of what is known as the ***limbic system,*** proposed by Papez as being structures that were intimately involved in the control of emotion. It is now clear that limbic system structures of the brain control many processes in addition to emotional responses, but that there are close linkages between emotion and these other physiological processes.

Hippocampus. This large horn-shaped bilateral structure curves laterally and ventrally into the tissue beneath each temporal lobe of the cerebral cortex. The primary connecting fibers between the hippocampus and other limbic structures are contained within the fornix. Established functions of hippocampus include the encoding of memory information as well as mediating some aspects of emotion. In regard to neuroendocrine regulation, electrical stimulation of the hippocampus can modify pituitary secretion of LH, and the morphology of the synaptic input from hippocampus to preoptic area differs between sexes in the rat. Neurons in the hippocampus also contain a very high density of receptors for adrenal steroids which implies that it is a target tissue for blood-borne adrenocortical hormones.

Amygdala. These circular structures are located at the lateral tip of each hippocampal horn beneath the temporal lobe. Each amygdala contains several subdivisions, the morphology of which are beyond the scope of this brief overview. The amygdala receives a sampling of the inputs from all the sensory modalities and is intimately involved with emotional states. Lesions of amygdala tissue in experimental animals can cause behavioral depression followed later by nonselective hyperactive behaviors (sexual and feeding). Electrical stimulation of the amygdala can also induce anxiety and fear, suggesting that this region is very important in the regulation of emotion. Experimental manipulation of the amygdala has also been shown to modulate the secretion of pituitary LH similar to the results described above for hippocampus. The amygdala also contains receptors for adrenal steroids (like hippocampus) and estradiol (unlike hippocampus). These two steroid receptors may mediate the effects of these blood-borne hormones on behavior and/or on feedback effects to influence the adenohypophysial secretion of ACTH and LH.

Other components of the limbic system include the septum, olfactory gray, and parts of the cerebral cortex (i.e., cingulate gyrus and hippocampal gyrus). The epithalamus and its associated pineal gland should probably be considered functionally part of the limbic system although they were not included in the original designation. Finally, the hypothalamus is the part of the limbic system most intimately involved with neuroendocrinology. Not only does it contain perikarya of neurons which participate in neuroendocrine regulation and integration, but it contains many tracts which connect with other parts of the limbic system. The preoptic area is included herein as a functional part of the hypothalamus although morphological classifications often list the preoptic area as a separate entity. The major fiber pathways connecting the hypothalamus with limbic and non-limbic structures are depicted in Figure 2-3 in a schematic and highly simplified diagram.

The following three tracts connect the hypothalamus with the midbrain and lower areas: mammillopeduncular tract (labeled H in Figure 2-3), mammillotegmental tract (G), dorsal longitudinal fasciculus (F). The mammillopeduncular tract and mammillotegmental tract both connect the mammillary bodies of the hypothalamus with the midbrain and lower regions. The hippocampus inputs into the hypothalamus pass by way of the fornix (E) to the preoptic area, arcuate nucleus and mammillary bodies of the hypothalamus. The thalamus connection with the mammillary bodies involves the mammillothalamic tract (I). The amygdala connections with the hypothalamus consist of (1) stria terminalis (C) which curves around in parallel with the fornix and (2) the shorter route from amygdala to hypothalamus called the direct amygdalohypothalamic tract (D). A major fiber tract passing through the hypothalamus is the medial forebrain bundle (B) which connects hypothalamus with the septum, rostral structures such as the olfactory gray, and structures caudal to the hypothalamus. The epithalamus sends input to the preoptic area of the hypothalamus via the stria medullaris (A).

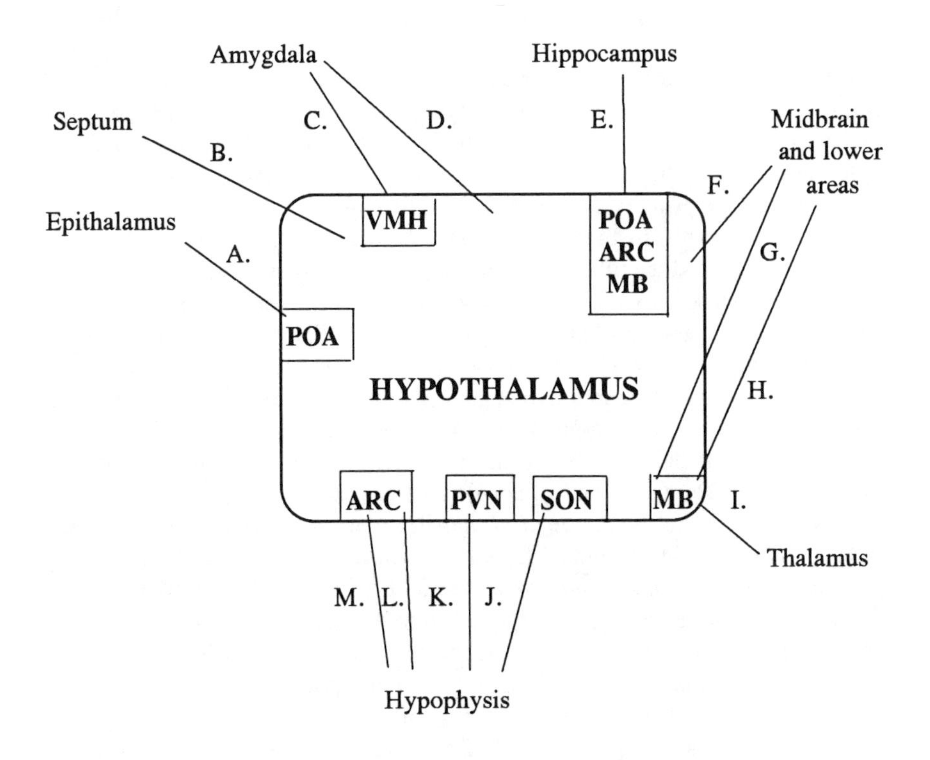

Figure 2-3. Fiber tract connections with the hypothalamus.

Schematic representation of neuroanatomic pathways connecting various hypothalamic areas with extrahypothalamic structures.

Key for hypothalamic areas: ARC = arcuate nucleus; MB = mammillary bodies; PVN = paraventricular nucleus; POA = preoptic area; SON = supraoptic nucleus; VMH = ventromedial hypothalamic nucleus.

Key for pathways: A = stria medullaris; B = medial forebrain bundle; C = stria terminalis; D = direct amygdalohypothalamic tract; E = fornix; F = dorsal longitudinal fasciculus; G = mammillotegmental tract; H = mammillopeduncular tract; I = mammillothalamic tract; J = supraoptic hypophysial tract; K = paraventricular hypophysial tract; L = tuberohypophysial tract; M = tuberoinfundibular tract.

Fiber tracts that connect the hypothalamus with the hypophysis are very important in neuroendocrinology. They include the paraventricular hypophysial tract (labeled K in Figure 2-3) from the paraventricular nucleus to the pars nervosa and the supraoptic hypophysial tract (J) from the supraoptic nucleus to the pars nervosa. These two tracts contain axons of the magnocellular neuronal perikarya, and they transport the neurohypophysial peptides to the pars nervosa for release into blood (like C in Figure 1-1). Another important tract is the tuberoinfundibular tract (M) containing axons that discharge their neuronal products into the primary capillaries of the hypophysial portal veins located in the median eminence of the pars eminens. A final tract from hypothalamus to hypophysis is called the tuberohypophysial tract (L), and its axons extend down the pituitary stalk into the pars nervosa along with axons of the supraoptic hypophysial and paraventricular hypophysial tracts. However, these axons of the tuberohypophysial tract do not originate from magnocellular perikarya.

Because of its intrinsic functions and its many connecting pathways illustrated in Figure 2-3, the hypothalamus is an important component of the limbic system. The hypothalamus also mediates the linkage between neuronal functions of the limbic system and the hormonal functions of the hypophysis.

Neuroendocrine Implications of the Blood-Brain Barrier. Most neuronal tissues of the brain are separated from the vascular compartment of the body by what is called the ***blood-brain barrier*** (Bradbury, 1979). This barrier restricts the movement of blood-borne compounds into the brain as well as the secretion of neuron-derived compounds into blood. The endothelium of brain capillaries may provide the morphological basis for this conceptual barrier. The ventricles of the brain appear to be on the brain side of the blood-brain barrier. Circumventricular organs (Chapter 4) lack the usual blood-brain barrier, and the endothelium of their capillaries is fenestrated. Therefore, circumventricular organs may be the functional link by which some blood-borne molecules affect the brain and by which neurohormones are released into blood (Johansson, 1990). This second function is especially true for the median eminence and pars nervosa where many neurons release their neurohormones into the blood. Perikarya of those neurosecretory neurons in which the axons have such access to the circulation may be localized by their uptake of a blood-borne dye which can later be visualized histochemically together with the neurohormone of interest (Witkin, 1990).

REFERENCES

Bicknell, R.J., S.M. Luckman, K. Inenaga, W.T. Mason, and G.I. Hatton. β-Adrenergic and opioid receptors on pituicytes cultured from adult rat neurohypophysis: regulation of cell morphology. *Brain Res. Bull.* (1989) 22:379-388.

Bradbury, M.W.B. *The Concept of a Blood-Brain Barrier.* (Wiley-Interscience, Chichester, UK, 1979) 465 pp.

Daniel, P.M., J.D.K. Dawes, and M.M.L. Prichard. Studies of the carotid rete and its associated arteries. *Philos. Trans. R. Soc.* (1953) 237B:173-208.

de Reviers, M.M., J.P. Ravault, Y. Tillet, and J. Pelletier. Melatonin binding sites in sheep pars tuberalis. *Neurosci. Lett.* (1989) 100:89-93.

Frawley, L.S., F.R. Brockfor, and J.P. Hoeffler. Indentifcation by plaque assays of a pituitary cell type that secretes both growth hormone and prolactin. *Endocrinology* (1985) 115:734-737.

Green, J.D. The comparative anatomy of the hypophysis, with special reference to its blood supply and innervation. *Am. J. Anat.* (1951) 88:225-312.

Gross, D.S. Hormone production in the hypophysial pars tuberalis of intact and hypophysectomized rats. *Endocrinology* (1983) 112:733-744.

Hatton, G.I., Q.Z. Yang, and K.G. Smithson. Synaptic inputs and electrical coupling among magnocellular neuroendocrine cells. *Brain Res. Bull.* (1988) 20:751-755.

Hayward, J.N., and M.A. Baker. A comparative study of the role of the cerebral arterial blood in regulation of brain temperature in five mammals. *Brain Res.* (1969) 16:417-440.

Kineman, R.D., W.J. Faught, and L.S. Frawley. Bovine pituitary cells exhibit a unique form of somatotrope secretory heterogeneity. *Endocrinology* (1990) 127:2229-2235.

Johansson, B.B. The physiology of the blood-brain barrier. In: Porter, J.C., and D. Jezova (eds). *Circulating Regulatory Factors and Neuroendocrine Function* (Plenum Press, N.Y., 1990), pg 25-39.

Laburn, H.P., D. Mitchell, G. Mitchell, and K. Saffy. Effects of tracheostomy breathing on brain and body temperatures in hyperthermic sheep. *J. Physiol.* (1988) 406:331-344.

Morrell, J.I. Application of in situ hybridization with radioactive nucleotide probes to detection of mRNA in the central nervous system. In: Bullock, G.R. and P. Petrusz (eds). *Techniques in Immunocytochemistry, Vol. 4,* (Academic Press, London, 1989), pg 127-146.

Murphy, D., A. Levy, S. Lightman, and D. Carter. Vasopressin RNA in the neural lobe of the pituitary: dramatic accumulation in response to salt loading. *Proc. Natl. Acad. Sci.* (1989) 86:9002-9005.

Nakazawa, K., U. Marubayashi, and S.M. McCann. Mediation of the short-loop negative feedback of luteinizing hormone (LH) on LH-releasing hormone release by melatonin-induced inhibition of LH release from the pars tuberalis. *Proc. Natl. Acad. Sci.* (1991) 88:7576-7579.

Oliver, C., R.S. Mical, and J.C. Porter. Hypothalamic-pituitary vasculature: evidence for retrograde flow in the pituitary stalk. *Endocrinology* (1977) 101:598-604.

Ordronneau, P., and P. Petrusz. Immunocytochemical demonstration of anterior pituitary hormones in the pars tuberalis of long-term hypophysectomized rats. *Am. J. Anat.* (1980) 158:491-506.

Pelletier, J., R. Counis, M.M. de Reviers, and Y. Tillet. Localization of luteinizing hormone beta-messenger RNA by in situ hybridization in the sheep pars tuberalis. *Cell. Tissue Res.* (1992) 267:301-306.

Polak, J.M., and S. Van Noorden. *Immunocytochemistry. Modern Methods and Applications, 2nd Ed.* (Wright & Sons, Bristol, UK, 1986) 703 pg.

Purves, H.D. Morphology of the hypophysis related to its function. In: Young, W.C. (ed). *Sex and Internal Secretions.* (Williams & Wilkins, Baltimore, 1961), pg 161-239.

Scarbrough, K., N.G. Weiland, G.H. Larson, M.A. Sortino, S. Chiu, A.N. Hirshfield, and P.M. Wise. Measurement of peptide secretion and gene expression in the same cell. *Mol. Endocrinol.* (1991) 5:134-142.

Witkin, J.W. Access of luteinizing hormone-releasing hormone neurons to the vasculature in the rat. *Neuroscience* (1990) 37:501-506.

Chapter 3

NEUROHYPOPHYSIS

Vasopressin (also known as ***antidiuretic hormone***) and oxytocin were the first secretory products of the neuroendocrine system to have their structures chemically identified (Du Vigneaud, 1954-55). This discovery was the first of many similar chemical identifications of neurohormones, all of which greatly advanced the discipline of Neuroendocrinology. The amino acids present in the neurohypophysial neurohormones are summarized in Table 3-1. There is a common structure of nine amino acids (i.e., ***nonapeptide***) with one disulfide bond between #1 and #6 completing a six-membered ring. The C-terminal glycine residue (#9) is amidated, and because of the ring structure there is no free N-terminus. The six neurohormones listed in Table 3-1 contain substitutions only at positions #3, #4 and #8. Glutamine at position #4 is present in both the vasopressinergic and oxytocinergic molecules. Only isotocin, found in the neurohypophysis of bony fishes, contains a serine at #4. At position #3, the two vasopressins (arginine and lysine) contain phenylalanine, whereas all the other neurohormones in Table 3-1 contain isoleucine at #3.

The difference at position #8 between arginine vasopressin and lysine vasopressin appears to reflect genetic diversity without physiological importance

Table 3-1. Amino acids in neurohypophysial peptides.

S-S (disulfide bond between #1 and #6)

CYS -	TYR -	-	-	ASN -	CYS -	PRO -	-	GLY(NH_2)
#1	#2	#3	#4	#5	#6	#7	#8	#9

	Substitutions at Various Positions		
	#3	**#4**	**#8**
Arginine vasopressin	PHE	GLN	ARG
Lysine vasopressin	PHE	GLN	LYS
Oxytocin	ILE	GLN	LEU
Arginine vasotocin	ILE	GLN	ARG
Mesotocin	ILE	GLN	ILE
Isotocin	ILE	SER	ILE

because both compounds have full vasopressinergic activity. The lysine vasopressin form is present in neurohypophysial tissues only in pigs and a few related species, whereas most mammals synthesize arginine vasopressin. It is likely that the gene for lysine vasopressin originated as a mutation of the gene for arginine vasopressin because ancestral pigs (e.g., warthogs) express one allele for each form of vasopressin (Ivell, 1990). The mammalian genome may actually contain two genes encoding the vasopressinergic compounds, with only one of these genes being expressed in neurohypophysial tissues. The other vasopressinergic gene appears to be expressed in gonadal tissue since testicular tissue from pigs contains both arginine and lysine vasopressin molecules (Nicholson et al., 1988).

Oxytocin contains isoleucine at position #3, whereas vasopressinergic molecules contain phenylalanine at that position. Although mammalian species secrete oxytocin for uterine-stimulating activity and milk ejection from the mammary glands, submammalian species appear to secrete mesotocin (avian and reptilian species) or isotocin (lower vertebrates) for uterine stimulation. As occurs for the vasopressin gene, the gene for oxytocin appears sometimes to be expressed in gonadal tissue. The corpus luteum of the ovary of selected species of mammals (e.g., ruminant ungulates such as sheep and cow) synthesizes very large quantities of oxytocin during specific stages of the reproductive cycle. During the period of maximum ovarian expression of the oxytocin gene, the bovine corpus luteum may contain 250-fold more oxytocin mRNA than the entire hypothalamus (Ivell and Richter, 1984). In sheep, each corpus luteum contained about 15-fold more oxytocin mRNA than the hypothalamus (Ivell et al., 1990).

Arginine vasotocin constitutes a molecular compromise, being vasopressin-like at position #8 and oxytocin-like at position #3 (Table 3-1). This molecule is found in neurohypophysial tissue of most submammalian vertebrates, but it also occurs in the neurohypophysis of mammalian fetuses and perhaps in the pineal gland of some adult mammals. The functional role of arginine vasotocin in mammals is unclear, but in bioassays the molecule has both oxytocinergic and vasopressinergic activity.

Biosynthesis and Release

Application of molecular biology techniques has contributed greatly to our understanding of the synthesis of oxytocin and vasopressin. Each nonapeptide neurohormone is synthesized in the rough endoplasmic reticulum of hypothalamic neurons as part of large precursors depicted in Figure 3-1. This synthesis is directed by three exons of each gene, with Exon 2 being identical for the oxytocin and vasopressin genes. Exon 1 directs the synthesis of the signal peptide, neurohormone and the N-terminal end of the neurophysin molecule that is specific to that neurohormone. Exon 2 directs the synthesis of the middle

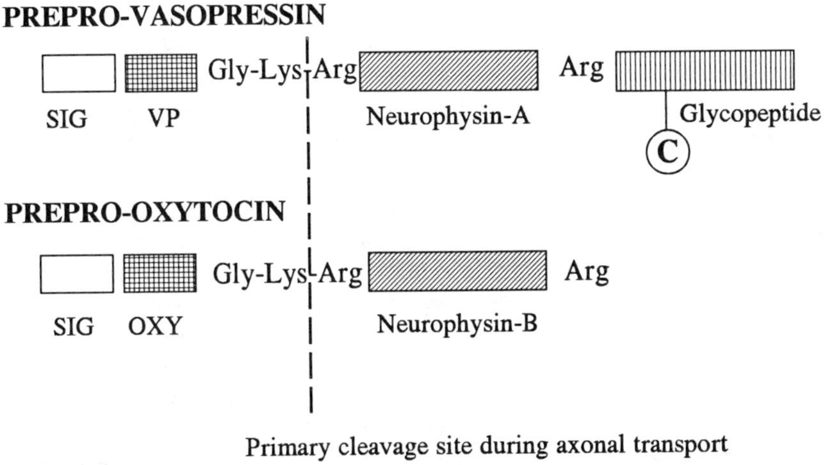

Figure 3-1. Biosynthetic precursor proteins.

Peptide sequences contained within the precursors of vasopressin (VP) and oxytocin (OXY). In each precursor, the signal sequence (SIG) is followed by the sequence of either VP or OXY. Each nonapeptide sequence is followed by glycine (Gly), a putative cleavage site (Lys-Arg), and the remainder of each precursor molecule.

portion of neurophysin (residues #22 through #88) that are identical in neurophysin-A (vasopressin-linked) and neurophysin-B (oxytocin-linked). Exon 3 directs synthesis of the C-terminal end of neurophysin (residues #89 through 107), and in the vasopressin gene this exon also directs synthesis of a 39-residue glycopeptide (Figure 3-1).

Synthesis and packaging into secretory vesicles occurs in the perikarya of magnocellular neurons (Figure 3-2). Following synthesis in the rough endoplasmic reticulum, the signal peptide (SIG in Figure 3-1) is cleaved and the resulting molecules (now called pro-vasopressin and pro-oxytocin) are packaged into dense core secretory vesicles by the Golgi apparatus. These vesicles are quite large (100-150 nm diameter), and they are transported down the axon in rodents at a rate of 1-3 mm/h. Although this rate of axoplasmic flow in magnocellular neurons that secrete oxytocin and vasopressin is much faster than that occurring in ordinary neurons, the time between synthesis and appearance in the axonal terminals of the pars nervosa is usually about 3 h. This time interval can be reduced to about 90 min if the stimuli for release of the neurohormone into blood are very strong.

During the hours of axonal transport of secretory vesicles, each molecule is cleaved at the point indicated in Figure 3-1 (between the Lys-Arg residues). This cleavage produces a 10-residue peptide with Gly at its C-terminus. This Gly residue serves as amide donor for amidation of residue #9 during axonal transport. The two cysteine residues (#1 and #6) of each amidated nonapeptide form a disulfide linkage to create the six-membered ring that is characteristic of these nonapeptide neurohormones. The fully formed nonapeptide now binds non-covalently to the neurophysin from which it was earlier cleaved. Although neurophysin-A and neurophysin-B differ at their N-terminal and C-terminal ends, the middle portions where binding occurs are identical. Therefore, purified neurophysin-A will bind equally well *in vitro* either to vasopressin, with which it binds *in vivo*, or to oxytocin. A similar situation exists for neurophysin-B. As indicated in Figure 3-1, neurophysin-A is linked to a 37-residue glycopeptide by an Arg residue. This glycopeptide is cleaved from neurophysin-A during axonal transport, but its function within the secretory vesicle is not known.

There has been extensive genetic analysis of prepro-vasopressin in a mutant strain of rats with diabetes insipidus (producing large volumes of dilute urine), also known as the Brattleboro strain. It is known that these rats lack the ability to synthesize vasopressin in neurohypophysial tissue. However, genetic analysis revealed that exon 1, which encodes the vasopressin molecule, is normal. The mutation was discovered in exon 2, which encodes the middle portion of neurophysin-A. The mutation apparently involves deletion of one nucleotide base which has drastic consequences because it shifts the reading frame of codon triplets. Although the nonapeptide and the N-terminus of neurophysin-A are properly encoded in Brattleboro rats, all residues on the C-terminal side of the deleted base are altered by the shift in reading frame. Moreover, the in-frame stop codon is not detected due to the shift, and translation proceeds into the intron region between exons 2 and 3. This abnormal form of prepro-vasopressin apparently cannot be processed to yield any vasopressin in neurohypophysial tissues. The vasopressin gene is also expressed in testicular tissue, but the mRNA transcripts differ from those in brain tissue (Foo et al., 1991; Lefebvre and Zingg, 1991). This testis-specific transcription may explain why Brattleboro rats possess the ability to synthesize vasopressin in peripheral tissues but lack that ability in the neurohypophysis.

The cleavage, amidation, and noncovalent binding of the nonapeptides occurs during the transport of the dense core secretory vesicles to the axon terminals located in the pars nervosa (Figure 3-2). Although the diagram shows a single terminal, the axon leaving each magnocellular neuron may give rise to as many as 2000 terminals from which the contents of the secretory vesicles can be released by exocytosis. There may also be dilated segments of the branched axons proximal to their terminals. These axonal dilations appear to function for storage and/or lysosomal degradation of the contents of secretory vesicles. Their discovery preceded biochemical knowledge of the system and was based upon their unique cytochemical properties. These structures were given the name "Herring bodies" at that time.

Nomenclature of the neurophysin molecules can be very confusing due to lack of uniformity among different authors. In this book, the molecule synthesized from the vasopressin gene is termed neurophysin-A while neurophysin-B is the one synthesized from the oxytocin gene. Some authors will use the terms

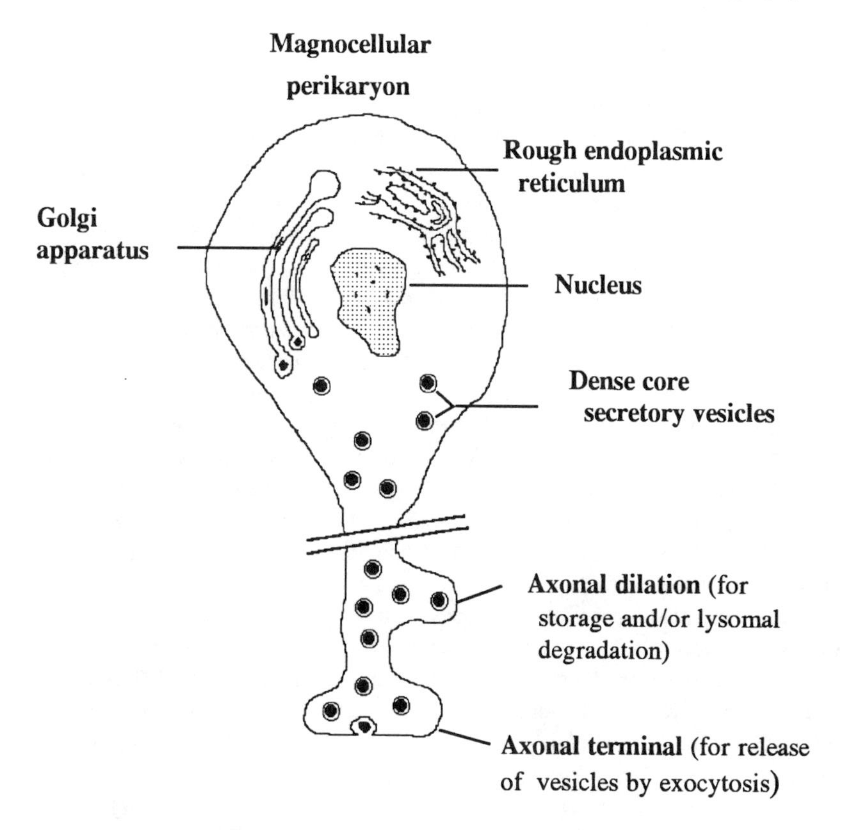

Figure 3-2. Magnocellular neuron.

This greatly simplified diagram illustrates the cellular elements of magnocellular neurons which are involved in the synthesis, packaging, intracellular translocation, and exocytotic release of hormones.

neurophysin-I and neurophysin-II, which are based upon electrophoretic migration of the purified molecules. In that nomenclature, neurophysin-I migrates further toward the anode than neurophysin-II. In most species, neurophysin-I is synthesized with vasopressin and neurophysin-II is synthesized with oxytocin. In contrast, neurophysin-I of sheep and cattle is associated with oxytocin. Recent evidence indicates considerable homology (93%) between sheep and cattle in the mRNA encoding for prepro-oxytocin, whereas human and rat prepro-oxytocin showed lower homology (66-74%) with the sheep sequence (Ivell et al., 1990). It may be noteworthy that ovarian expression of the oxytocin gene occurs to a much greater extent in sheep and cows than in other species. Another nomenclature for neurophysin is unique to humans. The term ***nicotine-stimulated*** is given to the neurophysin associated with vasopressin and ***estrogen-stimulated*** is given to the neurophysin associated with oxytocin. Assigning names based upon physiological stimuli provoking the selective release of each nonapeptide is no longer used, but this practice emphasizes that each neurophysin is released into blood along with the nonapeptide to which it is bound inside the secretory vesicle. This non-covalent bond breaks readily after exocytosis due to both dilution and elevation of the pH. Free circulating neurophysin can be measured immunologically in human blood, and this is the basis for the initial assignment of names to the neurophysin molecules.

The neuronal perikarya that synthesize oxytocin and vasopressin destined for transport to the pars nervosa are known as ***magnocellular*** neurons because of their large size (i.e., 25-30 μm diameter). Almost all of these magnocellular neurons are located in either the supraoptic nucleus or paraventricular nucleus, and the concentration of large readily stained perikarya make these nuclei histologically unique. In addition to these magnocellular neurons which release the nonapeptides into pars nervosa blood, there is another group of smaller neurons (10-15 μm diameter) called ***parvocellular*** neurons that synthesize vasopressin, project their axons to the median eminence, and release vasopressin into hypophysial portal blood. The perikarya of parvocellular neurons releasing vasopressin into blood are located in the paraventricular nucleus intermixed with magnocellular neurons that project to the pars nervosa. These parvocellular vasopressinergic neurosecretory neurons will be discussed more in Chapter 4 in relation to regulation of corticotropin release from the adenohypophysis. Other parvocellular neurons are located within the hypothalamus and synthesize either vasopressin or oxytocin. Probably they do not release the nonapeptides into blood, but they have axons that project to various CNS areas where oxytocin or vasopressin are released to modulate the firing of other neurons.

Each magnocellular neuron receives synaptic input on its perikaryon and dendrites from approximately 1000 other neurons by way of about 5000 synapses per magnocellular neuron. About 60% of these synapses are associated with axons that originate inside the individual nucleus containing the magnocellular perikaryon. Control of magnocellular neurons by neurotransmitters from the large number of different synaptic inputs is only partly understood (Crowley and Armstrong, 1992). Acetylcholine as well as glutamate and other excitatory amino acids are stimulatory to magnocellular neurons. Exogenous angiotensin II is also consistently stimulatory to oxytocin release, but the physiological role of endogenous angiotensin II is unclear. Norepinephrine appears to have dual effects on oxytocin perikarya with stimulation caused by α-adrenergic receptors and inhibition caused by β-adrenergic receptors. There is also evidence that oxytocin, acting as a neurotransmitter, can modulate the firing of oxytocin-containing magnocellular neurons. Although vasopressinergic magnocellular neurons are profoundly affected by osmotic stimuli, the vasopressin neuron is probably not especially osmosensitive. Rather, parvocellular neurons, located nearby or in distant sites, are osmosensitive and communicate with the magnocellular neuron using acetylcholine. Structures of the limbic system and the brain stem also communicate with magnocellular vasopressinergic neurons.

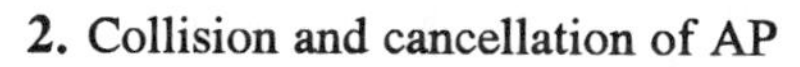

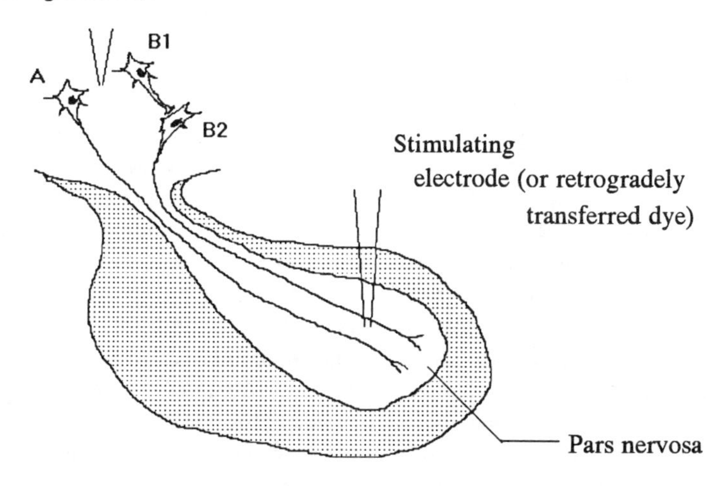

Figure 3-3. Antidromic identification.
This diagram illustrates the location of electrodes (stimulating and recording) and also lists the criteria for antidromic identification of those neurons which send their axons all the way to the pars nervosa (identified as A and B2 in this diagram). See text for the details of this procedure.

Electrophysiology of the Neurohypophysis

Current knowledge regarding the electrical characteristics of magnocellular neurons depends greatly upon recording from hypothalamic perikarya that have been antidromically identified as sending their axons to the pars nervosa (Figure 3-3). Electrical stimuli are applied to the pars nervosa as illustrated while simultaneously recording action potentials (AP) from neurons in the hypothalamus (either A or B1 in Figure 3-3). The stimulation-induced AP in the pars nervosa will ascend into the hypothalamus, and if neuron A typifies the one being recorded, the induced AP will be recorded with very short latency. If neuron B1 typifies the one being recorded, a short-latency induced AP will not be recorded in the hypothalamus because AP cannot cross synapses in the reverse direction. A second method for antidromic identification is to give simultaneous electrical stimuli in the hypothalamus and pars nervosa. If the same axon is activated at opposite ends, the induced AP will collide and cancel each other so that the AP induced in the pars nervosa will not be recorded in the hypothalamus. Satisfying these two criteria is considered sufficient to prove that the perikaryon being recorded in the hypothalamus projects, without an intervening synapse, to the pars nervosa (neuron A rather than B1 in Figure 3-3). Antidromic identification usually precedes the recording of electrophysiological data regarding the properties of the magnocellular neuron. After completion of data collection, dye is sometimes injected into axons of the pars nervosa so that it will be retrogradely transported to the magnocellular perikarya. Because the dye does not cross chemical synapses, this staining confirms that the dye-containing neurons projected without synapse to the pars nervosa.

One particular experimental model contributed greatly to our understanding of neurohypophysial electrophysiology. It was discovered by Lincoln and Wakerley (1975) that lactating rats, which are anesthetized and exposed continuously to suckling by hungry pups, release oxytocin only in periodic bursts. After about 15 min of suckling and at periodic intervals of 10-15 min thereafter, a bolus of oxytocin is released causing milk ejection and agitation among all the suckling pups. The milk ejection can be detected by measurement of intramammary pressure in a non-suckled mammary gland or alternatively by observing the simultaneous agitation of the suckling pups. Timing of the periodic discharges of oxytocin was not precisely regular and also seemed to depend upon the anesthetic used in the lactating rat. It was discovered later that oxytocin discharges only occurred when the suckled rat entered into a specific stage of sleep (i.e., slow-wave sleep). Entry into this sleep stage was not precisely regular, and it depended on the anesthetic used. It was also discovered later that oxytocin discharge in pigs and rabbits did not depend on entry into slow-wave sleep.

The consequences of knowing exactly when a bolus of oxytocin was being released allowed much new knowledge to be gained. Approximately 18 sec before each oxytocin discharge, some antidromically identified neurons increased their firing rate from 1-5 impulses/sec to 40-80 impulses/sec. This high-frequency burst of firing lasted only 1-3 sec and was followed by a few seconds in which very few spikes occurred. It was subsequently proven that those neurons that displayed these periodic high-frequency bursts just before milk ejection were in fact oxytocinergic magnocellular neurons. An example of the pattern of impulses from one such oxytocin neuron is illustrated in the upper left circle of Figure 3-4. The firing pattern of other antidromically identified magnocellular neurons was unrelated to the periodic milk ejections (lower left circle of Figure 3-4), and these were identified as vasopressin neurons. When either chronic or acute hyperosmotic stimuli were applied, a portion of the magnocellular neurons displayed a pattern of ***phasic*** firing (illustrated in lower right circle of Figure 3-4). This pattern consisted of recurring phases of firing at 6-15 spikes/sec, with each phase lasting 20-60 sec and followed by a silent period of similar length. Antidromically identified neurons displaying this phasic pattern of firing during osmotic stimulation are vasopressin-secreting neurons. In contrast to the firing of oxytocin neurons during suckling, there is no group synchrony of the phasic bursts of vasopressin neurons. It appears that the phasic firing of vasopressin neurons is an intrinsic property of individual neurons and is not controlled by other neurons as occurs in oxytocin neurons

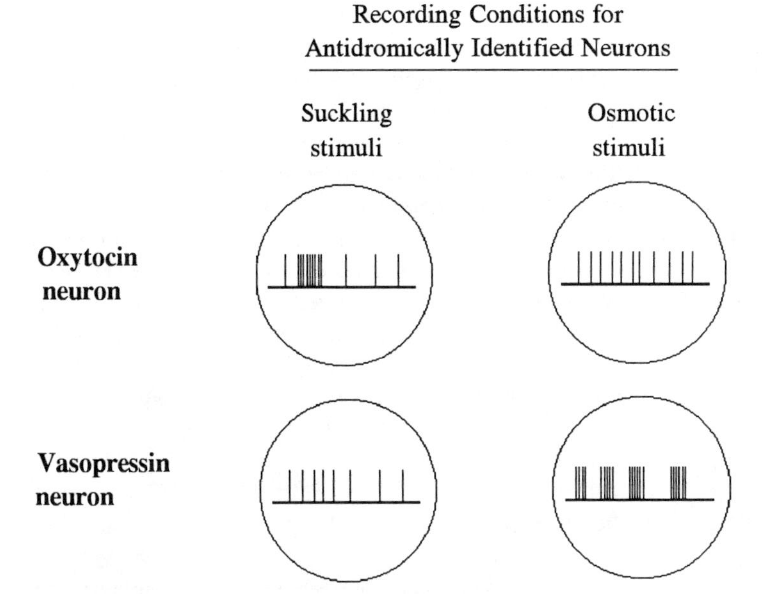

Figure 3-4. Firing patterns of magnocellular neurons in response to suckling or osmotic stimulation.

This figure contains representative firing patterns for antidromically identified oxytocin or vasopressin neurons during application of suckling or osmotic stimuli. High-frequency bursts of AP during suckling stimulation are observed only in oxytocin neurons whereas recurring middle-frequency bursts of AP during osmotic stimulation are observed only in vasopressin neurons.

(Leng and Bicknell, 1986). In contrast, almost all of the estimated 9000 magnocellular oxytocin neurons in the rat hypothalamus produce their firing burst at the same time regardless of whether they are located close together or far apart in distant nuclei, even in the contralateral side of the brain. Mechanisms responsible for this synchrony are not fully understood, but release of oxytocin at synaptic terminals may be involved because intrahypothalamic application of exogenous oxytocin can trigger the bursts of impulses in oxytocin neurons.

The paraventricular nuclei contain a mixture of magnocellular oxytocin and vasopressin neurons as well as parvocellular neurons containing vasopressin and a variety of other neuropeptides. The supraoptic nuclei consist almost exclusively of magnocellular neurons containing either vasopressin or oxytocin and their associated precursors. Some magnocellular neurons are also present outside of the paraventricular and supraoptic nuclei in what are sometimes called accessory neurosecretory cell groups. These vasopressin and oxytocin neurons are located both rostral and ventral to the paraventricular nuclei (Crowley and

Armstrong, 1992). It is thought that the magnocellular neurons in the accessory cell groups function the same as the majority of magnocellular neurons located in the paraventricular and supraoptic nuclei.

It has been generally accepted that an individual magnocellular neuron synthesizes either prepro-vasopressin or prepro-oxytocin but not both. However, *in situ* hybridization for the respective vasopressinergic and oxytocinergic mRNA in the supraoptic nuclei of lactating rats showed that 9-17% of the perikarya contained both forms of mRNA. This unexpected colocalization of vasopressin and oxytocin in a single neuron even extended to secretory vesicles where 21-24% of them contained both peptides by immunocytochemical analysis (Mezey and Kiss, 1991). These very unexpected results need to be confirmed before an established concept is modified, but they certainly raise questions regarding the idea that the vasopressin and oxytocin system are completely independent.

Neurohormone Release

As stated earlier, it takes several hours before newly synthesized vasopressin or oxytocin is ready for release into blood by the axon terminals. However, the process of neurohormone release must be instantly responsive to various stimuli. The ability of this neuroendocrine system to respond very rapidly and convert transient neural signals into neurohormone release for prolonged blood-borne action (i.e., ***neuroendocrine transduction***) is an essential element of the neurohypophysial system.

Events occurring at each axon terminal in the pars nervosa are illustrated in Figure 3-5, and these events begin with the arrival of a high-frequency burst of AP. As stated earlier, all oxytocinergic terminals in the pars nervosa simultaneously receive a high frequency (40-80 impulses/sec) burst of arriving AP lasting 1-3 sec. Vasopressinergic terminals of individual neurons under hyperosmotic conditions receive an elevated frequency of AP (6-15 impulses/sec) lasting between 20 and 60 sec. The amount of neurohormone released from the terminal per arriving AP is much greater at higher frequencies due to a phenomenon called ***frequency facilitation***. For example, a high frequency of AP can produce a 100- to 1000-fold increase in the quantity of oxytocin release per AP. *In vitro* studies of isolated pars nervosa tissue indicate that the high-frequency AP rates observed *in vivo* for each neurohormone are optimal for inducing release of that neurohormone. Moreover, release from each type of terminal requires a recovery period containing relatively few AP before optimum responses can be induced again.

Arrival of the AP burst depolarizes the terminal which allows intracellular entry of extracellular sodium ions followed by a similar entry of calcium ions (Figure 3-5). Intracellular calcium promotes the fusion of the dense core secretory vesicles to the inner surface of the plasma membrane. The area of fused membranes breaks down in a process called ***exocytosis*** to allow the contents of the dense core secretory vesicle to enter the extracellular space and ultimately the blood capillaries. Following exocytosis and somehow linked to it, there occurs a recapture of vesicle membrane (also called vesiculation). Although the ultrastructure of this recapture process can be readily observed, biochemical details of the process are not known. It is believed that the variable sized vesicles formed during recapture may ultimately be degraded in lysosomes present in the terminals. The final event illustrated in Figure 3-5 is the use of mitochondrial-generated ATP to repolarize the plasma membrane by pumping intracellular sodium to the outside of the cell in exchange for potassium. There may also be a role of cyclic AMP in the calcium-induced fusion of secretory vesicle and

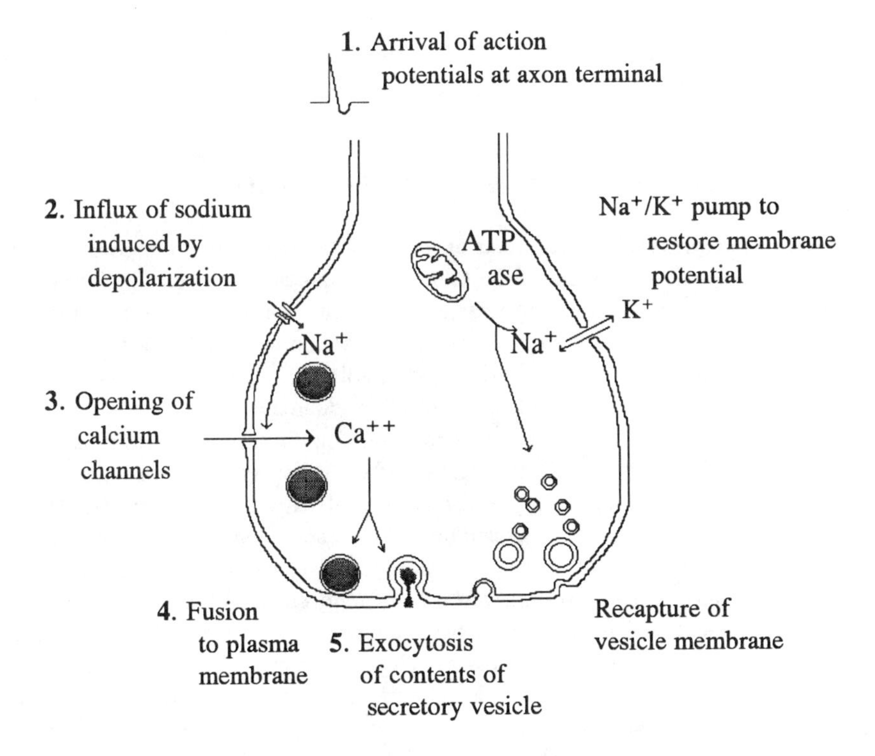

Figure 3-5. Release of neurohormone.

This diagram of an axon terminal illustrates morphological, biochemical, and membrane events associated with exocytosis of neuro-hormone. The temporal sequence of events starts in the upper left of the diagram with the arrival of a high-frequency burst of AP, progresses diagramatically left to right around the terminal, and ends in the upper right and lower right of the diagram with restoration of the membrane potential to normal and recapture of vesicle membrane, respectively.

plasma membrane, but this possibility is not illustrated in Figure 3-5. Disposal of excess intracellular calcium is also not illustrated, but this may occur by either incorporation into recaptured vesicles or direct extracellular transfer of calcium.

Although the arrival of high frequency AP in the pars nervosa is the primary regulator of neurohormone release, there may also be other factors that influence release from axon terminals. Opioid peptides such as dynorphin and enkephalin are released from terminals in the pars nervosa, and these neuropeptides are inhibitory to the release of oxytocin and vasopressin. The mechanisms by which opioid peptides could inhibit release from the terminals of magnocellular neurons are not fully understood, and the role of the pituicyte is especially unclear because these neuroglial cells have receptors for opioid peptides (Bicknell et al., 1989). Pituicytes surround the axon terminals in the pars nervosa, and during periods of increased release of vasopressin and/or oxytocin, the morphology of pituicytes changes to allow more contact of terminal membranes with the perivascular space (Hatton et al., 1988).

Endocrinology of Vasopressin and Oxytocin

Blood-borne vasopressin reduces diuresis by increasing reabsorption of water in the collecting ducts of the kidney, and this action represents its major endocrine action. The molecule was named vasopressin as well as ***antidiuretic hormone*** because of its initial pharmacological action to increase blood pressure by constricting arterioles. The primary physiological stimulus of vasopressin release is plasma osmolality, which is closely regulated by (1) retention of water by the kidneys and (2) thirst mechanisms promoting the intake of water. Osmosensitive neurons, distinct from magnocellular vasopressinergic neurons, continuously monitor the osmolality of the extracellular fluid. Whenever osmolality increases as little as 1%, vasopressin neurons are activated, probably by acetylcholine neurotransmission, to generate more impulses and discharge more vasopressin into blood. The resulting vasopressin-induced retention of water by the kidneys promotes the restoration of osmolality back to normal. Secondary and supplemental control over vasopressin release is provided by mechanisms to maintain a suitable blood volume. Monitoring mechanisms in the left atrium of the heart detect hypovolemia (i.e., low blood volume) and similar monitoring mechanisms in the aortic arch and carotid bodies detect hypotension (i.e., low blood pressure). Detection of either hypovolemia or hypotension decreases the inhibitory norepinephrine input to the vasopressin neurons, thereby allowing more vasopressin to be released to increase renal retention of water. These blood volume monitoring mechanisms require a much larger proportional change to affect vasopressin release than do the osmolality monitoring mechanisms, and they are useful mainly in emergency situations. Other short-term stimuli to release of vasopressin include pain, emotional stress, muscular exercise, and, in humans, nicotine administration.

Blood-borne oxytocin causes milk-ejection by stimulating the contraction of myoepithelial cells of the mammary alveoli. It also stimulates contractions of the smooth muscle in the myometrium of the uterus, and the name oxytocin (i.e., ***quick birth***) derives from this action in periparturient females. In males, oxytocin administration stimulates contractions of the smooth muscle lining the ducts of the genital tract, but the physiological relevance of this action is unclear. Sensory input from the nipple of the mammary gland constitutes a major physiological stimulus of oxytocin release into blood. However, psychological factors in anticipation of this sensory stimulation can also induce oxytocin release. Both of these mechanisms facilitate oxytocin-induced milk-ejection in suckled females, and this process is physiologically important for survival of the offspring.

Sensory stimuli from the genital tract of pregnant females experiencing parturition also stimulate the release of oxytocin into the blood and into the brain. Although it is not clear whether increased release of oxytocin into blood contributes to the initial stages of parturition, neurohypophysial oxytocin released in response to distention of uterine cervix and vagina (so-called ***Ferguson reflex***) may contribute to the completion of the birth process. From a therapeutic point of view, exogenous oxytocin facilitates the initiation of parturition, but the role of endogenous neurohypophysial oxytocin in the onset of spontaneous parturition remains to be established. However, the uterine endometrium near the end of pregnancy in rats may synthesize oxytocin to supplement the actions of blood-borne oxytocin on the myometrium (Lefebvre et al., 1992). The local release of oxytocin that occurs in the brain during the Ferguson reflex facilitates onset of maternal behavior of sheep toward their newborn offspring (see Chapter 14). To summarize the actions of oxytocin, circulating concentrations of the neurohormone released from the neurohypophysis in response to sensory input from the mammary gland, as well as the female genital tract, contribute to evacuation of the contents of both organs. Intracerebral release of oxytocin may also promote the appropriate and timely development of maternal behavior toward the offspring.

REFERENCES

Bicknell, R.J., S.M. Luckman, K. Inenaga, W.T. Mason, and G.I. Hatton. β-Adrenergic and opioid receptors on pituicytes cultured from adult rat neurohypophysis: regulation of cell morphology. *Brain Res. Bull.* (1989) 22:379-388.

Crowley, W.R., and W.E. Armstrong. Neurochemical regulation of oxytocin secretion in lactation. *Endocr. Rev.* (1992) 13:33-65.

Du Vigneaud, V. Hormones of the posterior pituitary gland: oxytocin and vasopressin. *Harvey Lect.* (1954-55) 50:1-26.

Foo, N.C., D. Carter, D. Murphy, and R. Ivell. Vasopressin and oxytocin gene expression in rat testis. *Endocrinology* (1991) 128:2118-2128.

Hatton, G.I., Q.Z. Yang, and K.G. Smithson. Synaptic inputs and electrical coupling among magnocellular neuroendocrine cells. *Brain Res. Bull.* (1988) 20:751-755.

Ivell, R., and D. Richter. The gene for hypothalamic peptide hormone oxytocin is highly expressed in the bovine corpus luteum: biosynthesis, structure and sequence analysis. *EMBO J.* (1984) 3:2351-2354.

Ivell, R., N. Hunt, N. Abend, B. Brackman, D. Nollmeyer, J.C. Lamsa, and J.A. McCracken. Structure and ovarian expression of the oxytocin gene in sheep. *Reprod. Fert. Devel.* (1990) 2:703-711.

Ivell, R. The structure and function of peptide hormone genes. In: Geldermann, H. and F. Ellendorff (eds). *Genome Analysis in Domestic Animals.* (VCH Verlagsgesellschaft, Weinheim, Germany, 1990), pg 185-201.

Lefebvre, D.L., A. Giaid, H. Bennett, R. Lariviere, and H.H. Zingg. Oxytocin gene expression in rat uterus. *Science* (1992) 256:1553-1555.

Lefebvre, D.L., and H.H. Zingg. Novel vasopressin gene-related transcripts in rat testis. *Molec. Endocrinol.* (1991) 5:645-652.

Leng, G., and R.J. Bicknell. The neurohypophysis. In: Lightman, S.L., and B. J. Everitt (eds). *Neuroendocrinology.* (Blackwell Scientific Publ., Oxford, UK, 1986) pg 177-196.

Lincoln, D.W., and J.B. Wakerley. Factors governing the periodic activation of supraoptic and paraventricular neurosecretory cells during suckling in the rat. *J. Physiol.* (1975) 250:443-461.

Mezey, C., and J.Z. Kiss. Coexpression of vasopressin and oxytocin in hypothalamic supraoptic neurons of lactating rats. *Endocrinology* (1991) 129:1814-1820.

Nicholson, H.D., A.J. Smith, S.D. Birkett, P.A. Denning-Kendall, and B.T. Pickering. Two vasopressin-like peptides in the pig testis? *J. Endocrinol.* (1988) 117:441-446.

Chapter 4

NEURAL CONTROL OF ADENOHYPOPHYSIS

There is no doubt today that the brain controls the release of hormones from the adenohypophysis. Moreover, synthetic analogues of the hypothalamic neurohormones that exercise this control over the adenohypophysis are readily available. These neurohormones, most of which are called ***releasing factors*** or ***releasing hormones***, stimulate the release of the appropriate pituitary hormone when injected into animals and man, and they also increase hormone release when applied directly to adenohypophysial tissue *in vitro*. In addition, antibodies can be raised against those releasing factors (RF) which are peptides, and endogenous blood-borne levels of a single RF can be immunoneutralized *in vivo*. The resulting deficits in circulating levels of the adenohypophysial hormone or in biological processes requiring that hormone provide conclusive evidence that blood-borne levels of that particular hypothalamic RF control the release of the adenohypophysial hormone.

It is sometimes difficult for today's student to appreciate the development of knowledge concerning the neural control of adenohypophysial secretions. The ease with which chemically pure RFs are obtained for administration often leads to the assumption that the injected material simulates completely the endogenous blood-borne CNS factors which cause release of the adenohypophysial hormone. How can we assume that the effects of synthetic RF preparations simulate physiology rather than represent pharmacology? Although much of this assumption is based upon the elegant chemical identification as well as the *in vivo* immunoneutralization research mentioned in the preceding paragraph, there is also a body of evidence that is independent of the chemical isolation and identification of hypothalamic neurohormones. This evidence led to and subsequently supported the concept originated by Harris (1948) that there is neurovascular control of adenohypophysial secretions.

As stated in Chapter 2, the long portal vessels linking the median eminence with the pars anterior of the adenohypophysis are always present in diverse vertebrate species, which is indicative of a functional importance. Direct neural control of the adenohypophysis via secretomotor innervation has been considered, but it was rejected for all the adenohypophysis except the pars intermedia.

In addition to these morphological observations, experimental manipulation of the hypothalamus-hypophysis relationship provided important physiological evidence. One of the earliest studies was conducted in female rabbits that only ovulate after copulation. However, electrical stimulation of the hypothalamus of unmated rabbits induced ovulation, which at that time was known to depend on the secretion of pituitary gonadotropins. Direct electrical stimulation of the adenohypophysis failed to induce ovulation suggesting that secretomotor innervation was not involved. The paradigm of electrical stimulation of brain tissue to provoke release of adenohypophysial hormones has been widely used in many species, especially after the advent of radioimmunoassays to quantify circulating hormones. Although differences exist regarding the most effective brain loci to modulate release of each hormone, the ability of localized electrical stimulation in the brain to stimulate or inhibit release of adenohypophysial hormones is well established.

The antithesis to electrical stimulation of the hypothalamus is to surgically eliminate the influence of the hypothalamus over the adenohypophysis. The earliest studies were performed in rats in which the hypophysial stalk was surgically transected. The resulting deficits in gonadal function were variable and inconclusive, but it was later shown that the variability depended on whether the hypophysial portal vessels regenerated spontaneously during the period following transection (Jacobsohn, 1966). When this regeneration was prevented by placement of a barrier at the time of stalk transection, gonadal deficits were severe and consistent. These results were especially important because they cleared up earlier inconsistencies and also emphasized portal blood vessels as the functional link between brain and adenohypophysis.

After it became possible to quantify adenohypophysial hormones in circulating blood, surgical transection of the hypophysial stalk was performed in many species. With the exception of prolactin (PRL), the release of hormones of the pars anterior (or pars distalis) is usually deficient after stalk transection. There may be basal secretion of some hormones into blood as evidenced by low but steady concentrations. However, stalk-transected animals never release large-amplitude pulsatile discharges of the hormones that occur in sham-operated controls. The effect of stalk transection on pituitary PRL is clearly different from that for other hormones. Figure 4-1 illustrates one such study from rats in which

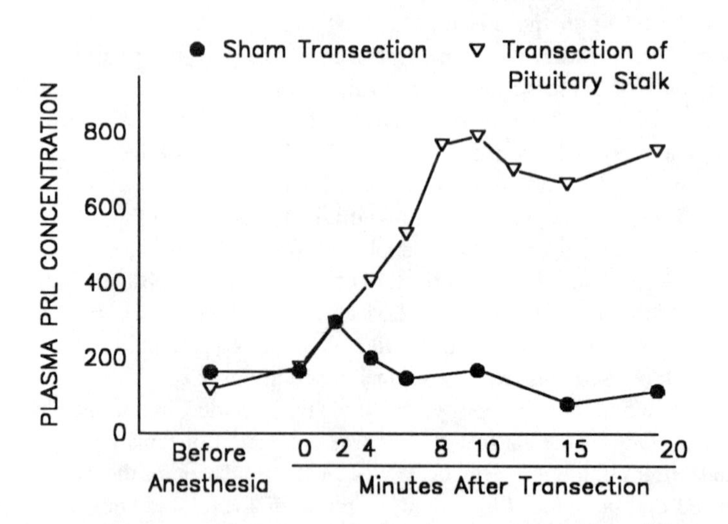

Figure 4-1. Prolactin increase after transection of pituitary stalk.
Redrawn data of Kanematsu et al. (1979). Plasma PRL (ng/ml) after surgical transection of the hypophysial stalk in female rats.

blood was drawn immediately after transection of the hypophysial stalk. Plasma concentrations of PRL began to rise above the control group as early as 4 min after transection, and plasma PRL reached as high as it would ultimately go in only 8 min (Kanematsu et al., 1979).

Experiments similar to that in Figure 4-1 have also been performed in domestic animals, but transection-induced increases in PRL release have not been as dramatic. In sheep, surgical disconnection of the hypothalamus and hypophysis by removing part of the median eminence allowed PRL secretion to increase dramatically over that in sham-operated controls, and a small part of this increase was sustained for several weeks (Thomas et al., 1986). In pigs and cattle, simple transection of the hypophysial stalk resulted in a small but consistent increase in circulating PRL which took several hours to develop and was maintained thereafter (Anderson et al., 1982; Benoit et al., 1989). These differences between rodents and domestic ungulates in the magnitude of PRL increase following surgical transection of the hypophysial stalk may be related to the relative importance, for each species, of hypothalamic neurohormones that stimulate the release of PRL and that would also be eliminated by transection. Accordingly, hypothalamic releasing factors that stimulate release of PRL may be relatively less important in rodents than in the domestic ungulates.

An alternative to transecting the hypophysial stalk is to surgically remove the hypophysis from its *in situ* location and transplant it to a distant site. Whenever tissue is transplanted to a site other than its usual location, it is called ***ectopic*** tissue. Rodents in which ectopic pituitary tissue was created, using either their own or donor tissue, were used widely for early neuroendocrine research. The ectopic site was usually the anterior chamber of the eye or beneath the capsule of the kidney. Although much of the transplanted tissue died, many cells received enough vascular supply to survive. Using biological and morphological criteria, the only cells in ectopic pituitaries that seem to maintain secretory activity are PRL-secreting mammotrophs. Using ultrastructural criteria, gonadotrophs and thyrotrophs can be identified in ectopic pituitary tissue, but they do not appear to have secretory activity. Although it is obvious from a modern perspective that the secretory deficits of ectopic pituitaries derive from a lack of hypothalamic neurohormones, physiological validation of this interpretation was enhanced greatly by the work of Nikitovitch-Winer and Everett (1958). These authors took ectopic pituitary tissue from the kidney capsule and transplanted it beneath the median eminence. Upon revascularization with hypophysial portal vessels, the twice-transplanted tissue resumed its gonadotrophic, thyrotrophic, and adrenocorticotrophic secretions. This result proved that the deficits in ectopic pituitary tissue were not permanent and could be restored by proximity to the hypothalamus.

Regardless of whether it is ectopic or stalk-sectioned, hypophysial tissue separated from its usual source of hypophysial portal blood appears unable to secrete normal quantities of most pars anterior hormones. However, some basal secretion of these hormones can be detected, especially when the volume of ectopic tissue is experimentally increased. In evaluating the results of such studies, it is important to remember that the secreted hormones may have been stimulated by hypothalamic RF that reached the hypophysial tissue via the general circulation. Alternatively, the detected hormones may have been secreted not by the isolated hypophysial tissue but by the pars tuberalis tissue that remained adhered to the median eminence. Only if both of these possibilities are excluded can one conclude that hormone secretion by the isolated hypophysial tissue is autonomous. Therefore, a better model in which to study autonomous secretion by pars anterior tissue involves the *in vitro* incubation or culture of pituitary tissue or cells. If synthesis of adenohypophysial hormones from radioactive precursors can be demonstrated *in vitro* without the addition of any stimulatory factors, the secretion can be viewed as autonomous. This criterion has been satisfied many times for secretion of PRL, but not for other hormones of the pars anterior. In short-term incubations, there may be release of these other hormones into the medium, but those released molecules were probably synthesized *in vivo* under the influence of hypothalamic neurohormones.

Another experimental approach helped establish that the hypothalamus regulates the adenohypophysis. It involves the creation of destructive lesions within the hypothalamus that result in deficits of hypophysial secretion. The early research used biological endpoints reflective of hypophysial secretion, but development of radioimmunoassay quantification has allowed specific adenohypophysial peptides to be measured in blood after lesions of the hypothalamus. In general, these results confirm the concepts derived from the studies described above. Destructive lesions within the hypothalamus create deficits in the secretion of most adenohypophysial hormones except for PRL. A special form of hypothalamic lesion that has yielded much valuable information is known as ***hypothalamic deafferentation***. In this procedure a special neurosurgical knife is lowered into the mediobasal hypothalamus and used to sever the connections of the mediobasal hypothalamus from rostral, dorsal, and caudal structures. Because only the ventral connections with the median eminence and hypophysial stalk remain intact, the model is sometimes called a hypothalamic island. Experimental variations in the location and completeness of the deafferentation can provide important information about specific neural inputs necessary for certain adenohypohysial secretions.

Valuable Experimental Techniques

Current knowledge about the neural regulation of the adenohypophysis depends greatly on an understanding of many experimental techniques. Knowledge of how to perform the techniques is not necessary, but it is important to know what inferences can be drawn from the results obtained with each technique. Therefore, a brief discussion of several techniques is presented here for those procedures not already covered in earlier chapters.

The first technique to be discussed is the Golgi-Cox staining of neuronal projections with silver ions. This technique allows microscopic visualization of individual neuronal perikarya and all their projections including terminal arborizations. Although very few neurons take up the silver ions applied to the tissue, those neurons that do acquire silver transport the ions throughout all the processes of that neuron. The resulting preparation consists of a very small proportion of neurons being stained in their entirety so that their connections with other stained neurons can be visualized. Application of Golgi-Cox silver staining to those areas of the hypothalamus that secrete neurohormones into the portal vessels has provided important cytoarchitectural information about neural regulation of the adenohypophysis.

Electron microscopy (EM) has also been very valuable for investigating this subject. Transmission EM has been used to determine the types and the diameters of secretory vesicles in parvocellular neuronal elements that influence the adenohypophysis. Scanning EM, which visualizes the morphological features of surfaces, has been useful for examining the inner ependymal lining of the third ventricle especially in the ventral region near the median eminence.

Electrophysiological techniques such as antidromic identification (Figure 3-3) and post-recording labeling of a neuron with dye have provided information on those parvocellular neurons that project to the median eminence. It is assumed that such neurons are neurosecretory neurons whose products influence the adenohypophysis. The peptidergic contents of these neurons can also be visualized by immunocytochemical staining such as described in Chapter 2 for hypophysial cells. Another technique for determining which neurons project their axons into contact with capillaries involves either the systemic or localized median eminence infusion of compounds (e.g., horseradish peroxidase or fluorogold) which are taken up by such axonal terminals. The compound taken up into the terminal is then transported retrogradely to the perikaryon where it can be visualized by appropriate techniques (Lechan et al., 1980; Witkin, 1990). In this way the precise location of perikarya of neurosecretory neurons can be ascertained. This technique can also be combined with immunocytochemical staining for specific peptide neurohormones.

Antibodies can be produced and used to stain (e.g., by immunocytochemistry) or quantify (e.g., by immunoassay) specific peptide neurohormones. *In situ* hybridization, as discussed in Chapters 2 and 3, can also assist in the localization of neurons that transcribe the gene for a particular neurohormone. The quantity of messenger RNA (mRNA) transcribed by a neurohormone gene can also be quantified by appropriate techniques. In general, these various peptidergic techniques can be used to generate the following types of information: (1) location of perikarya and axonal terminals, (2) close apposition and/or synaptic contact between two neurons each with different peptides, (3) colocalization of more than one peptide within a single neuron, and (4) quantification of peptide content or specific mRNA abundance in dissected neural tissue.

Besides hypothalamic peptides, a group of compounds known as ***monoamines*** participate directly or indirectly in the hypothalamic regulation of the adenohypophysis. Figure 4-2 shows the chemical structures of the three most important monoamines that regulate, directly or indirectly, the adenohypophysis. The structures of dopamine and norepinephrine are very similar, differing only by one hydroxyl group. Both compounds are known as catecholamines

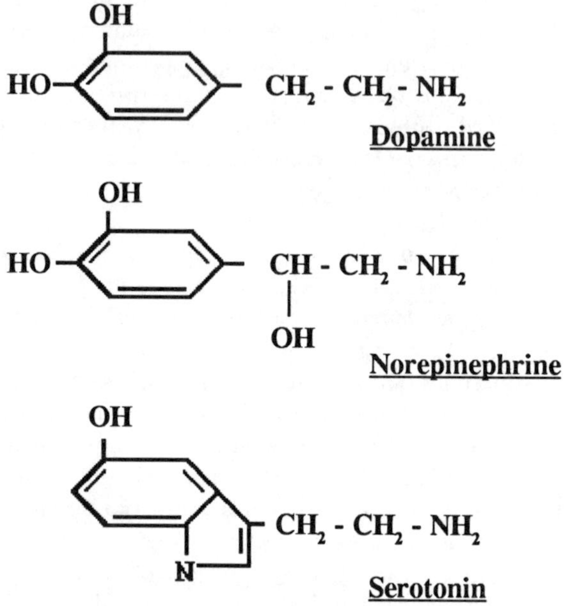

OH
HO— $CH_2 - CH_2 - NH_2$
Dopamine

OH
HO— $CH - CH_2 - NH_2$
OH
Norepinephrine

OH
$CH_2 - CH_2 - NH_2$
N
Serotonin

Figure 4-2. Monoamine compounds.
Chemical structures of monoamines which influence directly or indirectly the secretion of adenohypophysial hormones.

because of the six-membered catechol ring structure. Serotonin is known as an indolamine because of its unique double-ring structure. All three monoamines in Figure 4-2 share the single amine group from which the monoamine terminology is derived. Techniques to stain for monoamines actually preceded the immunocytochemical staining of peptides because the chemical nature of the hypothalamic RF had not yet been discovered. This staining was based on histochemical fluorescence of the catecholamines and indolamines after specific chemical treatments. These monoamines were also quantified in dissected tissues using chemical assays.

The final technique to be mentioned examines specific binding sites (i.e., receptors) for peripheral hormones or neuropeptides within the hypothalamus. Knowledge gained from such studies may improve our understanding of neuroendocrine integration (Figure 1-2). The precise location of such binding sites on neurons or neuroglia can be visualized by *in vitro* autoradiography. This technique involves the application of the binding ligand to tissue sections followed by an appropriate method to visualize the bound compound. Other aspects of ligand binding can be quantified from homogenates of dissected tissues. This approach permits one to quantify the affinity of binding as well as the number of binding sites per amount of dissected tissue or protein. Treatments that might alter either of these binding parameters can also be readily investigated using this technique. Immunocytochemical staining can also be combined with *in vitro* autoradiography in order to identify the peptide content of neurons that have specific binding sites for a particular ligand.

Functional Morphology of Median Eminence

The median eminence is that neural tissue located beneath the ventral floor of the third ventricle (3V). Alternative terms for this structure are infundibulum and tuber cinereum. Because the median eminence is continuous with the hypophysial stalk, the term stalk-median eminence (SME) is used by some authors. As indicated in Chapter 2, the primary capillaries of the portal vessels are located in the median eminence. Early descriptions of the morphology recognized an inner layer (or zone) and an outer layer. The inner layer contains axons of passage that continue down the stalk to the pars nervosa, whereas the outer layer contains axon terminals and primary portal capillaries. Adherent to the outer layer is the pars tuberalis tissue of the adenohypophysis (described in Chapter 2).

Morphological features of the median eminence are illustrated in Figure 4-3. Axonal input is represented as those which terminate in the median eminence and those which pass through the tissue. These later axons are mostly magnocellular axons, but there are also parvocellular axons of the tuberohypophysial tract (not shown). Figure 4-3 depicts four types of axonal terminations in the median eminence (numbered 1 through 4). Type #1 represents axo-axonal terminals in which one axon may synaptically influence another axon. Type #2 represents the neurosecretory terminal that discharges its neurohormones into a primary portal capillary. Type #3 typifies the axon terminals in contact with non-

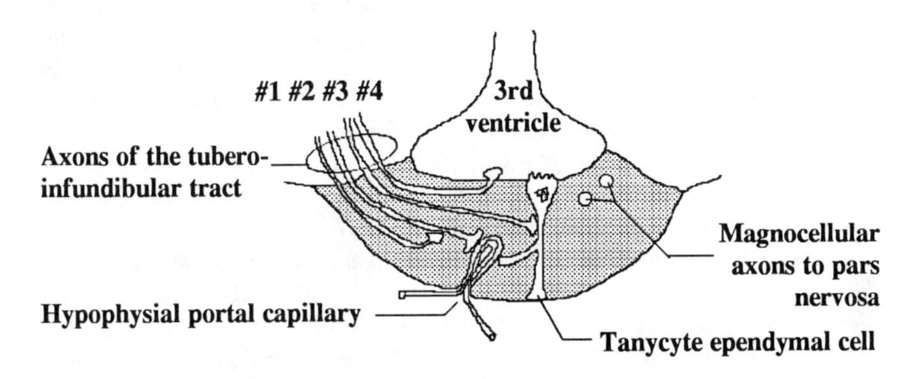

Figure 4-3. Axonal input to the median eminence.

Types of axonal terminations numbered 1 through 4 represent axo-axonal (#1), neurosecretory into portal capillary (#2), onto ependymal cell (#3), and into ventricular lumen (#4).

neuronal cells known as tanycyte ependymal cells. Ependymal cells line the walls of the cerebral ventricles, and tanycyte ependymal cells send processes from the ventricular wall to contact blood vessels or neurons. In the median eminence these cellular processes end mainly on portal capillaries, and they may also project to the ventral surface of the median eminence as illustrated in Figure 4-3. Type #4 of the axonal terminations represents those which reach into the lumen of the third ventricle. The functional roles for types #1 and #2 are well understood, but definitive conclusions about types #3 and #4 are not available. The functions of tanycyte ependymal cells are also not well understood. At one time, it was hypothesized that they transport neurohormones that regulate the adenohypophysis into portal blood. However, such a role for these cells is not generally accepted today.

The median eminence contains another type of ependymal cells not illustrated in Figure 4-3. These cells are called ***brick-like*** ependymal cells, and they are located exclusively in the floor of the third ventricle. Although not within the median eminence, there is a third type of ependymal cell lining the third ventricle in the hypothalamus. This ependymal cell is ***ciliated*** and occurs in areas dorsal and rostral to the median eminence. The ventricular surfaces of non-ciliated ependymal cells (tanycyte and brick-like) possess many microvilli indicative of secretory and/or absorptive functions. Gonadal steroids are important, at least in sheep, for maintenance of the extensive microvilli of these non-ciliated ependyma (Coates and Davis, 1979, 1982). The functional relevance of this relationship is not understood.

Tanycyte ependymal cells in the median eminence and basal hypothalamus project their processes onto primary portal capillaries, as illustrated in Figure 4-3, as well as onto ordinary capillaries and neuronal processes. The functions of these tanycyte projections onto neurons are not known. One other cell type found in the median eminence and not depicted in Figure 4-3 is the neuroglial cell. This cell is called a pituicyte when it is located in the pars nervosa, but that term is not used here. Except for a few sparse neuronal perikarya, neuroglial and tanycyte ependymal cells are the elements in the median eminence that contain ribosomes and rough endoplasmic reticulum (RER). This unique feature aids greatly in the interpretation of transmission electron micrographs. If a process contains RER, it can probably be identified as a non-neuronal process. Otherwise it would sometimes be difficult to distinguish ultrastructurally between axonal terminations and non-neuronal processes.

Although axon terminals in the median eminence lack ribosomes and RER, they possess a variety of secretory vesicles which aid in their ultrastructural identification. The magnocellular axons passing through the inner layer as they project to the pars nervosa contain large diameter (about 150 nm) secretory vesicles that are also electron dense. Other axon terminals usually located in the outer layer contain somewhat smaller dense core secretory vesicles with diameters of 100 nm or slightly less. Finally, there are terminals with a variety of smaller vesicles some of which are electron-dense and some of which are electron-lucent. It is generally thought that the intermediate size dense core secretory vesicles contain the neurohormones that are discharged into portal blood. The various other vesicles contain neuropeptides and monoamines which are probably not secreted into blood (types 1, 3, and 4 in Figure 4-3). Catecholamine-containing terminals contain electron-lucent vesicles of about 50 nm diameter or less.

Circumventricular Organs

The median eminence is just one of several unique structures in the brain known as ***circumventricular organs*** (CVO). These organs, which are also called neurohemal structures, share a common vascular and ependymal organization, which is different from the rest of the brain. As the name denotes, these organs are all located adjacent to some part of a cerebral ventricle. The capillaries in circumventricular organs have a characteristic fenestrated endothelium that probably accounts for the blood-brain barrier being less restrictive in these organs than in most brain tissue. Circumventricular organs are also unique in that their ependymal cells are non-ciliated, whereas ependymal cells in most other regions are ciliated. The diagram in Figure 4-4 shows the location of four different circumventricular organs including the median eminence. The organum vasculosum of the lamina terminalis (OVLT) is located around the rostral projection of the third ventricle above the optic chiasma. The subfornical organ is located on the midline beneath the descending fornix and in contact with the choroid plexus of the third ventricle. The subcommissural organ lines the roof of the third ventricle beneath the posterior commissure and habenula. The three circumventricular organs not illustrated in Figure 4-4 are pars nervosa, pineal gland, and area postrema. The first two of these are covered in detail in Chapters 3 and

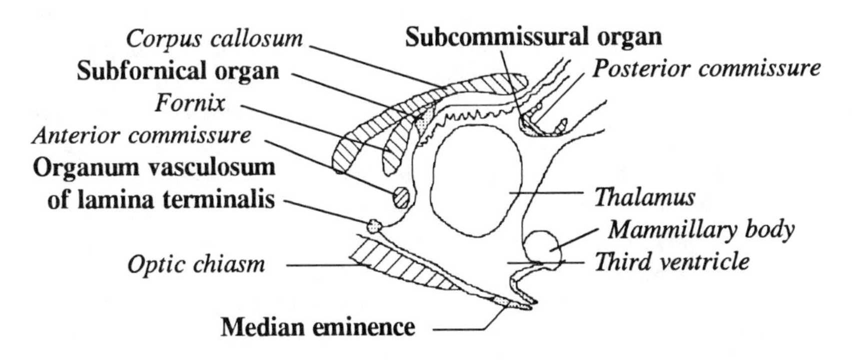

Figure 4-4. Circumventricular organs.
Sagittal section showing the location of four circumventricular organs which are in contact with the third ventricle. Each circumventricular organ is stippled and its label is in bold type.

10, respectively. The area postrema is located in the roof of the fourth ventricle caudal to the cerebellum.

Some functions of CVO are well understood while others are generally assumed, and all the functions are related to the unique location of each CVO between the blood vascular system and the cerebroventricular system of the brain. One well-established function for at least some CVOs is to secrete products of the brain into the blood as occurs in the median eminence, pars nervosa, and pineal organ. Other functions are to secrete substances into and/or absorb substances from the cerebrospinal fluid of the cerebral ventricles. Another potential function is to monitor chemical and hormonal changes in blood reaching the brain, especially since the blood-brain barrier is nonrestrictive in each CVO due to its fenestrated endothelium. Specific regulatory roles for those CVO which are less well understood include a chemoreceptive function for the area postrema in the emetic reflex and blood pressor responses. The subfornical organ is somehow involved in the control of drinking behavior because local infusion of angiotensin II causes this behavior.

Quantification of specific monoamines and neuropeptides in CVO and hypothalamic nuclei has been examined in control animals as well as in those which the mediobasal hypothalamus (MBH) has been surgically deafferented (Brownstein, 1977). This experimental approach allows conclusions to be drawn regarding the proportion of a measured compound that exists within the MBH in processes from perikarya located outside the MBH. Dopamine is present in very large quantities within the median eminence, and these quantities do not decline after surgical deafferentation of the MBH. Therefore, the dopaminergic perikarya are located within the MBH. Norepinephrine and serotonin are present in the median eminence but other hypothalamic structures contain higher concentrations. The concentration of both monoamines in the median eminence declines sharply after deafferentation of the MBH indicating that most of the perikarya are located outside the MBH.

After the peptide structures of certain hypothalamic RFs were discovered (to be covered in the next section), the concentration of these neuropeptides was quantified in CVO and specific hypothalamic nuclei (Brownstein, 1977). In addition to the median eminence, the OVLT was found to contain large amounts of LHRH. Although the OVLT contained more of various neuropeptide RFs than other CVO, concentrations of these neurohormones (other than LHRH) in OVLT were much lower than in median eminence. Among hypothalamic areas, the median eminence contains much higher concentrations of the various hypothalamic RFs than any other region, with the arcuate nucleus and paraventricular nucleus containing somewhat higher concentrations than other hypothalamic nuclei. Later research demonstrated that the immunoassays used in these early studies were often unable to detect the precursor forms in which the hypothalamic RF are synthesized in the hypothalamic perikarya. Therefore, some of the early investigations using microdissected CVO and hypothalamic regions were only able to detect mature releasable forms of RFs.

Hypophysiotrophic Neurohormones from the Hypothalamus

The physiological and morphological evidence suggesting that the hypothalamus regulates adenohypophysial secretions provided the rationale to extract hypothalamic tissue and search for such hypophysiotrophic bioactivity. The development of bioassays for pituitary hormone releasing activity allowed purification and ultimate chemical identification of hypothalamic compounds that stimulated or inhibited acutely the release of one or more adenohypophysial hormones. Advances in chemical analysis of peptides also contributed greatly to this field, as occurred previously in the identification of oxytocin and vasopressin (Chapter 3). The development of knowledge about hypothalamic factors (or hormones) that affect the adenohypophysis should be based on the following principles or criteria before a specific hypothalamic compound is accepted as an endogenous hormonal regulator of a particular adenohypophysial hormone. ***First***, the appropriate bioactivity must be extractable from hypothalamic tissue with proof that the bioactivity did not derive from compounds introduced during collection, storage, or extraction of the hypothalamic tissue. ***Second***, the extracted factor must alter the release of the appropriate adenohypophysial hormone in a bioassay system that has been proven to be free of nonspecific responses. ***Third***, the extracted factor must be able to release the appropriate hormone *in vivo* when applied directly to adenohypophysial tissue with proof that the induced release is not mediated by indirect effects of the factor on the hypothalamus. At this point in the development of knowledge, chemical identification of the purified activity (factor) is usually pursued, and detection of a single molecule that is responsible for most of the bioactivity is a ***fourth*** criterion supportive of a physiological role for the identified factor. The ***fifth*** criterion is that the identified hypophysiotrophic factor should be present in high concentrations in hypophysial portal blood and that concentrations should change appropriately in response to established physiological manipulations that affect release of the adenohypophysial hormone. The ***sixth*** and final criterion is that specific immunoneutralization of the identified factor in blood modifies the endogenous secretion of the adenohypophysial hormone. It should be noted that immunoneutralization can be investigated only when the hypothalamic factor is a peptide, but this is usually the case. When all six criteria have been satisfied, this author feels that the compound can be correctly called a releasing hormone. Unfortunately, the distinction between ***releasing factor*** and ***releasing hormone*** in the descriptive names of these compounds has historical and proprietary implications, but these should now be discarded for the sake of a uniform nomenclature. Therefore, the terms used throughout this book will be releasing (or inhibiting) hormone (RH) regardless of the term used by the lab that discovered the compound.

Thyrotropin-releasing hormone (TRH) and luteinizing hormone-releasing hormone (LHRH) were identified at about the same time by two competing laboratories. The structures of both TRH and LHRH are presented in Figure 4-5. Neither molecule has a free N-terminus due to having pyro-glutamine in

Thyrotropin-Releasing Hormone (TRH)

(pyro)Glu - His - Pro - NH_2
1 3

Luteinizing Hormone-Releasing Hormone (LHRH)

(pyro)Glu - His - Trp - Ser - Tyr - Gly - Leu - Arg - Pro - Gly - NH_2
1 10

Growth Hormone-Releasing Hormone (GHRH)

Tyr - Ala - Asp - [***residues # 4 through 40***] - Arg - Ala - Arg - Leu - NH_2
1 44

Corticotropin-Releasing Hormone (CRH)

Ser - Glu - Glu - [***residues #4 through 40***] - Ile - NH_2
1 41

Growth Hormone Inhibiting Hormone (GHIH or **Somatostatin)**

S - S
Ala - Gly - Cys - [***residues #4 through #10***] - Phe - Thr - Ser - Cys - OH
1 14

Figure 4-5. Partial amino acid sequences for peptidergic hypophysiotrophic hormones.

This figure presents amino acid sequences for peptides demonstrated to be hypophysiotrophic hormones. Only the N-terminus and C-terminus portion of GHRH, CRH, and GHIH are presented. All C-terminal ends are amidated except for GHIH.

that location. The C-terminus of both molecules is amidated, a feature found in many of these hypophysiotrophic hormones. TRH and LHRH are each synthesized as part of larger precursors with the LHRH precursor having a single copy of the decapeptide LHRH within its 68 residues (Chieffi et al., 1991). The 231-residue precursor of TRH contains five copies of the tripeptide sequence found in TRH, but only some of these sequences are likely to be sufficiently cleaved to yield the TRH tripeptide molecule (Wu et al., 1987).

Soon after LHRH was discovered, it was found to trigger the release of follicle-stimulating hormone (FSH) as well as LH. Because of the ability to stimulate release of both these gonadotropins, the LHRH molecule is sometimes called gonadotropin-releasing hormone (GnRH), but the term LHRH will be used throughout this book. Some research has focused on discovering a neurohormone, in addition to LHRH, that releases FSH perhaps even preferentially over the release of LH. However, no hypothalamic neurohormone with such activity has been chemically identified. With the availability of synthetic LHRH, specific binding sites (i.e., receptors) could also be investigated. As expected, adenohypophysial tissue contains many such receptors. Moreover, the biopotency of various LHRH agonists to release LH correlates very well with the affinity of each one for pituitary receptors of LHRH. Evolutionary studies of LHRH indicate that most non-mammalian species have two different molecular forms of LHRH, each encoded by a different gene. In birds, only one type of LHRH molecule seems to function as a hypophysiotrophic hormone (Sharp et al., 1990). Immunoneutralization of LHRH, whether achieved by active or passive immunization, consistently results in gonadal atrophy. There is also evidence from CNS administration of LHRH that endogenous levels of this neuropeptide may facilitate sexual behavior and thereby coordinate ovulation with sexual receptivity in females. However, the criterion of immunoneutralization is much more difficult to satisfy when the active neuropeptide is not blood-borne. In birds, it has been hypothesized that one form of LHRH may function as a central neuropeptide signal while the other LHRH molecule may function as a blood-borne neurohormone. Since mammals have only one form, perhaps the same molecule must serve both functions.

Because TRH is such a simple molecule, synthetic modifications were readily achieved soon after its discovery. As was observed for LHRH, the biopotency of these compounds to release thyrotropin (TSH) from the adenohypophysis correlated very well with the affinity of binding to TRH receptors in pituitary homogenates. An unexpected biological response to native TRH was that it released prolactin (PRL) as well as TSH from the adenohypophysis. Moreover, TRH-induced release of TSH could be antagonized by the hypothalamic neurohormone that inhibited the release of somatotropin (also named growth hormone and abbreviated as GH). An unexpected observation regarding TRH was its wide distribution in the spinal cord and brain areas other than the hypothalamus. Evolutionary studies also revealed a wide phylogenetic distribution in many diverse species and tissues of those species. The subsequent discovery of a large precursor containing five copies of the TRH sequence, only some of which are cleaved to the TRH tripeptide, helps explain the wide distribution of immunoreactive TRH and suggests that TRH-like molecules may have diverse functions.

Soon after the discovery of TRH and LHRH and while searching for GH-releasing activity, a molecule with GH-inhibiting activity was discovered and given the name ***somatostatin***. It was identified as a 14-residue peptide containing a disulfide bond between positions #3 and #14 (Figure 4-5). The disulfide linkage was not essential to bioactivity, and there was a free N-terminus as well as a non-amidated C-terminus, in contrast to TRH and LHRH. The bioactivity consisted of inhibiting the release of GH and also of antagonizing TRH-induced release of TSH. Although the original name somatostatin is widely used, the term GH-inhibiting hormone (GHIH) will be used in this book to maintain a consistent nomenclature with the other hypophysiotrophic hormones. The GHIH precursor was discovered some years later and found to contain a single copy of GHIH within its 116 residues. The precursor is not confined to the hypothalamus, and its cleavage products appear to function in brain, pancreas, and other tissues.

Hypothalamic bioactivity that stimulated the release of GH was subsequently identified, but in this case isolation and identification were facilitated by previous analysis of GH-releasing bioactivity from a tumor of the human pancreas. Hypothalamic activity was found to reside in a 44-residue linear peptide with an N-terminal tyrosine (Figure 4-5). There are relatively major differences among mammalian species in the amino acid sequences of the GH-releasing hormone (GHRH). Rat GHRH differs from the human GHRH in Figure 4-5 at 13 positions including the first six C-terminal residues. Porcine and bovine GHRH differ less from human GHRH than does the rat molecule. Rat GHRH is also unique in that the C-terminus is not amidated as occurs in most mammals (Guillemin et al., 1984). A precursor of GHRH was subsequently discovered to be a 108-residue peptide containing one copy of GHRH. The bioactivity of GHRH to stimulate release of GH is readily antagonized by concurrent administration of GHIH. Also, *in vivo* administration of GHRH often stimulates discharges of GH, but this effect is not consistent unless endogenous levels of GHIH have been previously immunoneutralized. It appears that episodic secretion of GH from the adenohypophysis is regulated by the opposing actions of GHRH and GHIH.

Although the search for corticotropin-releasing bioactivity was initiated even before the search for TRH and LHRH, it took much longer to identify the molecule responsible for this bioactivity (Vale et al., 1983). Both the ability of vasopressin to also stimulate corticotropin (ACTH) release and *in vivo* problems of bioassay specificity delayed the efforts to isolate, purify, and identify endogenous compounds with corticotropin-releasing bioactivity. Eventually, the 41-residue peptide depicted in Figure 4-5 was identified and named corticotropin-releasing factor. However, the term corticotropin-releasing hormone (CRH) will be used in this book in order to maintain a consistent nomenclature with other hypophysiotrophic hormones. The CRH molecule has a free N-terminus and an amidated C-terminus. Bioactivity of CRH includes the expected release of ACTH as well as stimulation of β-lipotropin and β-endorphin, which are synthesized from the same precursor as ACTH and sometimes released from the same adenohypophysial cell. As stated above, vasopressin also stimulates ACTH release and there is synergy between CRH and vasopressin to stimulate ACTH, although details of this synergy may differ between species (Madsen et al., 1991). CRH is synthesized as part of a larger precursor containing one copy of CRH. Besides stimulating ACTH release, CRH appears to function as a CNS neuropeptide signal for other neural systems. In one of these systems, CRH activates central mechanisms that stimulate sympathetic nervous outflow from the brain. This activation of the sympathetic system results in hyperglycemia, cardiovascular stimulation, hypertension, and increased blood-borne norepinephrine (Brown et al., 1982). The ability of endogenous CRH to activate the pituitary-adrenocortical axis as well as the sympathetic nervous system would appear to be beneficial in coordinating endogenous responses to stressful stimuli.

It has been known for a long time that the hypothalamus exerts an inhibitory influence over adenohypophysial secretion of PRL. Bioassay of this PRL-inhibiting activity was relatively easy because PRL secretion is well maintained *in vitro* and because hypothalamic extracts readily suppress it. The material in such bioactive extracts that was mainly responsible for inhibition of PRL release was found to be the catecholamine dopamine (Figure 4-2). There may have been some initial reluctance to accept such a non-peptide compound as the endogenous PRL-inhibiting hormone (PIH), but research results have satisfied all the criteria mentioned earlier in this chapter with the exception of immunoneutralization (Lamberts and Macleod, 1990). However, there is some evidence for an additional PIH bioactivity that is not dopamine and may be an unknown peptide.

In addition to PIH activity, hypothalamic extracts contain PRL-releasing bioactivity. Known hypothalamic compounds that possess this PRL-releasing hormone (PRH) activity include TRH, mentioned above, and vasoactive intestinal peptide (VIP). Both TRH and VIP appear to function in certain situations as a physiological PRH (Lamberts and MacLeod, 1990), but their relative roles may differ among mammalian species and endocrine states. The neurohypophysis also appears to contain an unidentified peptide with PRH bioactivity (Ben-Jonathan et al., 1989).

Figure 4-6 summarizes the hypophysiotrophic actions of chemically identified compounds from the hypothalamus, which appear to act physiologically on adenohypophysial cells to either stimulate or inhibit hormone release. There are two cases of cooperative action with two different compounds stimulating release

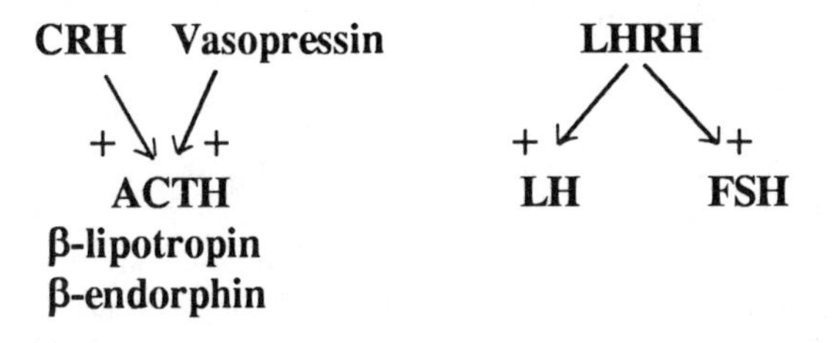

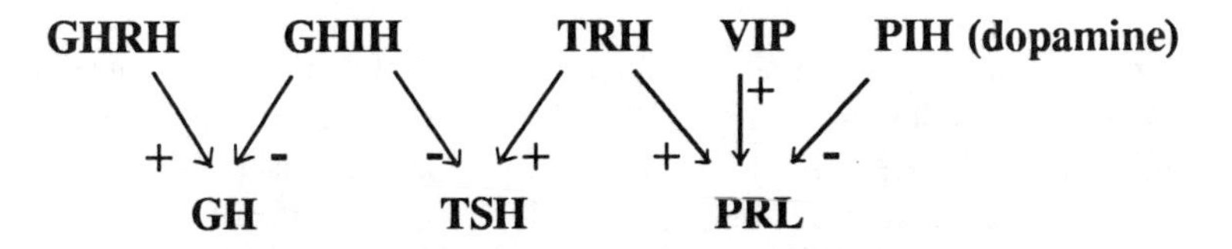

Figure 4-6. Cooperative and antagonistic control of adenohypophysial secretion by hypophysiotrophic hormones.

This diagram illustrates the interactive control of each adenohypohysial hormone by one or more neurohormones. The diagram also depicts stimulatory (+) as well as inhibitory (-) influences of each neurohormone.

from one type of adenohypophysial cell. Both CRH and vasopressin stimulate release of ACTH, β-lipotropin and β-endorphin. Both TRH and VIP stimulate the release of PRL. There are also two cases of a single RH stimulating the release of two hormones (LH and FSH by LHRH; TSH and PRL by TRH). There are three cases of dual stimulatory/inhibitory control of one hormone by at least two antagonistic compounds. GH is regulated by the opposing actions of GHRH and GHIH, whereas TSH is similarly regulated antagonistically by TRH and GHIH. PRL is regulated by the releasing actions of TRH and VIP as well as the inhibiting actions of dopamine.

The mode of action for hypothalamic neuropeptides that influence the adenohypophysis begins with specific binding to receptors on the cell surface. The ligand-receptor complex may then be internalized by endocytosis, although this step may not be required because second messenger systems involving cyclic AMP and cyclic GMP have been demonstrated to be involved. Calcium ions appear to be required for the action of hypophysiotropic releasing hormones, and potassium-induced depolarization of adenohypophysial cells can also discharge hormonal contents. The releasing and inhibiting actions of hypothalamic neurohormones primarily affect release from adenohypophysial cells, and alterations in hormone synthesis only occur secondarily to effects on release. However, research in which RH was administered chronically to hypophysectomized rats with ectopic adenohypophysial tissue clearly demonstrated that hormone synthesis can be stimulated and that synthesis must precede any induced release.

Location of Hypophysiotrophic Neurons

As a result of discovering the chemical identities of hypophysiotrophic hormones and later their precursors, it has been possible to study in detail the location of neuronal perikarya and axon terminals that contain each neurohormone. It has also become important to verify that a specific immunostained neuron projects its axon to the median eminence before it can be denoted as a hypophysiotrophic neuron. Figure 4-7 illustrates a hypophysiotrophic neuron. It has at least one axonal terminal in the outer layer of the median eminence in apposition to a portal capillary. It may have other axonal branches (collaterals) and terminals that make diverse synaptic connections inside and outside the mediobasal hypothalamus. Using antidromic identification techniques of electrophysiology, about 15% of perikarya located in the mediobasal hypothalamus and areas just rostral to it were demonstrated to have at least one axonal branch reaching the median eminence (Yagi and Sawaki, 1978).

The present consideration of neurohormone localization will begin with LHRH because it was studied first. Axon terminals that contact the endothelium of portal capillaries in the outer layer of the median eminence are readily stained for LHRH immunoreactivity. Some LHRH-stained terminals appear to terminate very near the ependymal lining of the third ventricle, but tanycyte ependymal cells do not stain for LHRH. Within a single LHRH-stained terminal in the

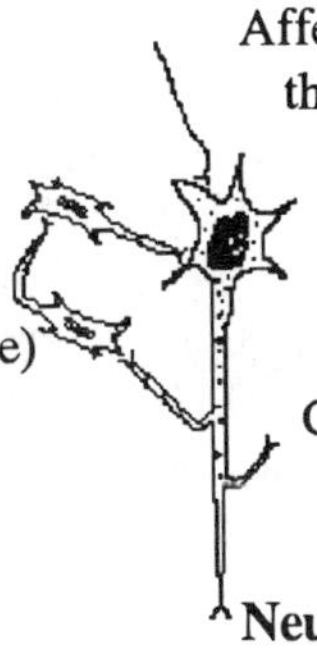

Afferent axons from areas outside the mediobasal hypothalamus

Parvocellular neuron which projects to the median eminence

Interneurons mediating the effects (inhibit/excite) of axonal collaterals

Collateral axonal branches to areas outside the mediobasal hypothalamus

Neurosecretory terminals in outer layer of median eminence

Figure 4-7. Hypophysiotrophic neuron.

Schematic diagram of a hypophysiotrophic neuron sending at least one axonal terminal to the median eminence. Other axonal collaterals are shown projecting to other structures. Synaptic input to the hypophysiotrophic neuron is also depicted in this highly simplified diagram.

median eminence, almost all of the secretory vesicles (75-95 nm diameter) stain for LHRH, which suggests that there is little colocalization of LHRH with other neurochemicals.

The distribution of perikarya that stain for LHRH differs among species. In early research, it was often necessary to inject a drug such as colchicine to block axoplasmic flow in order that antibodies directed against the mature neurohormone would stain neuronal perikarya. When antibodies are directed against the precursor of a neurohormone, the perikarya are more readily stained in untreated animals. In rodents and domestic ungulates, the mediobasal hypothalamus contains relatively few LHRH perikarya because most are located in rostral areas of preoptic area and even in areas rostral to the preoptic area such as the diagonal band of Broca. In primates and bats, a majority of LHRH perikarya are found in the mediobasal hypothalamus, with only a few in the rostral areas.

Two types of LHRH-containing perikarya were identified in the preoptic area and diagonal band of Broca of rats. One type contained numerous dendritic spines on its surface while the perikaryon of the other had a more smooth surface (Jennes et al., 1985). These LHRH perikarya contained dense core secretory vesicles with diameters of about 100 nm, but they contained another population of electron lucent secretory vesicles with diameters of only 30-40 nm. Application of retrogradely transported dye in the median eminence caused the dye to appear in both types of LHRH perikarya although only about 70% of the LHRH neurons were labeled by this procedure (Merchenthaler et al., 1989). LHRH neurons in the rostral areas of the brain display several differences between male and female rats. LHRH perikarya and dendrites of females receive more synaptic appositions than those of males (Chen et al., 1990). There is also a greater proportion (65% in females versus 20% in males) of LHRH neurons that also contain the neuropeptide galanin (Merchenthaler et al., 1991). In female monkeys, the elimination of ovarian hormones following ovariectomy decreased the synaptic appositions on LHRH neurons and increased neuroglial appositions (Witkin et al., 1991). In summary, the morphology of LHRH neurons differs between sexes and may be modified in adults by modifications of gonadal feedback. During embryonic development, LHRH neurons appear to originate in the epithelium of the vomernasal organ and adjacent olfactory structures, and then migrate caudally to their final location (Daikoku-Ishido et al., 1990). The distance of this migration would appear to vary among species, contributing to the variability in the final location of LHRH neuronal perikarya.

Determining the location of TRH neurons was greatly facilitated by the discovery of a precursor molecule containing five copies of the tripeptide TRH sequence. Antibodies against the precursor could be used for immunocytochemical localization without pretreatment to block axoplasmic flow (Lechan et al., 1986). Based on retrograde transport of dye applied to the median eminence, only a small proportion of the many neurons that are stained actually send axons to the portal capillaries. Those TRH neurons that do project to the median eminence are located in the paraventricular nucleus (PVN) of the hypothalamus and the periventricular area just ventral to the PVN (Ishikawa et al., 1988).

Among the many CNS neurons that contain CRH, those that send axons to the median eminence are located primarily in the PVN of the hypothalamus intermixed with TRH neurons (Swanson et al., 1987). The number of these CRH neurosecretory neurons has been estimated at about 2000 per individual rat, and a few of these are outside the PVN in adjacent periventricular areas. CRH neurons in the PVN contain many different neuropeptides, making these neurosecretory neurons somewhat unusual. In untreated rats, approximately 30% of CRH neurons also contain neurotensin, and 20% of CRH neurons contain enkephalin (Ceccatelli et al., 1989). This colocalization of different neuropeptides probably represents synthesis of both neuropeptides within a single cell because mRNA for both CRH and enkephalin are found in individual neurons (Pretel and Piekut, 1990). After inhibitory feedback of adrenocortical hormones was eliminated by adrenalectomy, a majority of CRH neurons began to synthesize vasopressin (Swanson et al., 1987) which can synergize with CRH to release ACTH from the adenohypophysis. The activation of vasopressin synthesis in CRH neurons deprived of adrenocortical feedback would be expected to increase even further the release of ACTH trying to restore homeostasis if the pituitary-adrenocortical axis were still intact.

Neurons that contain GHRH are concentrated in the arcuate nucleus of the hypothalamus and number about 2500 per rat. About 30% of these GHRH neurons also contain neurotensin which, as stated earlier, was also found in about 30% of CRH neurons in the PVN (Sawchenko et al., 1985). The neuropeptide galanin is also colocalized in GHRH neurons as it is in LHRH neurons (Niimi et al., 1990).

Immunostaining for GHIH (also known as somatostatin) in the median eminence was observed in about 30% of the terminals in the outer layer of the median eminence, and these terminals contained dense core secretory vesicles with diameters of about 100 nm. There was no staining of tanycyte ependymal cells. Neuronal perikarya staining for GHIH are distributed widely throughout the brain. Even within the hypothalamus, only a portion of these GHIH neurons project axons to the median eminence. Those perikarya located in the arcuate and ventromedial nuclei of the hypothalamus were not retrogradely labeled by dye applied to the median eminence (Ishikawa et al., 1987; Merchenthaler et al., 1989), whereas a majority of those perikarya located in the preoptic area, periventricular area, and even a few in the PVN of the hypothalamus were labeled. It would appear that neurosecretory neurons that make GHIH for adenohypophsyial action are dispersed more widely than are the neurosecretory neurons that produce most of the RH (i.e., TRH, CRH, GHRH).

Neurons that produce dopamine for secretion into portal capillaries and adenohypophysial action as a PRL-inhibiting hormone are concentrated in the arcuate nucleus. Their axons form a major part of what is known as the tuberoinfundibular tract to the median eminence. Other dopamine-containing axons project to the pars nervosa as the tuberohypophysial tract.

As indicated earlier and in Figure 4-6, VIP appears to be an endogenous releasing hormone for PRL. There are many VIP-staining neurons in the hypothalamus (including PVN and suprachiasmatic nucleus), but the location of those perikarya that send axons to the median eminence has not been determined. Within the PVN of sheep, most VIP-containing perikarya also contain CRH (Papadopoulos et al., 1990).

To summarize this chapter, we have advanced from the hypothesis of Harris (1948) through neurosurgical manipulations that altered adenohypophysial secretions. This advance culminated in the chemical identification of hypophysiotrophic hormones that were found to sometimes regulate more than one adenohypophysial hormone. It was subsequently discovered that these hypophysiotrophic hormones did more than just regulate the adenohypophysis. Only a portion of the neurons that contain these neurohormones send axons to the median eminence. Those neurons that do secrete their products into portal capillaries are located in very specific neuroanatomical locations, with some variation among species. Another interesting feature of these neurons is the colocalization of various neuropeptides with specific hypophysiotrophic hormones, but the implications of these colocalizations are not yet fully understood. However, it should be noted that, with the exception of CRH and vasopressin, neurosecretory neurons that regulate adenohypophysial secretion do not contain more than one type of hypophysiotrophic hormone. Such specificity would appear to be essential for selective regulation of individual hormones of the adenohypophysis for optimum biologic advantage.

REFERENCES

Anderson, L.L., J.G. Berardinelli, P.V. Malven, and J.J. Ford. Prolactin secretion after hypophysial stalk transection in pigs. *Endocrinology* (1982) 111:380-384.

Ben-Jonathan, N., L.A. Arbogast, and J.F. Hyde. Neuroendocrine regulation of prolactin release. *Prog. Neurobiol.* (1989) 33:399-477.

Benoit, A.M., J.R. Molina, D.L. Hard, and L.L. Anderson. Prolactin secretion after hypophysial stalk transection in prepubertal beef heifers. *Anim. Reprod. Sci.* (1989) 18:61-76.

Brown, M.R., L.A. Fisher, J. Rivier, J. Spiess, C. Rivier, and W. Vale. Cortiocotropin-releasing factor: effects on the sympathetic nervous system and oxytocin consumption. *Life Sci.* (1982) 30:207-210.

Brownstein, M. Neurotransmitters and hypothalamic hormones in the central nervous system. *Fed. Proc.* (1977) 36:1960-1963.

Ceccatelli, S., M. Eriksson, and T. Hokfelt. Distribution and coexistence of corticotropin-releasing factor, neurotensin, enkephalin, cholecystokinin, galanin and vasoactive intestinal peptide/peptide histidine isoleucine-like peptides in the parvocellular part of the paraventricular nucleus. *Neuroendocrinol*ogy (1989) 49:309-323.

Chen, W.P., J.W. Witkin, and A. J. Silverman. Sexual dimorphism in the synaptic input to gonadotropin releasing hormone neurons. *Endocrinology* (1990) 126:695-702.

Chieffi, G., R. Pierantoni, and S. Fasano. Immunoreactive GnRH in hypothalamic and extrahypothalamic areas. *Int. Rev. Cytol.* (1991) 127:1-55.

Coates, P.W., and S.L. Davis. The sheep third ventricle: Correlated scanning and transmission electron microscopy, and plasma luteinizing hormone in wethers and testosterone-treated wethers. *Scan. Elect. Micros.* (1979) 1979/III:497-505.

Coates, P.W., and S.L. Davis. Tanycytes in long-term ovariectomized ewes treated with estrogen exhibit ultrastructural features associated with increased cellular activity. *Anat. Rec.* (1982) 203:179-187.

Daikoku-Ishido, H., Y. Okamura, N. Yanaihara, and S. Daikoku. Development of the hypothalamic luteinizing hormone-releasing hormone-containing neuron system in the rat: in vivo and in transplantation studies. *Dev. Biol.* (1990) 140:374-387.

Guillemin, R., F. Zeytin, N. Ling, P. Bohlen, F. Esch, P. Brazeau, B. Bloch, and W.B. Wehrenberg. Growth-hormone-releasing factor: chemistry and physiology. *Proc. Soc. Exp. Biol. Med.* (1984) 175:407-413.

Harris, G.W. Neural control of the pituitary gland. *Physiol. Rev.* (1948) 28:139-179.

Ishikawa, K., Y. Taniguchi, K. Inoue, K. Kurosumi, and M. Suzuki. Immunocytochemical deliniation of thyrotrophic area: origin of thyrotropin-releasing hormone in the median eminence. *Neuroendocrinology* (1988) 47:384-388.

Ishikawa, K., Y. Taniguchi, K. Kurosumi, M. Suzuki, and M. Shinoda. Immunohistochemical identification of somatostatin-containing neurons projecting to the median eminence of the rat. *Endocrinology* (1987) 121:94-97.

Jacobsohn, D. The techniques and effects of hypophysectomy, pituitary stalk section and pituitary transplantation in experimental animals. In: Harris, G.W. and B.T. Donovan (eds). *The Pituitary Gland*, Vol. 2. (Univ. Calif. Press, Berkeley, 1966) pg 1-21.

Jennes, L., W.E. Stumpf, and M.E. Sheedy. Ultrastructural characterization of gonadotropin-releasing hormone (GnRH)-producing neurons. *J. Comp. Neurol.* (1985) 232:534-547.

Kanematsu, S., K. Kishi, and S. Mikami. Rise of plasma prolactin and changes in fine structure of the anterior hypophysis after pituitary stalk section in rats. *Endocrinology* (1979) 105:427-430.

Lamberts, S.W.J., and R.M. MacLeod. Regulation of prolactin secretion at the level of the lactotroph. *Physiol. Rev.* (1990) 70:279-318.

Lechan, R.M., J.L. Nestler, S. Jacobson, and S. Reichlin. The hypothalamic tuberoinfundibular system of the rat as demonstrated by horseradish peroxidase (HRP) microiontophoresis. *Brain Res.* (1980) 195:13-27.

Lechan, R.M., P. Wu, and I.M.D. Jackson. Immunolocalization of the thyrotropin-releasing hormone prohormone in the rat central nervous system. *Endocrinology* (1986) 119:1210-1216.

Madsen, G., E.C. Chan, J. Falconer, K.Y. Ho, and R. Smith. Reverse haemolytic plaque assay study of corticotropin-releasing hormone and aginine vasopressin interaction in ovine corticotropes. *J. Neuroendocrinol.* (1991) 3:193-197.

Merchenthaler, I., F. J. Lopez, D.E. Lennard, and A. Negro-Vilar. Sexual differences in the distribution of neurons coexpressing galanin and luteinizing hormone-releasing hormone in the rat brain. *Endocrinology* (1991) 129:1977-1986.

Merchenthaler, I., G. Setalo, C. Csontos, P. Petrusz, B. Flerko, and A. Negro-Vilar. Combined retrograde tracing and immunocytochemical identification of luteinizing hormone-releasing hormone and somatostatin-containing neurons projecting to the median eminence of the rat. *Endocrinology* (1989) 125:2812-2821.

Niimi, J., J. Takahara, M. Sato, and K. Kawanishi. Immunohistochemical identification of galanin and growth hormone-releasing factor-containing neurons projecting to the median eminence of the rat. *Neuroendocrinology* (1990) 51:572-575.

Nikitovitch-Winer, M., and J.W. Everett. Functional restitution of pituitary grafts re-transplanted from kidney to median eminence. *Endocrinology* (1958) 63:916-930.

Papadopoulos, G.C., J. Antonopoulos, A.N. Karamanlidis, and H. Michaloudi. Coexistence of neuropeptides in the hypothalamic paraventricular nucleus of the sheep. *Neuropeptides* (1990) 15:227-233.

Pretel, S., and D. Piekut. Coexistence of corticotropin-releasing factor and enkephalin in the paraventricular nucleus of the rat. *J. Comp. Neurol.* (1990) 294:192-201.

Sawchenko, P.E., L.W. Swanson, J. Rivier, and W.W. Vale. The distributon of growth-hormone-releasing factor (GRF) immunoreactivity in the central nervous system of the rat: an immunohistochemical study using antisera directed against rat hypothalamic GRF. *J. Comp. Neurol.* (1985) 237:100-115.

Sharp, P.J., I.C. Dunn, G.M. Main, R.J. Sterling, and R.T. Talbot. Gonadotropin-releasing hormones: Distribution and function. In: Wada, M., S. Ishii, and C.G. Scanes (eds). *Endocrinology of Birds: Molecular To Behavioral.* (Japan Sci. Soc. Press, Tokyo, 1990), pg 31-42.

Swanson, L.W., P.E. Sawchenko, R.W. Lind, and J.H. Rho. The CRH motoneuron: differential peptide regulation in neurons with the possible synaptic, paracrine, and endocrine inputs. *Ann. N. Y. Acad. Sci.* (1987) 512:12-23.

Thomas, G.B., J.T. Cummins, L. Cavanagh, and I.J. Clarke. Transient increase in prolactin secretion following hypothalamo-pituitary disconnection in ewes during anestrous and the breeding season. *J. Endocrinol.* (1986) 111:425-431.

Vale, W., C. Rivier, M.R. Brown, J. Spiess, G. Koob, L. Swanson, L. Bilezikjian, F. Bloom, and J. Rivier. Chemical and biological characterization of corticotropin releasing factor. *Rec. Prog. Horm. Res.* (1983) 39:245-270.

Witkin, J.W. Access of luteinizing hormone-releasing hormone neurons to the vasculature in the rat. *Neuroscience* (1990) 37:501-506.

Witkin, J.W., M. Ferin, S.J. Popilskis, and A.J. Silverman. Effects of gonadal steroids on the ultrastructure of GnRH neurons in the rhesus monkey: Synaptic input and glial apposition. *Endocrinology* (1991) 129:1083-1092.

Wu, P., R.M. Lechan, and I.M.D. Jackson. Identification and characterization of thyrotropin-releasing hormone precursor peptides in rat brain. *Endocrinology* (1987) 121:108-115.

Yagi, K., and Y. Sawaki. Electrophysiological characteristics of identified tuberoinfundibular neurons. *Neuroendocrinology* (1978) 26:50-64.

Chapter 5

MISCELLANEOUS NEUROPEPTIDES

Chemical identification of the neurohormones released by the pars nervosa as well as those secreted into portal vessels to regulate the adenohypophysis was a harbinger of new knowledge about other peptide products of the nervous system (i.e., neuropeptides). The present chapter will briefly summarize knowledge about various neuropeptides, with the greatest emphasis on the opioid neuropeptides. For many of these neuropeptides, it is doubtful that they play a role in neuroendocrine transduction (Figure 1-1), but it is highly likely that they are chemical messengers acting in a paracrine or neurocrine manner within the nervous system. In this latter role, many of the neuropeptides to be discussed herein probably participate in neuroendocrine integration (Figure 1-2). For this reason, and because new knowledge is rapidly emerging, this chapter will present current information about each of these neuropeptides.

Table 5-1 lists the various categories of neuropeptides to be covered in this chapter and gives examples of each type. Categories A and B represent neurohormones covered in Chapters 3 and 4, respectively, but in this chapter special consideration will be given to the ***non-endocrine*** actions of these compounds. The neurohormone vasopressin is known to produce endocrine (i.e., blood-borne) effects in the following two systems: (1) magnocellular neurons release vasopressin into systemic blood in the pars nervosa to promote renal retention of water, and (2) parvocellular neurons release vasopressin into portal blood in the median eminence to act as a releasing hormone for hypophysial ACTH. Concerning non-endocrine effects, vasopressin-containing axons project from the hypothalamus to various other CNS regions including the medulla, where it has been suggested that they influence cardiovascular regulation (Soffroniew and Schrell, 1981). As stated in Chapter 4, a majority of parvocellular CRH-containing neurons begin to produce vasopressin after adrenalectomy. These neurons project to diverse CNS locations, probably to activate the sympathetic nervous system when the organism is exposed to stress. Vasopressin may also play a role, together with CRH, in this sympathetic activation.

Besides magnocellular neurons, parvocellular neurons of the paraventricular nucleus also contain oxytocin. Some of these oxytocin-containing neurons also contain CRH or GHIF (Papadopoulos et al., 1990). Release of oxytocin in the CNS has been documented in sheep by intracerebral microdialysis during those events (parturition, vaginocervical stimulation, and suckling) when oxytocin would also be secreted into blood by the pars nervosa (Kendrick et al., 1988). Maternal behavior may be enhanced by the central release of oxytocin because both CNS administration of oxytocin and vaginocervical stimulation to provoke CNS release of endogenous oxytocin can promote such behavior in maternally experienced sheep (Keverne and Kendrick, 1991). The role played by CNS oxytocin in the initiation of maternal behavior is also discussed in Chapter 14.

Besides maternal behavior, there is evidence that CNS actions of oxytocin (either blood-borne or locally released) may promote sexual behavior. Administration of progesterone to estrogen-primed female rats induced within 30 min the appearance of receptors for oxytocin in a portion of the hypothalamic ventromedial nucleus where local infusion of oxytocin in such primed rats could cause sexual receptivity (Schumacher et al., 1990). Intracerebral administration of an antagonist of oxytocin receptors also interfered with female sexual behaviors induced by progesterone (Witt and Insel, 1991). In summary, neurocrine or paracrine actions of oxytocin within the CNS appear to facilitate both mating and maternal behaviors in females. Moreover, stimuli that enhance release of neurohypophysial oxytocin into blood may also stimulate central release of oxytocin that has neurocrine and/or paracrine actions.

Neuropeptides in category B of Table 5-1 are hypophysiotrophic hormones (also listed in Figure 4-5). Although initially discovered for their ability to stimulate or inhibit release of adenohypophysial hormones, these compounds were subsequently found to have other actions. Some of these actions seem to be synergistic with their adenohypophysial actions. For example, CNS actions of CRH as a neuropeptide activator of the sympathetic nervous system would cooperate with CRH-induced activation of the pituitary-adrenal axis to mobilize a coordinated response to stress. As mentioned in Chapter 4, LHRH appears to facilitate sexual behavior in rodents, and such an action might synchronize female receptivity with the preovulatory discharge of LH induced by hypophysiotrophic actions of LHRH. Notably, local immunoneutralization of LHRH in the midbrain region of female rats was able to interfere with some aspects of induced sexual receptivity (Sirinathsinghji et al., 1983; Pfaff, 1983). On the other hand, LHRH may also facilitate male copulatory behavior in rodents (Myers and Baum, 1980) which have no preovulatory discharge of LHRH and LH. Attempts to demonstrate an action of LHRH in facilitating female sexual behavior in domestic ungulates have been unsuccessful (Cook et al., 1986; Esbenshade and Huff, 1989).

The hypophysiotrophic neurohormone TRH is synthesized as part of a large precursor that contains five copies of the tripeptide sequence of TRH. However, the location of putative cleavage sites suggests that perhaps only two of these five copies will generate the tripeptide TRH (Wu et al., 1987). Areas of the telencephalon and brain stem were immunoreactive for the precursor but not for tripeptide TRH (Lechan et al., 1986), suggesting that these areas may process the precursor in ways that do not yield any tripeptide TRH. Administration of synthetic TRH can produce a variety of effects unrelated to adenohypophysial secretions. Therefore, it is hypothesized that tripeptide TRH and/or the larger TRH-containing peptides derived from the TRH precursor may exert various neurocrine or paracrine effects in the CNS.

Somatostatin (GHIH), discovered for its ability to inhibit adenohypophysial release of GH, was subsequently found to have very wide distribution and diverse actions. Immunoreactive GHIH is present in high concentrations in the spinal cord, cerebral cortex, hippocampus, and amygdala (Crawley, 1985). In

Table 5-1. Categories of neuropeptides that have neuroendocrine functions through paracrine or neurocrine actions.

A. Neurohypophysial neurohormones.

Vasopressin
Oxytocin

B. Hypophysiotrophic neurohormones.

CRH
LHRH
TRH
GHIH

C. Neuropeptides initially discovered as digestive tract hormones.

Cholecystokinin
Gastrin
Insulin
Glucagon
Secretin
Motilin

D. Miscellaneous neuropeptides.

Substance P
Neurotensin
Vasoactive intestinal peptide (also hypophysiotrophic neurohormone)
Pancreatic polypeptide
Neuropeptide Y
Galanin
Angiotensin II
Atrial natriuretic peptide (atriopeptin)
Calcitonin-gene related peptide

E. Neuropeptides derived from opioid precursors.

From proenkephalin:
Methionine-enkephalin (met-ENK); Leucine-enkephalin
Heptapeptide met-ENK; Octopeptide met-ENK
From pro-opiomelanocortin (POMC)
β-endorphin; ACTH; Melanotropin (MSH)
From prodynorphin:
Dynorphin-A; Dynorphin-B; Neo-endorphin

addition to neural locations, GHIH immunoreactivity is also found in secretory cells of the gastric mucosa and the pancreatic islets. It is likely that pancreatic GHIH acts in an endocrine manner to regulate metabolism (Unger et al., 1978), especially since blood from the hepatic portal vein contains enriched concentrations of GHIH (Pimstone et al., 1979).

As stated in Chapter 4, some GHIH-containing perikarya in the hypothalamus do not project their axons to the median eminence. Because the precursor of GHIH is known, it is also possible to localize neuronal perikarya which contain the mRNA for GHIH through *in situ* hybridization. As expected, the hypothalamus contains many such GHIH neurons, but the central amygdala also has a high density of perikarya that contain the mRNA for GHIH (Uhl et al., 1986). Projections from these perikarya in the amygdala may release GHIH in various CNS areas for neurocrine or paracrine action. Because of its dual localization in the CNS as well as in the pancreas and intestine, GHIH is also classified as a ***brain-gut peptide*** as are many of the neuropeptides to be discussed in this chapter.

Brain-Gut and Other Neuropeptides

Neuropeptides listed under category C in Table 5-1 were initially discovered as digestive tract hormones and only later found to be present in the CNS. These compounds were originally called brain-gut peptides, and further research revealed many other compounds for which this term is appropriate.

Cholecystokinin (CCK). This digestive hormone is present in high concentrations in many parts of the brain, including parts of the limbic system (hypothalamus, hippocampus, and amygdala) and the cerebral cortex (Crawley, 1985). The molecule of CCK found in the CNS contains only eight amino acid residues, which is much smaller than the hormonally active molecule in the digestive tract. As for other neuropeptides, synthesis occurs as a precursor (proCCK) that is subsequently cleaved, and this cleavage proceeds furthest in the brain (Dixon et al., 1987). Other evidence of tissue-specific processing of proCCK derives from adenohypophysial cells that contain unique molecular forms of CCK immunoreactivity, some of which have very limited bioactivity (Rehfeld, 1988).

CNS perikarya that contain CCK also contain various other putative modulatory compounds. These include parvocellular CRH-containing neurons in the paraventricular nucleus as well as magnocellular oxytocinergic neurons. Many CCK-containing neurons also contain dopamine or the enzyme tyrosine hydroxylase, which is necessary for catecholamine synthesis. In the spinal cord and midbrain region, CCK is colocalized with another neuropeptide named ***substance P,*** which will be discussed on the following page.

Because of its wide distribution, CCK is thought to exert many diverse neurocrine actions through specific neural receptors (Barrett et al., 1989). Modulation of feeding behavior is one of these actions that has been extensively studied and that is related to neuroendocrine integration. CCK acting within the brain is thought to provide a satiety signal that contributes to cessation of feeding behavior. Blood-borne CCK originating in the gut may contribute to satiety, but neurally produced CCK is probably a greater contributor. Although exogenous CCK readily causes satiety in many experimental animals, even stronger evidence of a functional role of endogenous CCK derives from the ability of a CCK receptor antagonist to promote feeding behavior (Ebenezer et al., 1990).

Gastrin. This protein is another digestive hormone that was subsequently discovered to be a neuropeptide. Gastrin is immunologically similar to CCK and derives from a related precursor. In early research, gastrin immunoreactivity present in neural tissue was probably due to CCK crossreacting in the gastrin immunoassays. Later research established that small amounts of authentic gastrin are present, especially in the neurohypophysis. Gastrin is also present in the adenohypophysis where it is found in the POMC-containing cells of the pars anterior and pars intermedia. Gastrin also has its own unique releasing factor known as ***gastrin-releasing peptide*** (GRP) that is similar to the amphibian peptide called bombesin. Although GRP was discovered in the stomach, GRP-containing neuronal perikarya also exist in the hypothalamus and administration of exogenous GRP in the hypothalamus modulated adenohypophysial release of PRL and GH (Kentroti et al., 1988).

Four other digestive tract hormones are listed in Table 5-1 as being neuropeptides. They are the glucose regulating hormones, ***insulin*** and ***glucagon***, as well as ***secretin*** and ***motilin***, which coordinate and promote digestive processes. When these digestive hormones circulate in blood at substantial concentrations, it is sometimes difficult to determine what proportion of the hormone present in neural tissue was taken up from the blood (Baskin et al., 1987).

Substance P. Table 5-1 contains a long list of miscellaneous neuropeptides (category D) and the discussion of them will begin with substance P. The bioactivity of this molecule, which causes smooth muscle to contract *in vitro,* was discovered over 50 years ago, and the sequence of 11 amino acids was determined in 1970 (Leeman, 1980). Substance P in the CNS was the first of the brain-gut peptides to be discovered due to its wide distribution, but it is now considered primarily as a neuropeptide without significant non-neural digestive functions. Substance P is member of the tachykinin family of neuropeptides which also contains neurokinin A that is cleaved from the same precursor as is substance P (Maggio, 1988).

Substance P is widely distributed in the spinal cord as well as in the submucosal and myenteric plexuses of the gut. In the spinal cord, substance P appears to be localized in sensory afferent neurons and to modulate transmission of sensory input. The brain also contains substance P in many regions, including the hypothalamus and striatum (Kumar et al., 1991). Pharmacological antagonism of endogenous opioids increased the level of substance P mRNA in the striatum, suggesting an interaction between substance P and opioid neuropeptides (Tempel et al., 1990). Within the hypothalamus, endogenous substance P may modulate release of those hypophysiotrophic neurohormones that regulate LH and PRL release from the adenohypophysis. Immunoneutralization of endogenous substance P near the third ventricle of ovariectomized rats decreased LH release and increased PRL release (Arisawa et al., 1990a, 1990b). In human females, substance P-producing neurons of the hypothalamus were hypertrophied and contained more mRNA for substance P in postmenopausal women than in premenopausal controls (Rance and Young, 1991). Finally, the pattern of substance P release in the preoptic area, as determined from *in vivo* microdialysis, was circadian in estrogen-primed ovariectomized rats, just as LH and PRL release were circadian in such rats (Jarry et al., 1988).

Neurotensin. The sequence of neurotensin has a N-terminal pyro-glutamine just like the hypophysiotrophic neurohormones LHRH and TRH (Figure 4-5). Neurotensin is found in nervous and non-nervous tissues of the intestinal wall as well as in the brain. It also occurs in LH and TSH cells of the rat adenohypophysis (Bello et al., 1992). Neurotensin-containing neurons of the arcuate nucleus and paraventricular nucleus accumulated a blood-borne fluorescent dye indicating that neurotensin occurs in neurosecretory neurons in which the axons have access to the circulation (Merchenthaler and Lennard, 1991). However, neurotensin has been colocalized with known hypophysiotrophic hormones such as dopamine, CRH, and GHRH. Therefore, the presence of neurotensin in neurosecretory neurons does not prove that this neuropeptide is a neurohormone. About 50% of the neurotensin-containing neurons in the preoptic area of the rat contain the estrogen receptor (Herbison and Theodosis, 1991). Because local immunoneutralization of neurotensin within the rat preoptic area disrupted the estrogen-induced surge of LH (Alexander et al., 1989), it seems possible that neurotensin may mediate part of the effect of estrogen on LHRH.

Vasoactive Intestinal Peptide (VIP). This neuropeptide also functions as a hypophysiotrophic neurohormone stimulatory to the release of PRL in mammals as well as turkeys (El Halawani, 1990). VIP is found in adenohypophysial, gut, and brain tissues. Brain locations in which VIP is found include the cerebral cortex and the supraoptic, paraventricular, and suprachiasmatic nuclei of the hypothalamus. Within the adenohypophysis, VIP is found within PRL-containing cells (Morel et al., 1982), which is very interesting because extracellular VIP can stimulate the release of PRL from these cells. Within the hypothalamus, VIP is colocalized with CRH as well as with acetylcholine. VIP appears to also have a function in the reproductive tracts of both males and females. It modulates contractions of the uterus and oviduct in females, and in male dogs VIP promotes vasodilation associated with penile erection (Fahrenkrug et al., 1988).

Neuropeptide Y (NPY). This neuropeptide is a member of the pancreatic polypeptide family which also contains a specific molecule known as pancreatic polypeptide that was discovered before NPY (Dixon et al., 1987). The pancreatic polypeptide molecule contains a C-terminal tyrosine residue, whereas NPY contains tyrosine residues at both C- and N-terminal ends. Because the single letter code for tyrosine is Y, the compound which predominated in neural tissue of the adrenal medulla was named NPY. Within the adrenal medulla, NPY is associated with norepinephrine, and NPY promotes vasoconstriction similar to norepinephrine. Within the brain, NPY is widely distributed with special concentrations in limbic structures, including the hypothalamus. NPY has been observed to be colocalized with GHIH as well as with galanin.

A variety of functional roles for NPY in the brain have been suggested (Mutt et al., 1989; Lehmann, 1990). There is very strong evidence that endogenous NPY promotes feeding behavior (Leibowitz, 1989). Neuronal production of NPY was increased in hungry animals and subsequent satiety led to less production of NPY. Besides promoting eating, administered NPY suppressed copulatory behavior. There is also evidence that NPY modulates release of the hypophysiotrophic neurohormones GHIH, CRH and LHRH into hypophysial portal blood. The specific effect of NPY on LHRH may depend on the amount of inhibitory feedback by gonadal hormones. When gonadal feedback is low and LHRH secretion is high, NPY appears to inhibit release of LHRH. In contrast, when gonadal feedback is strong and LHRH secretion is low, NPY appears to stimulate release of LHRH. NPY-containing neurons were shown to also contain receptors for estradiol which is consistent with a modulation by estrogen. In addition, NPY-containing neurons also have receptors for glucocorticoid that might be involved in the modulation of CRH release.

Galanin. This 29-residue neuropeptide was originally isolated from extracts of intestine where it was later shown to exist in enteric nerves (Rokaeus, 1987). Galanin is also present in CNS and adenohypophysial tissue. Within CNS perikarya, galanin has been colocalized with the following non-peptide neurotransmitters: dopamine, norepinephrine, serotonin, histamine, and acetylcholine. The neurophysiological significance of such diverse colocalizations is unclear. Galanin is also colocalized in the CNS with the following neuropeptides and neurohormones: NPY, CCK, GHRH, LHRH, and vasopressin. In addition to being colocalized within GHRH and LHRH neurons, galanin inputs to GHRH and LHRH neurons have also been observed. Moreover, there is evidence that galanin can modulate release of both GHRH and LHRH to influence the adenohypophysis. Female rats had more neuronal perikarya with colocalization between LHRH and galanin than did male rats (Merchenthaler et al., 1991). As stated above, galanin is also synthesized in the adenohypophysis, and it can be visualized in the secretory vesicles of PRL-containing cells of estrogen-treated rats (Hyde et al., 1991).

Angiotensin II. This compound functions as both a blood-borne hormone and a CNS-produced neuropeptide (Phillips, 1987). Within the CNS, angiotensin II is widely distributed, but it is especially concentrated within and adjacent to circumventricular organs. The peripheral actions of angiotensin II include stimulating the release of aldosterone from the adrenal gland and raising blood pressure. In the CNS, angiotensin II acts upon the subfornical organ discussed in Chapter 4 to stimulate drinking behavior. Angiotensin II is also present in LH-containing cells of the adenohypophysis, and CNS angiotensin II may also influence the release of LHRH. There is also evidence that CNS angiotensin II may modulate other hypophysiotrophic neurohormones to influence the release of PRL and ACTH (Ganong, 1989).

Atrial Natriuretic Peptide. This neuropeptide is also a blood-borne hormone originating in the atria of the heart. The name was derived from the original bioactivity consisting of enhanced urinary excretion of sodium (i.e., natriuresis). Multiple forms of atrial natriuretic peptide (also called ***atriopeptin***) have now been discovered, and CNS synthesis has been demonstrated, making atriopeptin the first brain-heart peptide (Goetz, 1988). Within the CNS, atriopeptin is concentrated in the preoptic area and is also present in the median eminence. Bioactivities associated with atriopeptin include inhibition of vasopressin release as well as antagonism of angiotensin II-induced drinking behavior.

Calcitonin Gene-Related Peptide. Calcitonin is a blood-borne hormone from the parafollicular C-cells of the thyroid gland. The gene that encodes calcitonin is also expressed in neural and hypophysial tissue, but the peptide product differs from that in the thyroid gland. This alternatively processed product of the calcitonin gene is therefore called calcitonin gene-related peptide (CGRP). This peptide is found in both peripheral and central nerves, and it is an extremely potent vasodilator of blood vessels (O'Halloran and Bloom, 1991). In sensory afferents from the periphery, CGRP is colocalized with substance P. CGRP is also synthesized in the LH-containing cells of the adenohypophysis (Gon et al., 1990).

As reviewed on the previous pages, the neuropeptides listed under category D in Table 5-1 represent extreme diversity in distribution and putative actions. However, the following generalizations seem warranted. Synthesis of these peptides is not restricted to one organ or even one type of tissue, including in many cases the gut. These miscellaneous neuropeptides are colocalized in neuronal perikarya with other peptides assigned to this category. Some of these neuropeptides are also secreted into blood and act in an endocrine manner. Many of the neuropeptides in category D can also be found in specific types of adenohypophysial cells, and it is often the same cell type which that particular neuropeptide appears to modulate by direct or indirect actions.

OPIOID NEUROPEPTIDES

Separate consideration of opioid neuropeptides is appropriate because the various compounds in category E of Table 5-1 share bioactivity with opioid drugs and because information about their biosynthesis, receptors, neurochemical actions, and antagonism is more advanced than for other neuropeptides. Development of knowledge began with the discovery of specific binding of opioid drugs to homogenates of brain tissue. This discovery raised the question of what might be the normal physiological role for such receptors in animals or humans that were not administered derivatives of the opium poppy. The obvious answer was that there might be endogenous compounds capable of interacting with these receptors. The first two of these compounds to be discovered were methionine-enkephalin and leucine-enkephalin. They were soon followed by the discovery of β-endorphin and several years later by the discovery of dynorphin. Each of these neuropeptides with opioid bioactivity is a member of a family of peptides, with all members of each family sharing a common precursor. This precursor can be cleaved in various ways in different tissues to yield different peptides, and more details about each opioid family will be presented later. The terms to be used when discussing bioactive opioid peptides can sometimes be confusing. The generic term ***endorphin*** was originally proposed to describe all endogenous opioids. However, this term is confusingly similar to β-endorphin, one of the specific peptides. The term ***opioid*** is sometimes used, but it must be prefaced with either endogenous or exogenous to distinguish drugs from endogenous compounds. The best generic term is ***endogenous opioid peptide,*** which is both descriptive and correct. Endogenous opioid peptide is a lengthy term often abbreviated as EOP, and this term will be used herein when making general statements. The name EOP will be reserved for compounds that possess bioactivity similar to opioid agonist drugs. Other drugs can antagonize this bioactivity, and they are called opioid antagonists. There may be endogenous compounds that have some antagonistic activity, and these will be called endogenous anti-opioids to distinguish them from exogenous opioid antagonists.

Opioid Receptors. The original discovery involved specific binding of radiolabeled naloxone to brain homogenates. Naloxone is a drug known to antagonize most types of opioid bioactivity, and its competitive binding to the same moiety as opioid drugs was not unexpected. It should be noted that many opioid drugs are not strictly agonists or antagonists, but rather produce some of both effects. Naloxone is relatively unique because it appears to lack any agonist bioactivity even at high dosages. Therefore, it is very useful for *in vivo* studies in drug-free subjects. Any parameter that changes as a result of naloxone treatment was probably being modulated by some EOP at the time of naloxone administration.

A large body of knowledge about opioid receptors has developed in the years since their discovery (Goldstein and Naidu, 1989). Based on comparisons of activity in various bioassays and binding affinities for many synthetic opioid-related drugs, a number of subtypes of the opioid receptor have been identified. Some of these subtypes may represent altered forms of a single receptor, or they may be different molecules (Wollemann, 1990). The selectivity of each subtype for any given EOP or antagonist is never absolute, so that selectivity of binding always depends on dosage (Goldstein and Naidu, 1989). For example, naloxone can antagonize most subtypes if the dosage is sufficiently high. The three primary subtypes are known as ***mu***, ***delta***, and ***kappa,*** with an additional one called ***epsilon***. In some of the early literature, there was another one called sigma, but that receptor is probably not opioid in nature. The mu subtype was named for the prototypic opioid drug, morphine, but an endogenous ligand specific for only mu opioid receptors has not been established. Subdivisions of the mu and kappa receptor subtypes, which will not be considered here, have also been characterized. The delta opioid receptor is preferentially selective for endogenous enkephalin-related EOP, although exogenous enkephalin-related compounds can be synthesized which react preferentially with mu receptors.

Kappa opioid receptors react preferentially with endogenous dynorphin-related EOP. The epsilon opioid receptor is defined as being specific for β-endorphin, but there is some question about the separate existence of epsilon receptors because β-endorphin binds readily to both mu and delta receptors.

The tissue distribution of opioid receptors and their subtypes is highly heterogeneous in the CNS (Mansour et al., 1987). They also occur in the gut and in the vas deferens of the male reproductive tract. In both these muscular tissues, opioid agonists inhibit contractile activity and both tissues have been used for opioid bioassays. Such bioassays were instrumental in the search for endogenous compounds with opioid agonist bioactivity.

Enkephalin Family of EOP. As stated above, the first EOPs to be discovered were the two pentapeptides methionine-enkephalin (met-ENK) and leucine-enkephalin (leu-ENK), and the sequences of both neuropeptides are presented in Table 5-2 (Hughes, 1984). A precursor was subsequently discovered and found to contain six replications of the met-ENK sequence and one sequence of leu-ENK. The precursor is called ***proenkephalin***, but some authors use the term proenkephalin A to distinguish it from proenkephalin B that is their name for the precursor of dynorphin-related peptides. The location of probable cleavage sites in proenkephalin (Figure 5-1) suggests that cleavage of the molecule will yield four molecules of met-ENK, one of leu-ENK, and two of C-terminally extended forms of met-ENK. The sequences of these latter two molecules are shown in Table 5-2. The one named heptapeptide met-enkephalin contains 7 residues, whereas the one named octapeptide met-enkephalin contains 8 residues.

Table 5-2. Enkephalin-related neuropeptides.

Proenkephalin [also called proenkephalin A] (1-267)

Methionine-enkephalin (1-5)

Tyr-Gly-Gly-Phe-Met-OH
1 5

Leucine-enkephalin (1-5)

Tyr-Gly-Gly-Phe-Leu-OH
1 5

Heptapeptide met-enkephalin (1-7)

Tyr-Gly-Gly-Phe-Met-Arg-Phe-OH
1 5 7

Octopeptide met-enkephalin (1-8)

Tyr-Gly-Gly-Phe-Met-Arg-Gly-Leu-OH
1 5 8

Note in Table 5-2 that octapeptide met-ENK also differs from heptapeptide met-ENK at position 7. Figure 5-1 illustrates the location of these various enkephalin-related sequences within the linear 267-residue sequence of proenkephalin. The pentapeptide sequences are numbered sequentially starting at the N-terminus, and Leu-ENK is the sixth one. Other fragments of proenkephalin have been studied and assigned various names. The only one of these to be mentioned here is ***Peptide E*** which contains the fifth and sixth pentapeptide sequences as well as the intervening 15 residues (Figure 5-1). Therefore, peptide E contains 25 residues with met-ENK at its N-terminus and leu-ENK at its C-terminus. These naturally occurring EOPs of the enkephalin family share higher affinities for the delta subtype of opioid receptors.

β-Endorphin Family of EOP. As stated above, the discovery of opioid receptors stimulated the search for opioid bioactivity in various tissue extracts. A protein called β-lipotropin based on its lipid-mobilizing bioactivity had been discovered some years earlier in pituitary extracts of domestic ungulates. When pituitary extracts from camels were examined, they lacked β-lipotropin but contained part of its sequence in a smaller peptide which was originally named C-fragment. This C-fragment derived from the camel pituitary was found to contain opioid bioactivity, and it was assigned a new name, ***β-endorphin.*** Although the camel is a physiologically unique species, it was subsequently learned that high levels of β-endorphin in its adenohypophysial tissue were due to postmortem degradation of β-lipotropin and not due to any unique physiology of the camel.

A large precursor for β-endorphin was subsequently identified and found to give rise to a variety of biologically active hormones and neuropeptides (Imura et al., 1982). Figure 5-2 illustrates this precursor, named pro-opiomelanocortin

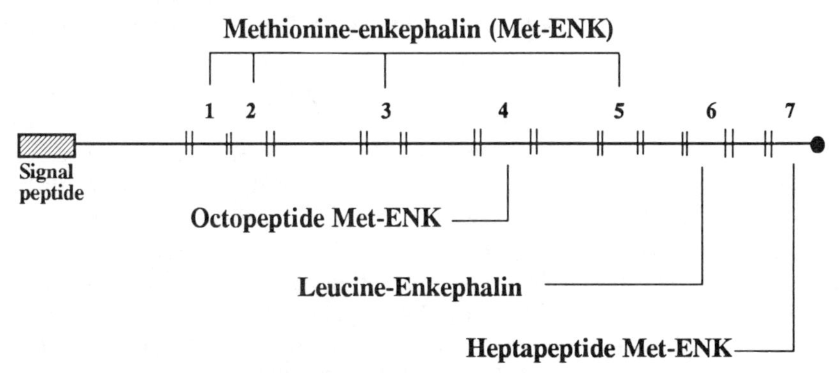

Figure 5-1. Pre-proenkephalin.

Multiple enkephalinergic sequences (sequentially numbered 1 through 7) contained within the proenkephalin precursor. Vertical double lines denote the position of probable cleavage sites in the 267-residue peptide that remains after removal of the signal sequence from the N-terminus of the precursor.

(POMC), and the many important sequences it contains. The C-terminal 91 residues of POMC constitute β-lipotropin, which is located adjacent to the 39 residues of the adenohypophysial hormone ACTH. The N-terminal end of ACTH contains the sequence for α-melanotropin (α-MSH), whereas the related sequences of β-MSH and γ-MSH are located in other parts of POMC (Figure 5-2). Because of sequence identity, the seven residues of ACTH (4-10) are also found in γ-MSH and β-MSH. Therefore, this seven-residue sequence recurs three times in the linear sequence of POMC.

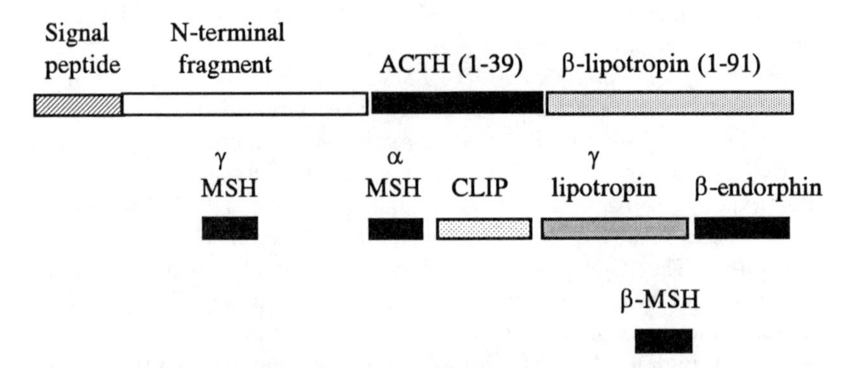

Figure 5-2. Prepro-opiomelanocortin (POMC).

Segments of the complete POMC molecule (top) are labeled with the names of peptides cleaved from POMC. Other cleavage products are depicted below that portion of POMC from which they originate.

The cleavage of POMC to constituent peptides is illustrated in Figure 5-3 and appears to differ among tissues (Hollt, 1986). The N-terminal fragment is cleaved from the N-terminus of ACTH, leaving a very transient compound sometimes called pro-opiocortin, which is rapidly cleaved into ACTH and β-lipotropin. These initial steps probably occur in all tissues. In adenohypophysial tissue of the pars anterior, the resulting ACTH is secreted into blood. About 80% of the β-lipotropin in pars anterior is also secreted, but about 20% of it is cleaved into β-endorphin (cleavage step labeled PA in Figure 5-3). Processing denoted by dashed lines in Figure 5-3 occurs in pars intermedia tissue of the adenohypophysis and probably in all neural tissues. This additional processing abolishes the bioactivity of ACTH by cleaving the molecule into α-MSH (1-13) and corticotropin-like intermediate peptide (abbreviated CLIP). The N-terminal fragment of POMC is cleaved to yield γ-MSH (1-27). The β-lipotropin molecules in neural and pars intermedia tissue are cleaved into γ-lipotropin (1-58) and β-endorphin (1-31). If the processing ended here, there would be large quantities of β-endorphin and γ-lipotropin in these tissues, but this does not occur. The two products of β-lipotropin are rapidly processed further. γ-Lipotropin yields β-MSH (1-18) while four C-terminal residues are cleaved from β-endorphin (1-31)

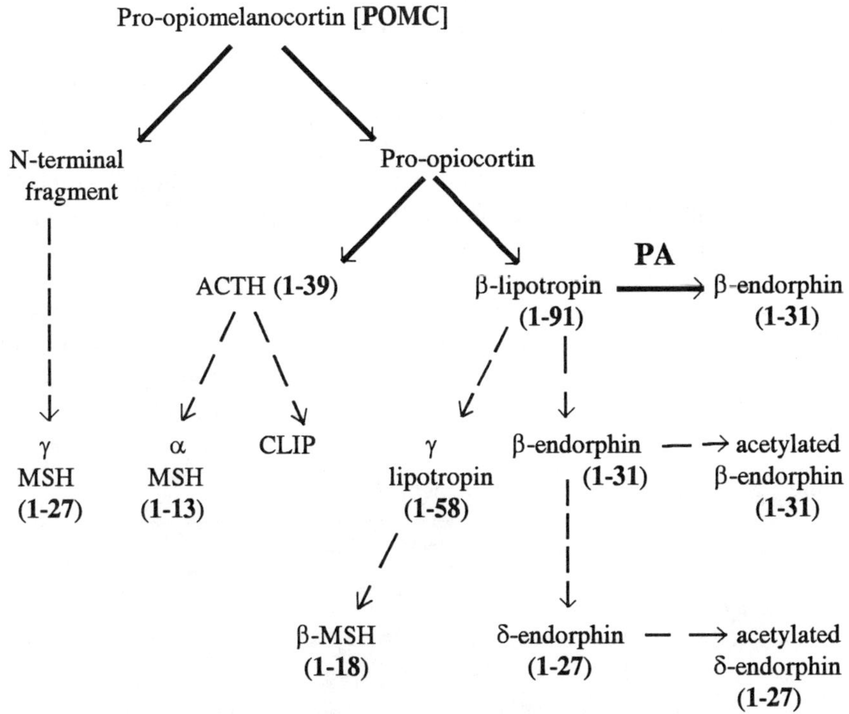

Figure 5-3. Processing of POMC.

Dashed lines denote cleavage processes occurring in neural tissue and in tissue of the pars intermedia. Solid lines denote cleavage processes occurring in all tissues expressing the POMC gene including the pars anterior (PA) of the adenohypophysis. Cleavage of β-lipotropin (1-91) into β-endorphin (1-31) which is denoted with PA is only about 20% complete in PA tissue.

to yield ***δ-endorphin*** (1-27), which is sometimes called β-endorphin (1-27). One additional C-terminal residue is sometimes cleaved to yield β-endorphin (1-26), which is not shown in Figure 5-3. Neural and pars intermedia tissues also N-acetylate the various forms of β-endorphin to inactivate any residual opioid bioactivity. Cleavage of C-terminal residues from β-endorphin also greatly decreases opioid bioactivity. It should be noted that the β-endorphin family of EOPs is unique because of the many neuropeptides with non-opioid bioactivity that are derived from POMC. More information about these other products of POMC will be presented in Chapter 6.

Dynorphin Family of EOP. For several years after the discoveries of met-ENK, leu-ENK and β-endorphin, there seemed to be some opioid bioactivity in pituitary extracts that could not be ascribed to any of the known peptides. Subsequent research demonstrated that this bioactivity resided in peptides which had leu-ENK at their N-terminus but which were larger than the leu-ENK molecule. Because of their extreme potency in some bioassays, the name ***dynorphin*** was given to the first identified peptide with this activity (Goldstein, 1984). The name was changed to dynorphin-A (1-17) when another form (dynorphin-B) was discovered by the same laboratory. A third form of dynorphin-related EOP was discovered by another laboratory, and it was given the name ***neo-endorphin*** (α and β forms). Table 5-3 summarizes the sequences of three dynorphin-related EOPs, all of which have been determined to be contained within the linear sequence of prodynorphin, a 256-residue precursor. This precursor is also called proenkephalin-B by some workers. As shown in Table 5-3, these dynorphin-related peptides share the N-terminal six residues consisting of the pentapeptide leu-ENK plus arginine at position #6. Dynorphin A (1-17) and dynorphin B (1-13) share the first seven residues. Due to different nomenclature in different labs, dynorphin B (1-13) is sometimes called rimorphin.

Table 5-3. Dynorphin-related neuropeptides.

Prodynorphin [also called proenkephalin B] (1-256)

Dynorphin-A (1-17)
Tyr-Gly-Gly-Phe-Leu--Arg-Arg-Ile-Arg-Pro-Lys-Leu-Lys-Tyr-Asp-Gln-OH
1 5 8 13 17

Dynorphin-B [also called rimorphin] (1-13)
Tyr-Gly-Gly-Phe-Leu--Arg-Arg-Gln-Phe-Lys-Val-Val-Thr-OH
1 5 8 13

β-neo-endorphin (1-9)
Tyr-Gly-Gly-Phe-Leu-Arg-Lys-Tyr-Pro-OH
1 5 9

Processing of prodynorphin to the smaller peptides has been studied and appears to differ somewhat among different CNS regions (Hollt, 1986). Figure 5-4 illustrates the generalized cleavage of prodynorphin. The molecule of dynorphin-A (1-17) comes from the middle portion of prodynorphin, and it is further cleaved to dynorphin-A (1-8) that has less bioactive potency and is more labile. Dynorphin-B (1-13) comes from the C-terminal end of prodynorphin, and it may be derived from the intermediate peptide called dynorphin-B (1-29) that was also named leumorphin by another laboratory. To emphasize the common origin in prodynorphin, this author prefers to use the name dynorphin-B to describe this peptide. The third dynorphin-related peptide derived from prodynorphin is β-***neo-endorphin*** that was discovered by yet another laboratory, which accounts for the different nomenclature. Only later, when its sequence was discovered in a middle portion of prodynorphin, was the association with dynorphin fully appreciated. As shown in Figure 5-4, α-neo-endorphin (1-10) is one residue

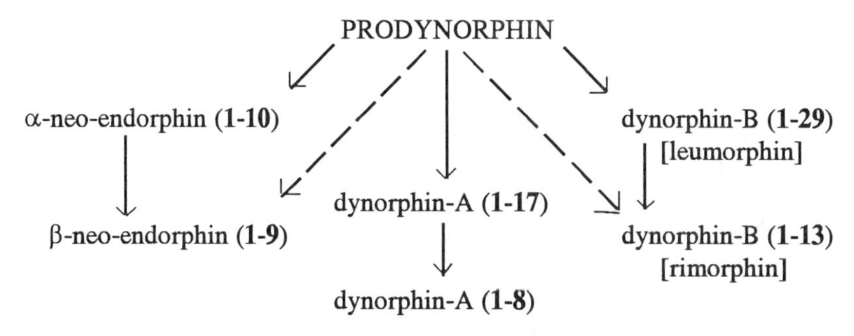

Figure 5-4. Cleavage of prodynorphin.

Solid lines denote probable cleavage of prodynorphin. Dashed lines denote possible production of some peptides without the intermediates shown by the solid lines. Details of prodynorphin cleavage may differ among CNS regions.

longer than β-neo-endorphin (1-9) and is probably its immediate precursor. Both α- and β-neo-endorphin have much less opioid bioactivity than the two forms of dynorphin. The dashed lines in Figure 5-4 denote the possibilities that β-neo-endorphin and dynorphin-B (1-13) may each originate directly from cleavage of prodynorphin without the intermediates shown by the solid lines. Besides sharing a common precursor in prodynorphin, bioactive peptides of this family of EOPs share high affinities for binding to the kappa subtype of opioid receptors (Hollt, 1986). It is interesting to note that the precursor for each EOP family contains between 239 and 267 residues and that each precursor contains three to six repeating sequences of five to seven amino acids.

CNS Locations of EOP Perikarya. Following the elucidation of the three EOP families, immunocytochemical approaches were used to locate neuronal perikarya specific to each family. These studies revealed highly heterogeneous distributions which differed among EOP families (Goldstein, 1984). The β-endorphin family was distributed in perikarya of only two CNS locations, the arcuate nucleus (ARC) of the hypothalamus and the nucleus of the solitary tract in the brain stem. Both of these regions were also found to contain perikarya staining for peptides of the enkephalin and dynorphin families. Peptides of the dynorphin family were also demonstrated in the following hypothalamic regions in addition to the ARC mentioned above: lateral hypothalamic area (LHA), paraventricular nucleus (PVN), and supraoptic nucleus (SON). Dynorphin-related peptides were also present in the hippocampus, amygdala, periaqueductal gray (PAG), as well as in a few other non-limbic regions. The prodynorphin gene appears also to be expressed in the LH-containing cells of the adenohypophysis, but the processing of the precursor does not proceed far enough to yield any small peptides with opioid bioactivity.

The enkephalin family of peptides is distributed more widely than the other families. Within the hypothalamus, stained perikarya have been demonstrated in ARC, PVN, and LHA where other opioids were present in other neurons. However, peptides of the enkephalin family are the only EOPs to be found in perikarya of the following hypothalamic regions: ventromedial nucleus, dorsomedial nucleus, and mammillary bodies. Stained perikarya of the enkephalin family are also found extensively in the amygdala, in the brain stem, including the PAG and other areas, and in many other CNS regions, including the basal ganglia. Enkephalin is also strongly expressed in the medulla of the adrenal gland, and much excellent biochemical information has been derived from this neural tissue.

In summary, individual perikarya of the arcuate nucleus and the nucleus of the solitary tract stained for peptides of all three EOP families. However, no individual neuron appears to express the gene for more than one EOP family. Nevertheless, EOP neurons often contain other non-opioid neuropeptides or receptors that are important in neuroendocrine integration. About 20% of the CRH-containing perikarya in the PVN also contain enkephalin. Some magnocellular oxytocin-containing neurons also stain for enkephalin. On the other hand, magnocellular vasopressin-containing neurons contain dynorphin-related peptides. Neurons of the striatum which expressed the proenkephalin gene also expressed the gene for the D2 dopamine receptor, suggesting that endogenous dopamine acts upon these enkephalinergic neurons. In the mediobasal hypothalamus, specific binding of estradiol has been reported to occur in perikarya of all three EOP families.

PHYSIOLOGICAL ROLE OF ENDOGENOUS OPIOIDS

Although the discoveries of opioid receptors and EOPs began as an effort to understand and perhaps treat or prevent drug addiction, the greatest advances have occurred in our understanding of possible roles for EOP in drug-free animals and humans. Information about these physiological roles for EOP depends very much on a single research approach, notably the administration of the opioid receptor antagonist naloxone to drug-free subjects. In this research paradigm, the investigator is searching for some measurable parameter (behavioral, endocrine or physiological) that changes as a result of the administration of naloxone. The observed change is usually very short-lived (less than 1 h) because naloxone-induced antagonism of opioid receptors is very short-lived. In some studies more prolonged antagonism of opioid receptors (several hours) is obtained by using a related compound called naltrexone. When a measurable parameter changes as a result of opioid antagonism and the animal was not being treated with exogenous opioid drugs, one can conclude that the parameter was being influenced by EOPs at the time of administration of the opioid receptor antagonist. This EOP influence will also probably resume as the effect of the exogenous antagonist wanes. This type of evidence usually provides the initial suggestion that EOPs modulate some parameter. Subsequent confirmatory evidence might be that brain areas that are known to control the parameter contain, perhaps in enriched concentrations, EOPs or their receptors. Moreover, experimental manipulations which modify the parameter also alter the concentration of EOPs or of opioid receptors in that brain area. A definitive, but less often achieved, result would be to immunoneutralize one or more EOPs in that brain area and produce changes in the parameter that are similar to those produced by opioid receptor antagonism. When experimental antagonism of opioid receptors fails to produce a change, the negative results are not definitive. It may have been that the dosage of antagonist and route of administration did not result in antagonism of the relevant subtype of opioid receptor in the tissue site where the EOP was acting to modulate the measured parameter.

The sensitivity of an organism to painful stimuli appears to be modulated by EOPs (Akil et al., 1984). The profound analgesic activity of exogenous opioid drugs is necessary but not sufficient evidence of a physiological role for EOPs. The enriched concentrations of opioid receptors as well as EOPs from all three families in the PAG of the brain stem are also suggestive. Localized electrical stimulation of the PAG promotes analgesia that can be antagonized by naloxone. The analgesia produced by suggestion and/or expectation (i.e., the ***placebo*** effect) is also antagonized by naloxone. Certain types of analgesia induced by stress or acupuncture can also be partly antagonized by naloxone. In summary, EOPs appear to be released in the CNS to decrease the perception of painful stimuli.

Another EOP action may be to modulate blood pressure, but only during the hypotension of circulatory shock (Holaday and Faden, 1981). Animals suffering from a variety of syndromes involving circulatory shock seem to be affected by naloxone administration. The consistent action of naloxone in these syndromes has been to elevate blood pressure from the abnormally low hypotensive state and to thereby improve survival. Because naloxone does not elevate blood pressure in animals that are not suffering from circulatory shock, it appears that EOPs are not affecting blood pressure in normotensive subjects. Therefore, EOPs appear to be released during induced hypotension and exacerbate the life-threatening hypotension. Such a physiological action of EOPs occurring during circulatory shock would certainly decrease the chances of survival and constitute the only established EOP action that is not beneficial in some way to the organism or its offspring.

Several types of behavior appear to be modulated by EOPs besides those that are related to the EOP-induced analgesia. Certain types of induced feeding behavior constitute one of these EOP-modulated behaviors (Gosnell, 1987). Administration of naloxone to experimental animals in which feeding has been enhanced by experimental means will decrease the consumption of food. Therefore, EOPs were contributing to the feeding behavior in these specific experimental situations. There are, however, many other neuropeptides that are known to modulate feeding behavior, and a major role for EOPs has not been established.

Another type of behavior that may be modulated by EOP is a presumably abnormal behavior called ***stereotypic*** behavior (Dantzer, 1986). This repetitive behavior is characterized as having no obvious benefit to the organism, and it is often observed in environments in which the organism has very little to do or nothing with which to interact. Therefore, some workers infer that the stereotypic behaviors reflect boredom in the organism. Animals which are isolated, physically confined, and/or hungry appear to display greater frequencies of stereotypic behavior. The ability of naloxone administration to decrease the quantity of stereotypic behavior suggests that EOPs may be contributing to them (Dodman et al., 1987). The repetitive nature of stereotypic behaviors suggests that the animal may be somewhat addicted to the performance of this behavior for which humans can see no benefit. There may be a similarity between such repetitive stereotypic behaviors and a syndrome of electrical stimulation of the brain where the animal controls its own rate of stimulation (i.e., ***self-stimulation***). It has been known for a long time that rates of self-stimulation are exceedingly high when the electrodes have been implanted in certain brain regions, notably the hypothalamus and parts of the limbic system. These regions have been called pleasure centers, and it is clear that their activation produces some sort of reward to the organism which contributes to a continuation of the self-stimulation (Olds, 1962). It is now known that opioid receptors and EOPs are present in brain regions in which self-stimulation rates are very high. Moreover, antagonism of EOP effects can sometimes decrease the rate of self-stimulation (Carr, 1990). Therefore, it seems possible that the actual purpose of stereotypic behaviors that animals perform is to release within selected brain regions unidentified EOPs which convey reward/pleasure to the organism.

The secretion of neurohypophysial and adenohypophysial hormones appears to be modulated by EOPs acting in various locations (Illes, 1989). In the pars nervosa, dynorphin-related peptides are probably released from vasopressinergic terminals, and they may suppress oxytocin release. The details of this EOP-mediated interaction between vasopressinergic and oxytocinergic neurons are not clear, but they may involve the pituicyte cells that have opioid receptors. Stressful stimuli provoke release of ACTH and PRL from the adenohypophysis, and EOPs may mediate some of these stimulatory effects because naloxone administration can decrease the amount of stress-induced release. In contrast, secretion of LH by the adenohypophysis may be decreased by EOPs in certain situations (Haynes et al., 1989). Those experimental situations in which naloxone administration in females will increase LH, presumably by increasing LHRH release, involve stress-induced suppression, suckling-induced anestrus, and gonadal feedback involving progesterone. Therefore, EOPs that inhibit LHRH may be activated during these states to prevent the secretion of LHRH and LH. Such a physiological role for EOPs would appear to suppress female reproduction when conditions are inappropriate for fertility.

As stated earlier, understanding drug addiction was the motivation for the early EOP studies. What conclusions can be drawn regarding the role of EOPs in addiction, tolerance, and the emotional states that contribute to the initial taking of opioid drugs as well as the voluntary return to them after opioid detoxification? There are a variety of theories that address these questions (Redmond and Krystal, 1984). Those straightforward theories that can be tested have not been convincingly supported. For example, self-administration of opioid drugs is probably not the result of inadequate production of EOP or the excessive requirements of opioid receptors in the CNS. However, self-administration of opioid drugs can depress CNS levels of EOPs and desensitize opioid receptor mechanisms that may contribute to the development of pharmacological tolerance and dependence. In attempting to understand the initial self-administration of opioid drugs, the existence of endogenous compounds with anti-opioid activity has been suggested (Galina and Kastin, 1986). Although not well characterized, the possibility of endogenous compounds with naloxone-like actions provides another theory to explain the predisposition for taking exogenous opioids. Perhaps such individuals synthesize so many anti-opioids that their EOP production is inadequate to maintain the needed balance between EOPs and endogenous anti-opioid compounds. More research will be necessary before this possibility can be confirmed or rejected.

One other point needs to be made about addiction. It is sometimes said that individuals are addicted to a particular activity or behavior such as exercise or perhaps one of several excitement-generating activities. This so-called addiction is quite different from pharmacologic addiction to exogenous opioids because the consequences of not performing the activity are much less severe than withdrawal from opioid drugs. It is often stated in non-scientific publications that EOP mediate the compulsion of individuals to repeat certain behaviors or activities, but definitive evidence is lacking. It has also been suggested that defects in EOP systems contribute to mental illness such as various psychoses. However, when psychotic patients were administered naloxone, therapeutic improvements were not consistently observed.

It is quite clear from the preceding discussion that EOPs play a physiological role in the CNS regulation of many processes. Do EOPs act in an endocrine, paracrine, or neurocrine manner to regulate these processes? Evidence is weakest for an endocrine action. Molecules of β-endorphin originating in the adenohypophysis circulate in peripheral blood. Although blood-borne β-endorphin fluctuates due to pulsatile secretion, the concentrations are quite low and may not be sufficient to affect receptors in peripheral tissues. Entry of blood-borne β-endorphin into the CNS can occur, but the efficiency of such transfer is low. Hypophysial portal blood contains enriched concentrations of β-endorphin, but evidence for any EOP as a hypophysiotrophic neurohormone is equivocal. In summary, β-endorphin might be a hormone and act in an endocrine manner, but definitive proof is lacking. EOPs of the enkephalin and dynorphin families do not appear to act via the blood. On the other hand, EOPs of all three families are produced locally in the vicinity of their putative receptors, but it is not clear whether they always act via synapses. Only a few β-endorphin containing presynaptic terminals have been observed in the CNS. Although a neurocrine action via such synapses does probably occur, the major action of EOPs is thought to be through paracrine actions which modulate adjacent neuronal structures that possess opioid receptors. It should be noted that the axonal projection which releases the EOP for paracrine action can be quite long and similar to axonal projections that end in presynaptic terminals. As stated earlier in this chapter, β-endorphin perikarya are found in only two brain regions, but their axonal projections distribute β-endorphin throughout the brain.

REFERENCES

Akil, H., S.J. Watson, E. Young, M.E. Lewis, H. Khachaturian, and J.M. Walker. Endogenous opioids: biology and function. *Annu. Rev. Neurosci.* (1984) 7:223-255.

Alexander, M.J., P.D. Mahoney, C.F. Ferris, R.E. Carraway, and S.E. Leeman. Evidence that neurotensin participates in the central regulation of the preovulatory surge of luteinizing hormone in the rat. *Endocrinology* (1989) 124:783-788.

Arisawa, M., L. De Palatis, R. Ho, G.D. Snyder, W.H. Yu, G. Pan, and S.M. McCann. Stimulatory role of substance P on gonadotropin release in ovariectomized rats. *Neuroendocrinology* (1990a) 51:523-529.

Arisawa, M., G.D. Snyder, W.H. Yu, L.R. De Palatis, R.H. Ho, and S.M. McCann. Physiologically significant inhibitory hypothamic action of substance P on prolactin release in the male rat. *Neuroendocrinology* (1990b) 52:22-27.

Barrett, R.W., M.E. Steffey, and C.A. Wolfram. Type-A cholecystokinin binding sites in cow brain: characterization using (-)-[^{3}H]L364718 membrane binding assays. *Mol. Pharmacol.* (1989) 36:285-290.

Baskin, D.G., D.P. Figlewicz, S.C. Woods, D. Porte, and D.M. Dorsa. Insulin in the brain. *Annu. Rev. Physiol.* (1987) 49:335-347.

Bello, A.R., P. Dubourg, O. Kah, and G. Tramu. Identification of neurotensin-immunoreactive cells in the anterior pituitary or normal and castrated rats. *Neuroendocrinology* (1992) 55:714-723.

Carr, K.D. Effects of antibodies to dynorphin-A and β-endorphin on lateral hypothalamic self-stimulation in ad libitum fed and food-deprived rats. *Brain Res.* (1990) 534:8-14.

Cook, D.L., T.A. Winters, L.A. Horstman, and R.D. Allrich. Induction of estrus in ovariectomized cows and heifers: Effects of estradiol benzoate and gonadotropin-releasing hormone. *J. Anim. Sci.* (1986) 63:546-550.

Crawley, J.N. Comparative distribution of cholecystokinin and other neuropeptides. Why is this peptide different from all other peptides? In: Vanderhaegan, J.J., and J.N. Crawley (eds). *Neuronal Cholecystokinin.* (*Ann. N. Y. Acad. Sci.* Vol. 448, 1985 (1985) pg 1-8.

Dantzer, R. Behavioral, physiological and functional aspects of stereotyped behavior: a review and a re-interpretation. *J. Anim. Sci.* (1986) 62:1776-1786.

Dixon, J.E., R.S. Haun, C.D. Minth, and R. Nichols. Neuropeptide gene families. In: Turner, A.J. (ed). *Neuropeptides and Their Peptidases.* (VCH Publishers, Weinheim, FRG, 1987), pg 9-30.

Dodman, N.H., L. Shuster, M.H. Court, and R. Dixon. Investigation into the use of narcotic antagonists in the treatment of a stereotypic behavior pattern (crib-biting) in the horse. *Am. J. Vet. Res.* (1987) 48:311-319.

Ebenezer, I.S., C. de la Riva, and B.A. Baldwin. Effects of the CCK receptor antagonist MK-329 on food intake in pigs. *Physiol. Behav.* (1990) 47:145-148.

El Halawani, M.E., O.M. Youngren, L.J. Mauro, and R.E. Phillips. Incubation and the control of prolactin secretion in the turkey. In: Wada, M., S. Ishii, and C.G. Scanes (eds). *Endocrinology of Birds: Molecular To Behavioral.* (Japan Sci. Soc. Press, Tokyo, 1990), pg 287-295.

Esbenshade, K.L., and B.G. Huff. Involvement of hypothalamic compounds in the expression of estrus in the pig. *Anim. Reprod. Sci.* (1989) 18:51-59.

Fahrenkrug, J., B. Ottesen, and C. Palle. Vasoactive intestinal polypeptide and the reproductive system. In: Said, S.I., and V. Mutt (eds). *Vasoactive Intestinal Peptide and Related Peptides* (*Ann. N. Y. Acad. Sci.* Vol. 257, 1988), pg 393-404.

Galina, Z.H., and A.J. Kastin. Existence of antiopiate systems illustrated by MIF-1/TYR-MIF-1. *Life Sci.* (1986) 39:2153-2159.

Ganong, W.F. Angiotensin II in the brain and pituitary: contrasting roles in the regulation of adenohypophysial secretion. *Horm. Res.* (1989) 31:24-31.

Goetz, K.L. Physiology and pathophysiology of atrial peptides. *Am. J. Physiol.* (1988) 254:E1-E15.

Goldstein, A. Biology and chemistry of the dynorphin peptides. In: Udenfriend, S., and J. Meienhofer (eds). *The Peptides. Vol. 6. Opioid Peptides: Biology, Chemistry and Genetics* (Academic Press, Orlando, 1984), pg 95-145.

Goldstein, A., and A. Naidu. Multiple opioid receptors: ligand selectivity profiles and binding site signatures. *Mol. Pharmacol.* (1989) 36:265-272.

Gon, G., A. Giaid, J.H. Steel, D.J. O'Halloran, S. Van Noorden, M.A. Ghatei, P.M. Jones, S.G. Amara, H. Ishikawa, S.R. Bloom, and J.M. Polak. Localization of immunoreactivity for calcitonin gene-related peptide in the rat anterior pituitary during ontogeny and gonadal steroid manipulations and detection of its messenger ribonucleic acid. *Endocrinology* (1990) 127:2618-2629.

Gosnell, B.A. Central structures involved in opioid induced feeding. *Fed. Proc.* (1987) 46:163-167.

Haynes, N.B., G.E. Lamming, K.P. Yang, A.N. Brooks, and A.D. Finnie. Endogenous opioid peptides and farm animal reproduction. *Oxford Rev. Reprod. Biol.* (1989) 11:111-145.

Herbison, A.E., and D.T. Theodosis. Neurotensin-immunoreactive neurons in the rat medial preoptic area are oestrogen-receptive. *J. Neuroendocrinol.* (1991) 3:587-589.

Holaday, J.W., and A.I. Faden. Naloxone reverses the pathophysiology of shock through an antagonism of endorphin systems. In: Martin, J.B., S. Reichlin, and K.L. Bick (eds). *Neurosecretion and Brain Peptides* (Raven Press, New York, 1981), pg 421-434.

Hollt, V. Opioid peptide processing and receptor selectivity. *Annu. Rev. Pharmacol. Toxicol.* (1986) 26:59-77.

Hughes, J. Reflections of opioid peptides. In: Hughes, J., H.O.J. Collier, M.J. Rance, and M.B. Tyers (eds), *Opioids, Past, Present and Future* (Taylor & Francis, London, 1984), pg 9-19.

Hyde, J.F., M.G. Engle, and B.E. Maley. Colocalization of galanin and prolactin within secretory granules of anterior pituitary cells in estrogen-treated Fischer 344 rats. *Endocrinology* (1991) 129:270-276.

Illes, P. Modulation of transmitter and hormone release by multiple neuronal opioid receptors. *Rev. Physiol. Biochem. Pharmacol.* (1989) 112:139-233.

Imura, H., Y. Nakai, K. Nakao, S. Oki, and I. Tanaka. Control of biosynthesis and secretion of ACTH, endorphins and related peptides. In: Muller, E.E., and R.M. MacLeod (eds). *Neuroendocrine Perspectives, Vol. I* (Elsevier, Amsterdam, 1982), pg 137-167.

Jarry, H., A. Perschl, H. Meissner, and W. Wuttke. In vivo release rate of substance P in the preoptic/anterior hypothalamic area of ovariectomized and ovariectomized estrogen-primed rats: Correlation with luteinizing hormone and prolactin. *Neurosci. Lett.* (1988) 88:189-194.

Kendrick, K.M., E.B. Keverne, C. Chapman, and B.A. Baldwin. Microdialysis measurement of oxytocin, aspartate, gamma-aminobutyric acid and glutamate release from the olfactory bulb of sheep during vaginocervical stimulation. *Brain Res.* (1988) 442:171-174.

Kentroti, S., W.L. Dees, and S.M. McCann. Evidence for a physiological role of hypothalamic gastrin-releasing peptide to suppress growth hormone and prolactin release in the rat. *Proc. Natl. Acad. Sci.* (1988) 85:953-957.

Keverne, E.B., and K.M. Kendrick. Morphine and corticotropin-releasing factor potentiate maternal acceptance in multiparous ewes after vaginocervical stimulation. *Brain Res.* (1991) 540:55-62.

Kumar, M.S.A., T. Becker, and K. Ebert. Distribution of substance-P, GnRH, met-enkephalin in the central nervous system of the pig. *Brain Res. Bull.* (1991) 26:511-514.

Lechan, R.M., P. Wu, and I.M.D. Jackson. Immunolocalization of the thyrotropin-releasing hormone prohormone in the rat central nervous system. *Endocrinology* (1986) 119:1210-1216.

Leeman, S.E. Substance P and neurotensin: discovery, isolation, chemical characterization and physiological studies. *J. Exp. Biol.* (1980) 89:193-200.

Lehmann, J. Neuropeptide-Y: an overview. *Drug Devel. Res.* (1990) 19:329-351.

Leibowitz, S.F. Hypothalamic neuropeptide Y, galanin, and amines. Concepts of coexistence in relation to feeding behavior. In: Schneider, L.H., S.J. Cooper, and K.A. Halmi (eds). *The Psychology of Human Eating Disorders: Preclinical and Clinical Perspectives.* (*Ann. N. Y. Acad. Sci.* Vol. 575, 1989) pg 221-233.

Maggio, J.E. Tachykinins. *Annu. Rev. Neurosci.* (1988) 11:13-28.

Mansour, A., H. Khachaturian, M.E. Lewis, H. Akil, and S.J. Watson. Autoradiographic differentiation of mu, delta, and kappa opioid receptors in the rat forebrain and midbrain. *J. Neurosci.* (1987) 7:2445-2464.

Merchenthaler, I., and D.E. Lennard. The hypophysiotrophic neurotensin-immunoreactive neuronal system of the rat brain. *Endocrinology* (1991) 129:2875-2880.

Merchenthaler, I., F.J. Lopez, D.E. Lennard, and A. Negro-Vilar. Sexual differences in the distribution of neurons coexpressing galanin and luteinizing hormone-releasing hormone in the rat brain. *Endocrinology* (1991) 129:1977-1986.

Morel, G., J. Besson, G. Rosselin, and P.M. Dubois. Ultrastructural evidence for endogenous vasoactive intestinal peptide-like immunoreactivity in the pituitary gland. *Neuroendocrinol.* (1982) 34:85-89.

Mutt, V., K. Fuxe, T. Hokfelt, and J.M. Lundberg. *Neuropeptide Y.* (Raven Press, New York, 1989), pp 353.

Myers, B.M., and M.J. Baum. Facilitation of copulatory performance in male rats by naloxone: effects of hypophysectomy, 17α-estradiol and luteinizing hormone releasing hormone. *Pharm. Biochem. Behav.* (1980) 12:365-370.

O'Halloran, D.J., and S.R. Bloom. Calcitonin gene related peptide - a major neuropeptide and the most powerful vasodilator known. *Br. Med. J.* (1991) 302:739-740.

Olds, J. Hypothalamic substrates of reward. *Physiol. Rev.* (1962) 42:554-604.

Papadopoulos, G.C., J. Antonopoulos, A.N. Karamanlidis, and H. Michaloudi. Coexistence of neuropeptides in the hypothalamic paraventricular nucleus of the sheep. *Neuropeptides* (1990) 15:227-233.

Pfaff, D.W. Impact of estrogens on hypothalamic nerve cells: ultrastructural, chemical and electrical effects. *Rec. Prog. Horm. Res.* (1983) 39:127-175.

Phillips, M.I. Functions of angiotensin in the central nervous system. *Annu. Rev. Physiol.* (1987) 49:413-435.

Pimstone, B.L., M. Sheppard, B. Shapiro, S. Kronheim, A. Hudson, S. Hendricks, and K. Waligora. Localization in and release of somatostatin from brain and gut. *Fed. Proc.* (1979) 38:2330-2332.

Rance, N.E., and W.S. Young. Hypertrophy and increased gene expression of neurons containing neurokinin-B, substance-P messenger ribonucleic acids in the hypothalami of postmenopausal women. *Endocrinology* (1991) 128:2239-2247.

Redmond, D.E., and J.H. Krystal. Multiple mechanisms of withdrawal from opioid drugs. *Annu. Rev. Neurosci.* (1984) 7:443-478.

Rehfeld, J.F. The expression of progastrin, procholecystokinin and their hormonal products in pituitary cells. *J. Mol. Endocrinol.* (1988) 1:87-94.

Rokaeus, A. Galanin: a newly isolated biologically active neuropeptide. *Trends Neurosci.* (1987) 10:158-164.

Schumacher, M., H. Coirini, D.W. Pfaff, and B.S. McEwen. Behavioral effects of progesterone associated with rapid modulation of oxytocin receptors. *Science* (1990) 250:691-694.

Sirinathsinghji, D.J.S., P.E. Whittington, A. Audsley, and H.M. Fraser. β-Endorphin regulates lordosis in female rats by modulating LH-RH release. *Nature* (1983) 301:62-64.

Sofroniew, M.V., and U. Schrell. Evidence for direct projection from oxytocin and vasopressin neurons in the hypothalamic paraventricular nucleus to the medulla oblongata: Immunohistochemical visualization of both the horseradish peroxidase transported and the peptide produced by the same neuron. *Neurosci. Lett.* (1981) 22:211-217.

Tempel, A., J.A. Kessler, and R.S. Zukin. Chronic naltrexone treatment increases expression of preproenkephalin and preprotachykinin messenger RNA in discrete brain regions. *J. Neurosci.* (1990) 10:741-747.

Uhl, G.R., J. Evans, M. Parta, C. Walworth, K. Hill, C. Sasek, M. Voigt, and S. Reppert. Vasopressin and somatostatin mRNA in situ hybridization. In: Uhl, G.D. (ed). *In Situ Hybridization in Brain* (Plenum Press, New York, 1986), pg 21-47.

Unger, R.H., R.E. Dobbs, and L. Orci. Insulin, glucagon, and somatostatin secretion in the regulation of metabolism. *Annu. Rev. Physiol.* (1978) 40:307-343.

Witt, D.M., and T.R. Insel. A selective oxytocin antagonist attenutates progesterone facilitation of female sexual behavior. *Endocrinology* (1991) 128:3269-3276.

Wollemann, M. Recent developments in the research of opioid receptor subtype molecular characterization. *J. Neurochem.* (1990) 54:1095-1101.

Wu, P., R.M. Lechan, and I.M.D. Jackson. Identification and characterization of thyrotropin-releasing hormone precursor peptides in rat brain. *Endocrinology* (1987) 121:108-115.

Chapter 6

CORTICOTROPIN AND MELANOTROPIN

The present chapter will discuss the neuroendocrinology of the pars intermedia as well as the endocrinology of the corticotroph cells located in the pars anterior and pars distalis. As stated in Chapter 2, the correct name for the corticotroph-containing part of the adenohypophysis differs among species. However, in this discussion the name pars anterior will be used for all species, including those in which pars distalis is the more appropriate term. The present discussions of pars intermedia cells and adenohypophysial corticotrophs have been combined into one chapter because their secretory products are all derived from a common precursor, pro-opiomelanocortin (see Figure 5-2).

The hormonal secretions of the pars intermedia have been studied most extensively in submammalian species, especially fish, amphibians, and reptiles. It has been known for many years that the pars intermedia of these species secretes melanocyte-stimulating hormone which, for purposes of uniform nomenclature in this book, will be called ***melanotropin*** and abbreviated MSH. The bioactivity of MSH involves activation of melanocyte cells in the dermis of the skin. MSH acts to disperse intracellular granules of melanin within melanocytes, which results in a darkening of the skin.

After the discovery of pro-opiomelanocortin (POMC), it was recognized that there existed three different forms of MSH. α-MSH contains 13 residues and it is derived from the N-terminus of ACTH (Figure 6-1). After cleavage from ACTH, the N-terminus of ACTH (1-13) is acetylated and the C-terminus is amidated to yield α-MSH. The acetylation of the N-terminus is essential for melanocyte-stimulating bioactivity. This enhancement of bioactivity by acetylation of α-MSH is opposite to that occurring when the β-endorphin molecule is N-acetylated because most of the opioid bioactivity in that molecule is lost after acetylation. The cleavage of ACTH (1-39) to form α-MSH also generates a molecule known as corticotropin-like intermediate peptide (CLIP) that contains 22 residues corresponding to ACTH (18-39). A definitive function for CLIP has not yet been discovered.

A second form of MSH, known as β-MSH (1-18), is derived from the sequential cleavage of POMC to β-lipotropin and then to γ-lipotropin as illustrated previously in Figure 5-3. The third form of MSH, known as γ-MSH (1-27), was discovered in the N-terminal fragment of POMC. All three forms of MSH share a common sequence of seven amino acids at positions #4 through #10 of each molecule, and this common core is identified in Figure 6-1 at positions #4 through #10 of ACTH. The N-terminus of all three forms of MSH appears to be acetylated. Although endogenous α-MSH exerts endocrine actions on melanocytes in submammalian species and exogenous β-MSH or γ-MSH imitates these effects, it is not clear whether endogenous β-MSH and γ-MSH act via the blood. Therefore, α-MSH is a hormone produced by adenohypophysial cells of

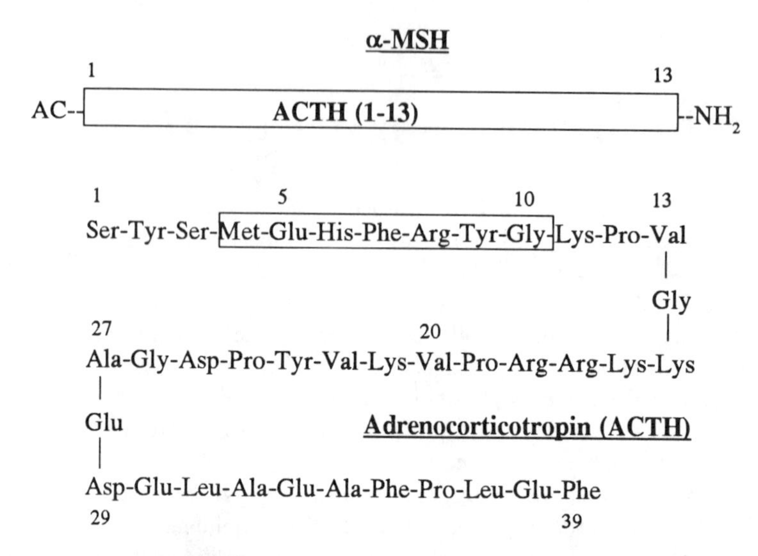

Figure 6-1. Amino acid sequences of α-MSH and ACTH.

Box drawn around residues #4 through #10 of ACTH denotes a common sequence of seven amino acids also found in γ-MSH and β-MSH. The sequence of 13 amino acids in α-MSH is identical to residues #1 through #13 of ACTH from which it is cleaved. The N-terminus of α-MSH is acetylated (AC) and the C-terminus is amidated (NH_2).

the pars intermedia as well as a neuropeptide produced by neurons of the arcuate nucleus. β-MSH and γ-MSH should be considered only as neuropeptides at the present time.

Blood-borne α-MSH originates in the pars intermedia and, as stated in Chapter 2, many species such as primates lack a defined pars intermedia. However, such defined tissue is present during fetal development in humans, but it produces a form of α-MSH which lacks the acetylation required for bioactivity (Tilders et al., 1981).

Hormonal secretion from the pars intermedia (primarily α-MSH and acetylated forms of β-endorphin) is under the inhibitory influence of thc hypothalamus mediated by secretomotor innervation involving dopamine and perhaps other aminergic neurotransmitters (Jackson and Lowry, 1983). Surgical disconnection of the hypothalamus and hypophysis in sheep increased the quantities of MSH and acetylated β-endorphin in the neurointermediate lobe which included the pars intermedia. Administration of dopamine agonist drugs to such operated sheep partially reversed the effects of hypothalamic disconnection (Smith et al., 1989). Many of the dopaminergic axons that innervate the pars intermedia originate in the periventricular area of the hypothalamus (Goudreau et al., 1992). In summary, synthesis and secretion of POMC-derived peptides in the pars intermedia appears to be tonically inhibited by hypothalamic axons that release dopamine at their terminals. Secretomotor innervation to regulate these cells is consistent with the poorly vascularized nature of the pars intermedia tissue relative to the extreme vascularity of the pars anterior tissue. Moreover, adrenocortical hormones from the adrenal gland which feedback and inhibit corticotroph cells of the pars anterior do not regulate the POMC cells of the pars intermedia.

Recent evidence suggests that α-MSH secreted by the pars intermedia may act upon the pars anterior to recruit additional lactotroph cells to release stored PRL. Immunoneutralization of α-MSH increased the magnitude of PRL secretion induced by stress and also increased the proportion of lactotrophs that secreted PRL in estrogen-treated rats (Khorram et al., 1985; Ellerkmann et al., 1992).

As noted in Figure 5-3, pars intermedia cells cleave β-endorphin (1-31) from β-lipotropin, but these cells rapidly process this potent molecule into the biologically weak molecules of acetylated β-endorphin (1-31), δ-endorphin (1-27) and other additionally processed molecules. These weak opioid peptides may be secreted into the blood and quantified by some radioimmunoassay procedures. However, there is no convincing evidence that they exert biological effects in the periphery.

In horses that possess tumors of the pars intermedia and a clinical diagnosis of Cushing's disease (excessive adrenocortical secretion), there is increased production of POMC combined with reduced processing of its cleavage products. Therefore, large quantities of ACTH, β-lipotropin and bioactive β-endorphin can be detected in the tumorous tissue of the pars intermedia and in the blood of such horses (Wilson et al., 1982; Millington et al., 1988). By contrast in normal pars intermedia tissue of horses and other mammals, ACTH would be cleaved into α-MSH and CLIP, and the β-endorphin would be processed into various molecules lacking very much opioid bioactivity.

Neurogenic Actions of MSH-Related Peptides. In addition to effects on skin pigmentation in submammalian species, there is evidence that MSH-related peptides may influence higher nervous function in mammals (De Weid, 1980). Administration of the seven amino acid sequence common to ACTH (#4 through #10; Figure 6-1) or any one of the three forms of MSH can profoundly modify learning and retention of avoidance behaviors in rodents (see Chapter 14). There are also reports of learning benefits in human subjects administered this seven-residue peptide. If these effects are related to putative effects of endogenous peptides that contain this sequence, the endogenous molecules are probably not blood-borne because increased blood levels of the endogenous peptides do not correlate with performance of these behaviors. Of course, POMC and its products are also produced in hypothalamic neurons that project to many CNS areas, and effects on higher brain function may reflect neuropeptide effects of MSH-related peptides. Another POMC-derived peptide, des-tyrosine-β-endorphin, has been postulated to also modify higher brain functions (De Wied, 1980). This peptide is formed after cleavage of the N-terminal tyrosine of β-endorphin or its processed products, and small quantities of these des-tyrosine peptides have been detected *in vivo*. Administration of des-tyrosine forms of β-endorphin can produce neurogenic effects, but it is not known if these effects reflect physiological events.

Secretion of POMC Products by Pars Anterior

Corticotropin (ACTH), β-lipotropin, and β-endorphin (1-31) are the three principal bioactive products derived from POMC in pars anterior tissue. As shown in Figure 5-3, corticotroph cells of the pars anterior cleave POMC into all three peptides. The full 39-amino acid sequence of ACTH is depicted in Figure 6-1, but it should be noted that only residues #1 through #24 are required for bioactivity. It is well documented that physiological and unstressed levels of blood-borne ACTH regulate, on a minute-to-minute basis, the secretion of glucocorticoids from the adrenal cortex. The major glucocorticoid secreted into blood varies among species, with cortisol being the most prevalent in humans and corticosterone being primary in rats. Other species secrete a mixture of cortisol and corticosterone. The term ***glucocorticoid*** will be used herein to denote the primary hormonal secretion of the adrenal cortex under the influence of ACTH. Because the output of glucocorticoid is so closely tied to ACTH, the secretory profiles of glucocorticoid can often be used as an index of ACTH secretory profiles.

Patterns of ACTH Release. Basal nonstress profiles of ACTH are episodic with both regular and irregular elements. The regular episodes of secretion can recur at intervals of about 24 h (i.e., circadian) or at intervals less than 24 h (i.e., ultradian). A wide variety of stressors can provoke immediate irregular discharges of ACTH (stress-related secretion).

Circadian increases of ACTH (or glucocorticoids) are invariably composed of many transient increases occurring with greater frequency and/or amplitude during the period of the observed circadian increase. In some cases, these circadian increases may be entrained by photoperiod, especially in those species such as rodents, in which transitions from light to darkness greatly modify behavior. Maximum glucocorticoid secretion occurs in rats just before the onset of darkness, the period in which nocturnal species such as rats are behaviorally most active. In humans, maximum glucocorticoid secretion is observed during the later portion of the sleep period, again shortly before initiation of behavioral activity. In humans deprived of time cues for several days and allowed to express free-running rhythms, maximum secretion of glucocorticoid coincided with periods of voluntary sleep (Weitzman, 1980). In other animal species, such as domestic ungulates, circadian patterns of glucocorticoid (and presumably ACTH) are often statistically significant but usually less robust than those observed in rodents or primates. It should be noted that the environments in which these domestic ungulates were studied often had environmental elements (i.e., feeding, milking, and other human-animal interactions) that recurred regularly at circadian intervals. Therefore, one cannot be certain whether these species possess a free-running circadian rhythm of glucocorticoid secretion or one that is entrained by human-animal or other environmental influences recurring at about the same time each day. When ultradian rhythms of glucocorticoid are observed (Holaday et al., 1977; Fulkerson and Tang, 1979), it seems more likely that the rhythms are free-running. The period of such ultradian rhythms of adrenocortical output usually ranges from 45 to 90 min, and it is unlikely that any environmental cues would recur at these short intervals.

Neural mechanisms that entrain circadian rhythms of pituitary-adrenal function as well as other parameters appear to involve the suprachiasmatic nucleus (SCN) of the hypothalamus. Lesions of the SCN abolish the circadian regularity of many parameters. Such observations do not, however, prove that the timing mechanisms reside in the SCN. An experiment that addressed this important question involved the discovery and investigation of a mutant strain (called the ***tau*** strain) of hamsters in which the period of the free-running circadian rhythm was 20 h rather than 24 h. Transplantation of fetal tissue containing the SCN could restore circadian rhythms to hamsters from either strain in which SCN ablation had abolished them. Moreover, the period of the restored free-running rhythm corresponded to that of the donor rather than the recipient strain (Ralph et al., 1990). Therefore, the SCN appears to contain mechanisms that govern the exact timing of free-running circadian rhythms.

A wide variety of stressful stimuli (i.e., stressors) acutely increase the release of ACTH and thereby increase glucocorticoid secretion. The ability of acute stressors to increase ACTH is so general that it is sometimes an operational definition of the stress response. Besides the stress-induced release of ACTH, serum concentrations of β-endorphin (1-31) and perhaps β-lipotropin also increase abruptly after stress (Guillemin et al., 1977; Engler et al., 1988). This result is not surprising because adenohypophysial corticotroph cells process the POMC precursor into all three of these peptides. The hypothalamic neurohormones (CRH and vasopressin) which may mediate the stress-related activation of the corticotroph cells will be discussed later. While the effects of acute stressors on the corticotroph cells are well understood, the effects of chronically applied stressors are more complex. Presumably, chronic exposure of corticotrophs to stimulatory hypothalamic neurohormones leads to a hypersecretory state, but the profile of peptides secreted by the hypophysis may change (DeSouza and VanLoon, 1985; Young and Akil, 1985).

Genetically Stress-Susceptible Pigs. A syndrome called porcine stress-susceptibility (PSS) exists in certain strains of pigs. Severely affected animals that are exposed to relatively mild environmental stressors may experience a muscular rigidity and hyperthermia that can sometimes result in death. For this reason, the PSS syndrome is also known as porcine malignant hyperthermia, and animals can be screened for the trait by exposure to the anesthetic gas halothane. Another consequence of PSS occurs when less severely affected pigs are slaughtered for meat. Due to unique conditions existing in the skeletal muscle after death, the meat undergoes a series of postmortem changes that decrease its consumer appeal (Mitchell and Heffron, 1982).

Because of the economic consequences both in terms of death losses and reduced value of the meat, much research has been conducted on the PSS syndrome. Chronic overstimulation of the pituitary-adrenocortical axis was hypothesized, but extensive quantification of glucocorticoid secretion failed to support the hypothesis (Lister et al., 1972; Marple and Cassens, 1973). Affected pigs do seem to have greater activation of the sympathetic nervous system as reflected in basal urinary concentrations of catecholamines and a stress-induced depletion of catecholamines in the caudate nucleus (Hallberg et al., 1983). However, these indices of sympathetic nervous system function may be consequences of the PSS syndrome rather than causes of it. A recent discovery has shifted attention from the brain-pituitary-adrenal axis to the skeletal muscle. Several strains of pigs that have a high incidence of malignant hyperthermia were found to have a single nucleotide mutation in the gene that encodes the skeletal muscle receptor for ryanodine (Fujii et al., 1991; Otsu et al., 1991). Because ryanodine controls intracellular calcium, it seems possible that a calcium-related defect in skeletal muscle may be responsible for secondary stress-like effects in the brain-pituitary-adrenal axis. Similar mutations in the human gene that encodes ryanodine appear to be responsible for the occurrence of malignant hyperthermia in anesthetized humans (MacLennan and Phillips, 1992). However, genetically affected humans do not have increased susceptibility to stress.

Patterns of β-Endorphin Release. Concentrations of immunoreactive β-endorphin fluctuate episodically in the peripheral circulation. In almost all cases, there is coincidence between the secretory pulses of β-endorphin and those of either ACTH and/or glucocorticoids (DeSouza and VanLoon, 1985; Engler et al., 1988; Iranmanesh et al., 1989). Such coincidence is to be expected since exogenous CRH provokes the release of both ACTH and β-endorphin in a variety of experimental models in rats (Rivier et al., 1982).

It should be noted that only a portion of the immunoreactive β-endorphin quantified in peripheral blood is the fully bioactive form (β-endorphin 1-31). It also remains to be determined whether this bioactive β-endorphin (1-31) exerts an endocrine action (i.e., via blood) or is only a secretory by-product of the release of ACTH by corticotrophs of the pars anterior. For example, experimental electroacupuncture (EA) of two different body regions in sheep elevated circulating β-endorphin and caused cutaneous analgesia (Bossut et al., 1986). Although the induced elevation of β-endorphin was similar after EA of the two regions, only the analgesia caused by EA in one region could be antagonized by intravenous naloxone. This result showed that increases in blood-borne β-endorphin could not have been generally responsible for the EA-induced analgesia because its opioid effects would have been antagonized by naloxone equally well in both situations. This paradigm illustrates what must be established to prove that blood-borne β-endorphin exerts an endocrine action. When an event hypothesized to have been caused by the blood-borne β-endorphin occurs concurrently with an induced elevation in β-endorphin secretion, the event should be capable of being blocked by systemic administration of naloxone to antagonize those opioid receptors to which blood-borne β-endorphin has access. This criterion is rarely satisfied and, in the opinion of this author, an endocrine action of β-endorphin has yet to be definitively established. However, the molecule is clearly a neuropeptide acting locally in a paracrine and/or neurocrine manner (see Chapter 5).

Neuroendocrine Integration of ACTH Release. As stated in Chapter 5 and illustrated in Figure 6-2, there are two hypothalamic neurohormones that stimulate release of ACTH from the pars anterior. Both CRH and vasopressin are secreted into hypophysial portal blood by axons which originate primarily from parvocellular perikarya in the PVN and adjacent periventricular structures. Under normal conditions, there are separate perikarya for CRH and vasopressin, but after adrenalectomy to remove glucocorticoid feedback, both neurohormones can be produced in the same perikaryon. There are receptors for both CRH and vasopressin in the pars anterior, and both peptides stimulate release of ACTH, although their relative potencies differ between rats and sheep (Shen et al., 1990). CRH was relatively more potent than vasopressin at stimulating ACTH in pars anterior cells from rats as compared to sheep. Notably, all the dispersed corticotrophs from rats that responded to vasopressin also responded to CRH, whereas some ACTH-containing cells of the rat pars anterior responded to CRH but not to vasopressin (Jia et al., 1991). Immunoneutralization of blood-borne vasopressin in sheep decreased stress-activated release of ACTH but not basal secretion (Guillaume et al., 1992). Concentrations of both CRH and vasopressin in the hypophysial portal blood of rats and sheep increased in experimental situations involving hypersecretion of ACTH (Fink et al., 1988; Canny et al., 1989). However, the stimulus of acute exercise in horses appears to increase the secretion of vasopressin selectively over that of CRH (Alexander et al., 1991). However, blood collection in exercised horses involved hypophysial venous blood rather than hypophysial portal blood, and this difference may complicate the interpretation.

Figure 6-2 denotes neural influences of stress (physical and psychogenic) and the circadian oscillator on CRH and vasopressin neurons. However, aminergic and peptidergic neurotransmitters that influence the release of CRH and vasopressin into hypophysial portal blood are only partially understood. Acetylcholine is thought to be stimulatory to neural elements that secrete neurohormones stimulatory to ACTH. A similar action of angiotensin II has also been observed. The effect of norepinephrine is probably indirectly inhibitory to ACTH, but other effects have also been observed. In summary, the neural inputs to the CRH and vasopressin neurons illustrated in Figure 6-2 are very complex, and more work is needed before they can be fully understood.

Anatomical locations of brain structures that mediate stress-induced release of ACTH are only partly known. Early research showed clearly that destruction of the median eminence blocked stress-related stimulation of ACTH release. Because all hypothalamic influences on ACTH must pass through the median eminence, this result does not really provide any information on localization. Complete neurosurgical deafferentation of the mediobasal hypothalamus in rats

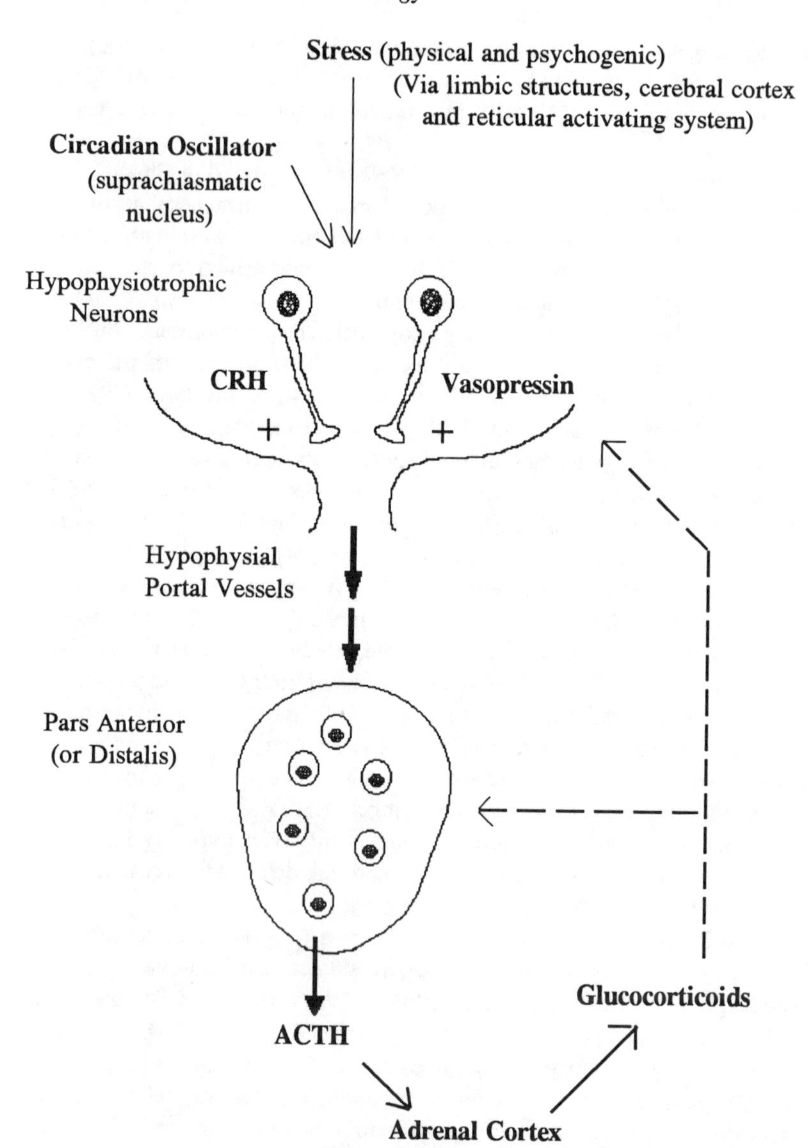

Figure 6-2. Neuroendocrine integration of ACTH release.

This simplified diagram illustrates functional interactions among hypothalamic hypophysiotrophic neurons, hypophysial portal vessels, pars anterior (or distalis) hormones, systemic circulation, and peripheral endocrine glands which are target tissues for adenohypophysial hormones. The diagram depicts schematically mechanisms involved in the neuroendocrine integration of ACTH release. Subsequent chapters of this book contain similar diagrams for each adenohypophysial hormone.

blocked glucocorticoid responses to stress (Markara et al., 1980). Notably, the neurosurgical block was not effective if inputs via the basolateral retrochiasmatic area remained intact.

Inhibitory Feedback from the Adrenal Cortex. The ability of circulating glucocorticoids to feedback in an inhibitory manner on the brain-ACTH axis has been recognized for a very long time. The precise site of this inhibitory feedback has been thoroughly investigated, and evidence exists to support both neural and pituitary sites of action as illustrated in Figure 6-2. Glucocorticoids can act directly on corticotroph cells to antagonize the ability of hypothalamic extracts or purified CRH to stimulate ACTH release either *in vivo* or *in vitro*. Early studies in which glucocorticoids were implanted locally into various brain regions revealed that feedback at a CNS site was also possible. Systemically administered glucocorticoids also decreased the concentrations of CRH and vasopressin in hypophysial portal blood, indicating a neural site of action. Immunocytochemical staining for the glucocorticoid receptor has recently become possible, and this approach demonstrated that most CRH-containing cells of the PVN also stain for the glucocorticoid receptor (Cintra et al., 1987). In addition, hypothalamic perikarya located in the ARC nucleus, and which contain GHRH or NPY, also stain for this same glucocorticoid receptor (Hisano et al., 1988). The potential role of these GHRH and NPY neurons possessing glucocorticoid receptors in control of the ACTH-adrenal axis remains to be determined. However, not all neurons with receptors for glucocorticoid are necessarily part of the feedback system governing ACTH release. For example, there are many neurons having glucocorticoid receptors in the hippocampus and amygdala, but glucocorticoid hormones could exert other effects on such neurons.

Blood-borne glucocorticoids can interact with either of two receptors designated, respectively, Type I (also called high-affinity or mineralocorticoid) and Type II (also called low-affinity or glucocorticoid). Low-affinity Type II receptors are highly specific for glucocorticoid molecules and produce maximal biological responses, but activation of these low-affinity receptors requires high levels of glucocorticoid such as occur only during stress or maximal circadian release. High-affinity Type I receptors bind mineralocorticoids (e.g., aldosterone) as well as glucocorticoids, but ligand-activated Type I receptors produce only modest biological responses (Arriza et al., 1988). Because of their high affinity and lack of specificity, Type I receptors are activated by low concentrations of circulating glucocorticoid such as occur during basal, nonstress conditions. Because low levels of glucocorticoid can activate Type I receptors, whereas high levels are required to activate Type II receptors, the range of concentrations over which inhibitory feedback of ACTH release can occur is greatly increased (Bradbury et al., 1991).

The molecular structures of Type I and Type II receptors for glucocortioids are known, and both molecules are members of a *receptor superfamily* which includes receptors for thyroid hormones, estrogen, androgen, and progesterone (Evans, 1988). The morphological distributions of each glucocorticoid receptor subtype are clearly different (Reul and de Kloet, 1986). Low-affinity Type II receptors have been demonstrated in both CRH and vasopressin perikarya of the hypothalamus (Uht et al., 1988). In contrast, high-affinity Type I receptors are found only in areas of the limbic system outside the hypothalamus (Reul and de Kloet, 1986). Feedback at the level of the adenohypophysis appears to require high levels of glucocorticoid acting via low-affinity Type II receptors (Bradbury et al., 1991).

One other feedback mechanism that is not illustrated in Figure 6-2 involves direct neural inputs from the adrenal gland to the CNS. After one adrenal gland was surgically removed, it was possible to observe morphological changes in the hypothalamus that differed between the left and right halves of the brain. In addition, alterations in hypothalamic mechanisms regulating pituitary ACTH occurred so rapidly after bilateral adrenalectomy that circulating glucocorticoids had not yet declined (Engeland et al., 1980). The functional importance of direct neural signals from adrenal gland to CNS has not yet been determined.

Another possible feedback mechanism not illustrated in Figure 6-2 is the action of secreted ACTH on neural structures that govern the release of CRH and vasopressin into hypophysial portal blood. This mechanism is called ***short feedback*** to distinguish it from the feedback of hormones secreted by target organs of the hypophysial hormones. Molecules of ACTH that exert this short feedback could be delivered to the hypothalamus in the general circulation or possibly ascend the hypophysial stalk by retrograde delivery (see Chapter 2). To experimentally demonstrate that short feedback can suppress the brain-ACTH axis, it is necessary to eliminate feedback of glucocorticoids secreted in response to exogenous ACTH. When this criterion is satisfied, the ability of exogenous ACTH to suppress CRH or vasopressin and endogenous ACTH can sometimes be demonstrated. However, many days or even weeks of ACTH administration are usually required before such an effect can be observed. Therefore, short feedback of ACTH to indirectly suppress its own secretion cannot explain the rapid episodic fluctuations of ACTH release observed *in vivo*. In fact, the physiological role of short feedback remains to be established.

Adrenocortical Actions of ACTH. Acting through receptors located in the plasma membrane of glucocorticoid-secreting cells, ACTH regulates the minute-to-minute release of cortisol, corticosterone and other glucocorticoids. There is some evidence that the neuropeptides VIP and CGRP, discussed in Chapter 5, may synergize with ACTH to enhance glucocorticoid secretion from the bovine adrenal cortex (Bloom et al., 1987, 1989). Electrical stimulation of the splanchnic nerves to the adrenal medulla can also enhance the ability of ACTH to stimulate the release of glucocorticoid. Whether this mechanism is physiological and whether it occurs in species other than cattle is not known.

In contrast to glucocorticoid secretion, where ACTH is the major regulator, aldosterone secreted by the glomerular zone of the adrenal cortex is primarily regulated by other factors such as blood-borne angiotensin II. ACTH can directly stimulate aldosterone-secreting cells, but this action, even if physiological, is relatively minor.

REFERENCES

Alexander, S.L., C.H.G. Irvine, M.J. Ellis, and R.A. Donald. The effect of acute exercise on the secretion of corticotropin-releasing factor, arginine vasopressin, and adrenocorticotropin as measured in pituitary venous blood from the horse. *Endocrinology* (1991) 128:65-72.

Arriza, J.L., R.B. Simerly, L.W. Swanson, and R.M. Evans. The neuronal mineralocorticoid receptor as a mediator of glucocorticoid response. *Neuron* (1988) 1:887-900.

Bloom, S.R., A.V. Edwards, and C.T. Jones. Adrenal cortical responses to vasoactive intestinal peptide in conscious hypophysectomized calves. *J. Physiol.* (1987) 391:441-450.

Bloom, S.R., A.V. Edwards, and C.T. Jones. Adrenal responses to calcitonin gene-related peptide in conscious hypophysectomized calves. *J. Physiol.* (1989) 409:29-41.

Bossut, D.F.B., M.W. Stromberg, and P.V. Malven. Electroacupuncture-induced anaglesia in sheep: Measurement of cutaneous pain thresholds and plasma concentrations of prolactin and β-endorphin immunoreactivity. *Am. J. Vet. Res.* (1986) 47:669-676.

Bradbury, M.J., S.F. Akana, C.S. Cascio, N. Levin, L. Jacobson, and M.F. Dallman. Regulation of basal ACTH secretion by corticosterone is mediated by both type I (MR) and type II (GR) receptors in rat brain. *J. Steroid Biochem. Molec. Biol.* (1991) 40:133-142.

Canny, B.J., J.W. Funder, and I.J. Clarke. Glucocorticoids regulate ovine hypophysial portal levels of corticotropin-releasing factor and arginine vasopresin in a stress-specific manner. *Endocrinology* (1989) 125:2532-2539.

Cintra, A., K. Fuxe, A. Harfstrand, L.F. Agnati, A.C. Wikstrom, S. Okret, W. Vale, and J.A. Gustafsson. Presence of glucocorticoid receptor immunoreactivity in corticotropin releasing factor and in growth hormone releasing factor neurons of the rat di- and telencephalon. *Neurosci. Lett.* (1987) 77:25-30.

De Souza, E.B., and G.R. Van Loon. Differential plasma β-endorphin, β-lipotropin, and adrenocorticotropin responses to stress in rats. *Endocrinology* (1985) 116:1577-1586.

De Wied, D. Hormonal influences on motivation, learning, memory, and psychosis. In: Krieger,D.T., and J.C. Hughes (eds). *Neuroendocrinology* (Sinauer Associates Inc., Sunderland, MA, 1980), pg 194-204.

Ellerkmann, E., G.M. Nagy, and L.S. Frawley. Alpha-melanocyte-stimulating hormone is a mammotrophic factor released by intermediate lobe cells after estrogen treatment. *Endocrinology* (1992) 130:133-138.

Engeland, W.C., F. Siedenburg, C.W. Wilkinson, J. Shinsako, and M.F. Dallman. Stimulus-induced corticotropin-releasing factor content and adrenocorticotropin release are augmented after unilateral adrenalectomy independently of circulating corticosteroid levels. *Endocrinology* (1980) 106:1410-1415.

Engler, D., T. Pham, M.J. Fullerton, J.W. Funder, and I.J. Clarke. Studies of the regulation of the hypothalamic-pituitary adrenal axis in sheep with hypothalamic-pituitary disconnection. I. Effect of an audiovisual stimulus and insulin-induced hypoglycemia. *Neuroendocrinology* (1988) 48:551-560.

Evans, R.M. The steroid and thyroid hormone receptor superfamily. *Science* (1988) 240:889-895.

Fink, G., I.C.A.F. Robinson, and L.A. Tannahill. Effects of adrenalectomy and glucocorticoids on the peptides CRF-41, AVP and oxytocin in rat hypophysial portal blood. *J. Physiol.* (1988) 401:329-345.

Fujii, J., K. Otsu, F. Zorazato, S. De Leon, V.K. Khanna, J.E. Weiler, P.J. O'Brien, and D.H. MacLennan. Identification of a mutation in porcine ryanodine receptor associated with malignant hyperthermia. *Science* (1991) 253:448-451.

Fulkerson, W.J., and B.Y. Tang. Ultradian and circadian rhythms in the plasma concentration of cortisol in sheep. *J. Endocrinol.* (1979) 81:135-141.

Goudreau, J.L., S.E. Lindley, K.J. Lookingland, and K.E. Moore. Evidence that hypothalamic periventricular dopamine neurons innervate the intermediate lobe of the rat pituitary. *Neuroendocrinology* (1992) 56:100-105.

Guillaume, V., B. Conte-Devolx, E. Magnan, F. Boudouresque, M. Grino, M. Cataldi, L. Muret, A. Priou, J.C. Figaroli, and C. Oliver. Effect of chronic active immunization with antiarginine vasopressin on pituitary-adrenal function in sheep. *Endocrinology* (1992) 130:3007-3014.

Guillemin, R., T. Vargo, J. Rossier, S. Minick, N. Ling, C. Rivier, W. Vale, and F. Bloom. β-Endorphin and adrenocrticotropin are secreted concomitantly by the pituitary gland. *Science* (1977) 197:1367-1369.

Hallberg, J.W., D.D. Draper, D.G. Topel, and D.M. Altrogge. Neural catecholamine deficiencies in the porcine stress syndrome. *Am. J. Vet. Res.* (1983) 44:368-371.

Hisano, S., Y. Kagotani, Y. Tsuruo, S. Daikoku, K. Chihara, and M.H. Whitnall. Localization of glucocorticoid receptor in neuropeptide Y-containing neurons in arcuate nucleus of the rat hypothalamus. *Neurosci. Lett.* (1988) 95:13-18.

Holaday, J.W., H.M. Martinez, and B.H. Natelson. Synchronized ultradian cortisol rhythms in monkeys: persistence during corticotropin infusion. *Science* (1977) 198:56-58.

Iranmanesh, A., G. Lizarralde, M.L. Johnson, and J.D. Veldhuis. Circadian, ultradian, and episodic release of β-endorphin in men, and its temporal coupling with cortisol. *J. Clin. Endocrinol. Metab.* (1989) 68:1019-1026.

Jackson, S., and P.J. Lowry. Secretion of pro-opiocortin peptides from isolated perfused rat pars intermedia cells. *Neuroendocrinology* (1983) 37:248-257.

Jia, L.G., B.J. Canny, D.N. Orth, and D.A. Leong. Distinct classes of corticotropes mediate corticotropin-releasing hormone- and arginine vasopressin-stimulated adrenocorticotropin release. *Endocrinology* (1991) 128:197-203.

Khorram, O., J.C. Bedran de Castro, and S.M. McCann. Stress-induced secretion of alpha-melanocyte-stimulating hormone and its physiological role in modulating the secretion of prolactin and luteinizing hormone in the female rat. *Endocrinology* (1985) 117:2483-2489.

Lister, D., J.N. Lucke, and B.N. Perry. Investigation of the hypothalamic-pituitary-adrenal axis in mesomorphic types of pig. *J. Endocrinol.* (1972) 53:505-506.

MacLennan, D.H., and M.S. Phillips. Malignant hyperthermia. *Science* (1992) 256:789-794.

Markara, G.B., E. Stark, and M. Palkovits. Reevaluation of the pituitary-adrenal response to ether in rats with various cuts around the medial basal hypothalamus. *Neuroendocrinology* (1980) 30:38-44.

Marple, D.N., and R.G. Cassens. A mechanism for stress-susceptibility in swine. *J. Anim. Sci.* (1973) 37:546-549.

Millington, W.R., N.O. Dybdal, R. Dawson, C. Manzini, and G.P. Mueller. Equine Cushing disease: differential regulation of β-endorphin processing in tumors of the intermediate pituitary. *Endocrinology* (1988) 123:1598-1604.

Mitchell, G., and J.J.A. Heffron. Porcine stress syndromes. *Adv. Food Res.* (1982) 28:167-230.

Otsu, K., V.K. Khanna, A.L. Archibald, and D.H. MacLennan. Cosegregation of porcine malignant hyperthermia and a probable casual mutation in the skeletal muscle ryanodine receptor gene in backcross families. *Genomics* (1991) 11:744-750.

Ralph, M.R., R.G. Foster, F.C. Davis, and M. Menaker. Transplanted suprachiasmatic nucleus determines circadian period. *Science* (1990) 247:975-978.

Reul, J.M.H.M., and E.R. de Kloet. Anatomical resolution of two types of corticosterone receptor sites in rat brain with in vitro autoradiography and computerized image analysis. *J. Steroid Biochem.* (1986) 24:269-272.

Rivier, C., M. Brownstein, J. Spiess, J. Rivier, and W. Vale. In vivo corticotropin-releasing factor-induced secretion of adrenocorticotropin, β-endorphin, and corticosterone. *Endocrinology* (1982) 110:272-278.

Shen, P.J., I.J. Clarke, B.J. Canny, J.W. Funder, and A.I. Smith. Arginine vasopressin and corticotropin releasing factor: binding to ovine anterior pituitary membranes. *Endocrinology* (1990) 127:2085-2089.

Smith, A.I., C.A. Wallace, I.J. Clarke, and J.W. Funder. Dopaminergic agents differentially regulate both procesing and content of alpha-N-acetylated endorphin and alpha-MSH in the ovine pituitary intermediate lobe. *Neuroendocrinology* (1989) 49:545-550.

Tilders, F.J.H., C.R. Parker, A. Barnea, and J.C. Porter. The major immunoreactive alpha-melanocyte-stimulating hormone (alpha-MSH)-like substance found in human fetal pituitary tissue is not alpha-MSH but may be desacetyl-alpha-MSH (adrenocorticotropin$_{1-13}$NH$_2$). *J. Clin. Endocrinol. Metab.* (1981) 52:319-323.

Uht, R.M., J.F. McKelvy, R.W. Harrison, and M.C. Bohn. Demonstration of glucocorticoid receptor-like immunoreactivity in glucocorticoid-sensitive vasopressin and corticotropin-releasing factor neurons in the hypothalamic paraventricular nucleus. *J. Neurosci. Res.* (1988) 19:405-411.

Weitzman, E.D. Biologic rhythms and hormone secretion patterns. In: Krieger, D.T., and J.C. Hughes (eds). *Neuroendocrinology* (Sinauer Associates Inc., Sunderland, MA, 1980), pg 85-92.

Wilson, M.G., W.E. Nicholson, M.A. Holscher, B.J. Sherrell, C.D. Mount, and D.N. Orth. Proopiolipomelanocortin peptides in normal pituitary, pituitary tumor, and plasma of normal and Cushing's horses. *Endocrinology* (1982) 110:941-954.

Young, E.A., and H. Akil. Corticotropin-releasing factor stimulation of adrenocorticotropin and β-endorphin release: effects of acute and chronic stress. *Endocrinology* (1985) 117:23-30.

Chapter 7

THYROTROPIN

A population of cells of the pars anterior (or pars distalis) that stain with basic dyes (i.e., cells are basophilic) synthesize and secrete thyrotropin into blood for delivery to the thyroid gland. For this reason, thyrotropin is sometimes called thyroid-stimulating hormone and abbreviated TSH. The present chapter will focus on the neural control of TSH release by adenohypophysial cells as well as the neuroendocrine integration involved in regulating the brain-pituitary-thyroid axis.

Thyrotropin is a glycosylated protein of approximately 28,000 daltons (Da), and it consists of two dissimilar subunits that are covalently linked to form the bioactive dimer illustrated in Figure 7-1. The subunits have been designated alpha-TSH and beta-TSH, respectively. The alpha subunit of TSH is identical to the alpha subunit of LH and of FSH as denoted in Figure 7-1. However, the beta subunits of TSH, LH and FSH differ and confer the bioactive specificity to each dimer (Pierce and Parson, 1981).

The actions of TSH on the cells of the thyroid gland are to promote the cellular uptake of iodide ions and the production/release of the iodide-containing thyroid hormones. The primary secretory product of the thyroid gland is ***thyroxine***, also known as tetra-iodothyronine and abbreviated T_4. Although the thyroid gland may also secrete some ***tri-iodothyronine*** (T_3), the major portion

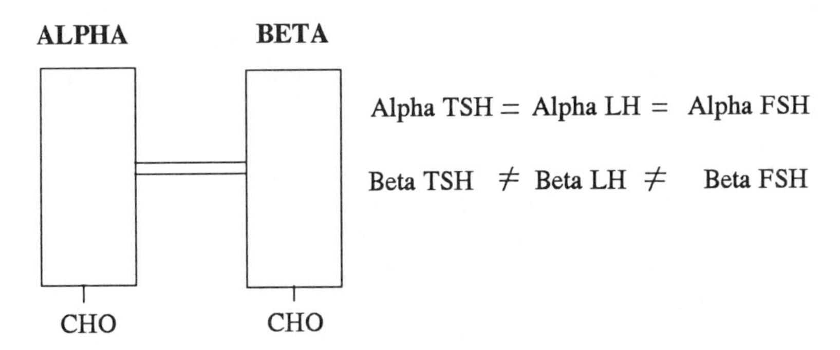

Figure 7-1. Glycosylated subunits of TSH and the gonadotropins, LH and FSH.

The rectangles denoted alpha and beta represent the individual subunits of the adenohypophysial glycoprotein hormones. The double lines between the two rectangles denote the covalent bonds that link the subunits. The CHO symbols attached to each rectangle denote the carbohydrates attached to each subunit during post-translational glycosylation.

of T_3 is formed outside the thyroid in blood and target tissues by cleavage of one iodide molecule from secreted T_4. The bioactive potency of T_3 is much greater than that of T_4 but the duration of its action is much shorter. It is thought that T_3 is probably the form that acts on cellular receptors to produce the stimulation of cellular metabolism for which both T_3 and T_4 are known. The biological actions of thyroid hormones (T_3/T_4) involve a general stimulation of cellular metabolism and oxygen utilization by a majority of the cells in adult animals. Although cells of the adult brain are not generally stimulated by T_3/T_4, cells of the fetal and neonatal brain do respond to T_3/T_4 during development. All the cellular responses to T_3/T_4 are mediated by intracellular receptors that are very similar, if not identical, to the cellular oncogene known as ***erbA***. These intracellular receptors for T_3/T_4 occur in two different forms (alpha and beta), with the beta form expressed in fewer tissues than the alpha form (Bradley et al., 1989). The T_3/T_4 receptor is a member of the ***receptor superfamily*** which also includes receptors for glucocorticoid, estrogen, androgen, and progesterone (Evans, 1988). As occurs for all members of this receptor superfamily, binding of the appropriate ligand, in this case probably T_3, initiates a series of events leading to alteration of gene transcription and translation (Parker, 1988).

The secretory profile of TSH is spontaneously episodic as occurs for all adenohypophysial hormones, and environmental factors that require homeostatic adjustment of general body metabolism can profoundly modify the spontaneous pattern. The secretion of adenohypophysial TSH is also regulated by a closed-loop feedback system involving blood-borne T_3/T_4.

As presented in Chapter 4, secretion of TSH is regulated by the antagonistic actions of GHIH (also known as somatostatin) and TRH (see Figures 4-5 and 4-6). The administration of TRH stimulates TSH release while GHIH administration decreases TSH release, but more definitive evidence of a physiological role for endogenous TRH and GHIH comes from immunoneutralization studies. When endogenous blood-borne TRH was immunoneutralized, serum concentrations of TSH declined in normal rats and in rats exposed acutely to cold temperatures to activate the pituitary-thyroid axis (Koch et al., 1977; Szabo and Frohman, 1977). Immunoneutralization of blood-borne GHIH in rats increased both basal and induced secretion of TSH, which suggests a physiological inhibitory role for endogenous GHIH (Chihara et al., 1978). When the effects of endogenous TRH were also immunoneutralized, there was no longer any effect of GHIH immunoneutralization on TSH secretion, suggesting a functional antagonism between endogenous levels of GHIH and TRH. As expected, experimental immunoneutralization of GHIH markedly increased the secretion of somatotropin but did not affect the release of prolactin. By contrast, immunoneutralization of TRH decreased the release of prolactin as well as of TSH. Exposure of rats to stressors decreased both TSH and somatotropin, but immunoneutralization of GHIH blocked the stress-induced decrease in both hormones. In summary, the pattern of selective and non-selective responses to immunoneutralization of TRH and of GHIH provides strong evidence for the dual control mechanisms illustrated previously in Figure 4-6.

As discussed in Chapter 4, TRH-containing perikarya that project axons to the median eminence are located in the paraventricular nucleus (PVN) and in periventricular areas just ventral to the PVN. Perikarya that contain GHIH, and that also project to the median eminence, are located in the following regions: preoptic area, periventricular parts of the anterior hypothalamus and a few in the PVN. The hypophysiotrophic neurons that release these two TSH-regulating neurohormones into hypophysial portal blood are illustrated diagrammatically in Figure 7-2 as one TRH neuron and one GHIH neuron.

Although not illustrated in Figure 7-2 (or previously in Figure 4-6), there is evidence that another hypothalamic compound, dopamine, acts directly on adenohypophysial TSH-containing cells. Dopamine secreted into hypophysial portal blood is generally regarded as an endogenous prolactin-inhibiting hormone, but dopamine also inhibits the release of TSH in adenohypophysial tissue from normal rats as well as rats made hypothyroid (Foord et al., 1986). These authors also suggest that thyrotroph cells may have autoreceptors for secreted TSH and that activation of these autoreceptors enhances the action of dopamine to suppress further release of TSH. The proponents of this dopamine-mediated inhibitory autofeedback by secreted TSH on the thyrotroph cells acknowledge that this control system probably acts only to regulate fine adjustments in the secretion of TSH.

Neuroendocrine integration of TSH release (Figure 7-2) involves a variety of neurally mediated input to the TRH and GHIH neurons. An animal's perception of the external ambient temperature is a major factor controlling the pituitary-thyroid axis. Peripheral temperature sensors provide neural input that passes through the preoptic area and other CNS regions (Bligh, 1966). If the input from peripheral temperature sensors should prove inadequate to regulate metabolism (via TSH-induced changes in T_3/T_4) or neurogenic thermoregulation, core body temperature may either increase or decrease. Because the hypothalamus is also directly thermosensitive, these marked increases or decreases in core body temperature can activate/deactivate the heat-generation and heat-loss mechanisms of the organism, including modulation of TSH release. As discussed above, acute exposure of an animal to cold rapidly increases TRH release and in turn secretion of TSH and T_3/T_4. Chronic exposure to cold temperatures seems to involve more complex adaptations of many hormonal systems in addition to thyroid hormones. In rodents, acute exposure to stressful stimuli decreases both TSH and GH, presumably through stimulation of GHIH release. Several CNS structures outside the hypothalamus have been implicated in the control of adenohypophysial TSH. These include limbic system structures such as the epithalamus that appear to be inhibitory to the release of TSH, but only in situations that require hypersecretion of TSH (i.e., cold exposure or surgical removal of thyroid gland). Other areas of the limbic system have been implicated in the regulation of TSH through lesion and electrical stimulation studies. Lesions in some areas enhanced the release of TSH, indicative of a tonic inhibition while electrical stimulation of other limbic structures enhanced release

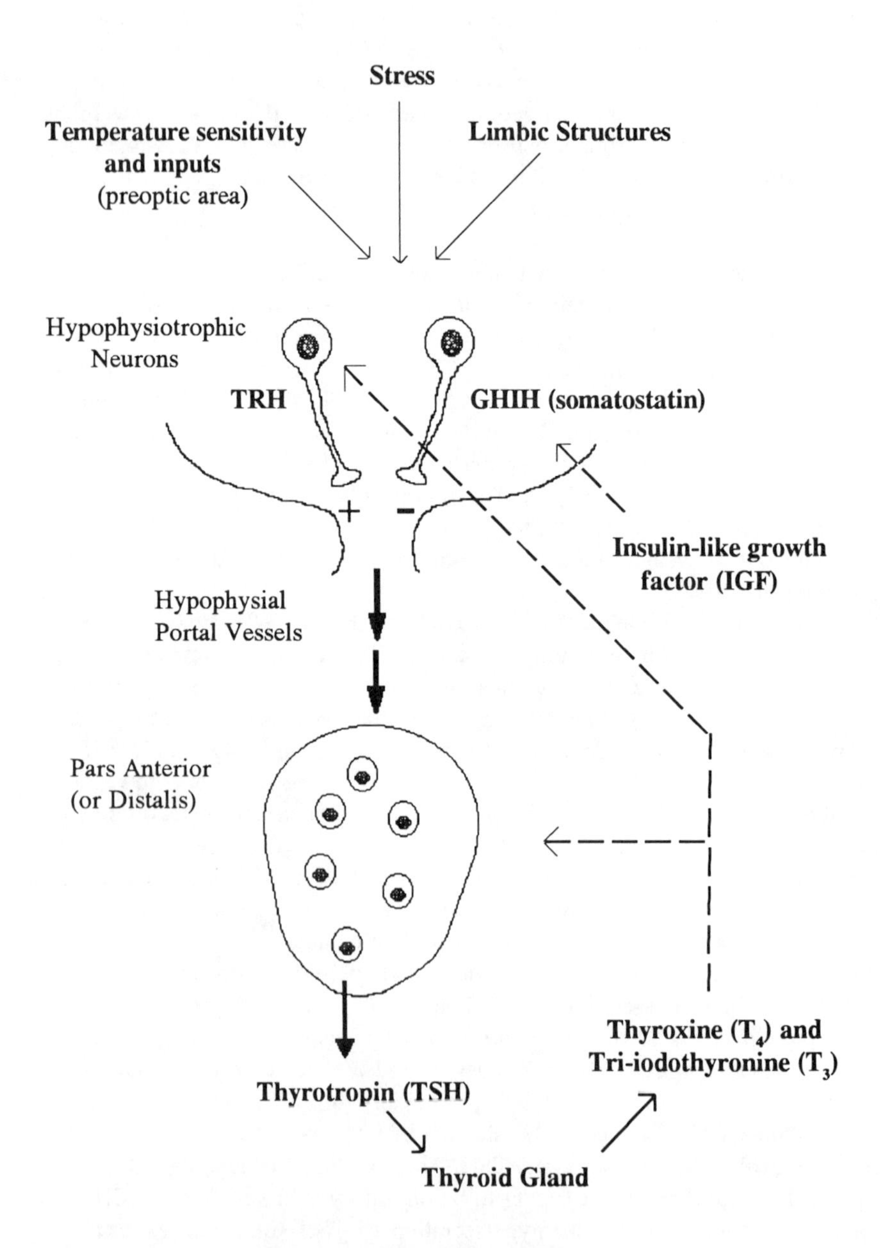

Figure 7-2. Neuroendocrine integration of TSH release.

This simplifed diagram illustrates mechanisms involved in the neuroendocrine integration of TSH release.

of TSH indicative of a stimulatory potential. There may indeed be dual limbic control over release of TSH, but the details and neuroanatomical substrates of such control remain to be determined.

Inhibitory Feedback from the Thyroid Gland. Secretory products of the thyroid gland exert a closed-loop inhibitory feedback that adjusts the release of TSH into blood to maintain homeostasis (Figure 7-2). Although T_4 is the major secretory product, the primary feedback molecule is probably T_3. There is abundant evidence for T_3 feedback at both adenohypophysial and neural sites. One established action of T_3/T_4 on the thyrotroph cells of the adenohypophysis is to antagonize TRH-induced release of TSH, and this action occurs at concentrations that are physiologically relevant.

Definitive evidence for T_3/T_4 feedback on neural structures was obtained only after the precursor of TRH was discovered and its sequence determined. Experimentally induced hypothyroidism was shown to increase hypothalamic transcription of the gene that encodes the TRH precursor (Segerson et al., 1987). Localized intrahypothalamic administration of T_3 was also able to selectively inhibit transcription of this gene in TRH-containing parvocellular perikarya of the PVN (Dyess et al., 1988). These same TRH-containing neurons are known to project their axons to the median eminence (see Chapter 4). Conversion of blood-borne T_4 into T_3 is minimal in the rat PVN, and administration of physiological concentrations of T_3 could not fully suppress PVN transcription of the TRH gene (Kakucska et al., 1992). Therefore, it is likely that blood-borne T_4 contributes to the physiological feedback on TRH synthesis.

As stated earlier, cellular receptors for T_3/T_4 exist in either an alpha or beta form. *In situ* hybridization for each receptor type in the brain and hypophysis revealed that the beta form was less widely distributed than was the alpha form. However, the beta form was highly expressed in pars anterior and in paraventricular nucleus, the two primary sites for inhibitory feedback to regulate TSH release (Bradley et al., 1989). Moreover, the pars anterior appeared to express the alpha form to only a small extent. In summary, T_3/T_4 acts on TRH neurons to decrease their synthesis of TRH and on adenohypophysial thyrotrophs to antagonize TRH-induced release of TSH. This combination of sites and mechanisms of inhibitory feedback probably contributes to the highly effective feedback regulation of the pituitary-thyroid axis.

One other feedback system illustrated in Figure 7-2 involves insulin-like growth factor (IGF), which is also known as somatomedin. The secretion of liver-derived IGF into blood depends in large part on the concentration of somatotropin. As part of the feedback control of somatotropin, IGF appears to stimulate GHIH neurons. This activation of GHIH neural elements will in turn inhibit somatotropin release and eventually lead to lower levels of IGF in blood. Because endogenous GHIH also inhibits the release of TSH, the feedback of IGF is also part of the neuroendocrine integration of TSH release.

Another feedback mechanism, which is not illustrated in Figure 7-2, involves short feedback of TSH onto neural structures. It remains possible that blood-borne TSH acts on the CNS to decrease the synthesis or release of TRH.

However, definitive evidence for this mechanism has not yet been obtained. The multiple mechanisms for inhibitory feedback of T_3/T_4 might make it unnecessary to have any direct inhibitory feedback of TSH.

Transfer of Maternal Thyroid Hormones to Neonates. Milk obtained from cows, rats, and humans has been demonstrated to contain T_4 and/or T_3 (Strbak et al., 1974, 1976; Sato and Suzuki, 1979). These milk-borne hormones appear to exert biological effects in the neonates that ingest this milk. Thyroidectomy of lactating rats decreased blood-borne T_4 in suckling neonates and subsequently activated the neonatal thyroid gland (Strbak et al., 1974). Conversely, administration of T_3 to lactating rats increased circulating T_3 and decreased serum TSH in the suckling neonates (Koldovsky et al., 1980). Serum T_4 was low in fasted neonatal rats, and suckling promptly increased both T_4 and T_3 in the neonate, whereas oral administration of a milk substitute failed to increase these thyroid hormones (Oberkotter, 1988). In summary, the pituitary-thyroid axis of a nursing mother may influence metabolism as well as the pituitary-thyroid axis of her neonate.

REFERENCES

Bligh, J. The thermosensitivity of the hypothalamus and thermoregulation in mammals. *Biol. Rev. Camb. Philos. Soc.* (1966) 41:317-367.

Bradley, D.J., W.S. Young, and C. Weinberger. Differential expression of alpha and beta thyroid hormone genes in rat brain and pituitary. *Proc. Natl .Acad. Sci.* (1989) 86:7250-7254.

Chihara, K., A. Arimura, M. Chihara, and A.V. Schally. Studies on the mechanism of growth hormone and thyrotropin responses to somatostatin antiserum in anesthetized rats. *Endocrinology* (1978) 103:1916-1923.

Dyess, E.M., T.P. Segerson, Z. Liposits, W.K. Paull, M.M. Kaplan, P. Wu, I.M.D. Jackson, and R.M. Lechan. Triiodothyronine exerts direct cell-specific regulation of thyrotropin-releasing hormone gene expression in the hypothalamic paraventricular nucleus. *Endocrinology* (1988) 123:2291-2297.

Evans, R.M. The steroid and thyroid hormone receptor superfamily. *Science* (1988) 240:889-895.

Foord, S.M., J.R. Peters, C. Dieguez, M.D. Lewis, B.M. Lewis, R. Hall, and M.F. Scanlon. Thyroid stimulating hormone. In: Lightman, S.L, and B.J. Everitt (eds). *Neuroendocrinology* (Blackwell Scientific Publ., Oxford, UK, 1986), pg 450-471.

Kakucska, I., W. Rand, and R.M. Lechan. Thyrotropin-releasing hormone gene expression in the hypothalamic paraventricular nucleus is dependent upon feedback regulation by both triiodothyronine and thyroxine. *Endocrinology* (1992) 130:2845-2850.

Koch, Y., G. Goldhaber, I. Fireman, U. Zor, J. Shani, and E. Tal. Suppression of prolactin and thyrotropin secretion in the rat by antiserum to thyrotropin-releasing hormone. *Endocrinology* (1977) 100:1476-1478.

Koldovsky, O., L. Krulich, A. Tenore, J. Jumawan, C. Horowitz, and H. Lau. Effect of triiodothyronine injection on levels of triiodothyronine and thyroid-stimulating hormone in sera and milk of lactating rats and in sera of their sucklings; precocious development of jejunal alpha-disaccharidases in the sucklings. *Biol. Neonate* (1980) 37:103-108.

Oberkotter, L.V. Suckling, but not formula feeding, induces a transient hyperthyroxinemia in rat pups. *Endocrinology* (1988) 123:127-133.

Parker, M.G. Gene regulation by steroid hormones. In: Cook, B.A., R.J.B. King, and H.J. van der Molen (eds). *Hormones and Their Actions. Part I* (Elsevier, Amsterdam, 1988), pg 39-59.

Pierce, J.G., and T.F. Parsons. Glycoprotein hormones: structure and function. *Annu. Rev. Biochem.* (1981) 50:465-495.

Sato, T., and Y. Suzuki. Presence of triiodothyronine, no detectable thyroxine and reverse triiodothyronine in human milk. *Endocrinol. Jpn.* (1979) 26:507-513

Segerson, T.P., J. Kauer, H.C. Wolfe, H. Mobtaker, P. Wu., I.M.D. Jackson, and R.M. Lechan. Thyroid hormone regulates TRH biosynthesis in the paraventricxular nucleus of the rat hypothalamus. *Science* (1987) 238:78-80.

Strbak, V., L. Macho, J. Knopp, and L. Struharova. Thyroxine content in mother milk and regulation of thyroid function of suckling rats. *Endocrinol. Exp.* (1974) 8:59-69.

Strbak, V., L. Macho, R. Kovac, M. Skultetyova, and J. Michalickova. Thyroxine (by competitive protein binding analysis) in human and cow milk and in infant formulas. *Endocrinol. Exp.* (1976) 10:167-174.

Szabo, M., and L.A. Frohman. Suppression of cold-stimulated thyrotropin secretion by antiserum to thyrotropin-releasing hormone. *Endocrinology* (1977) 101:1023-1033.

Chapter 8

SOMATOTROPIN

A population of cells in the pars anterior (or pars distalis) that stain with acidic dyes (i.e., cells are acidophilic) synthesize and secrete somatotropin for delivery to its receptors located in several tissues of the body. The first known action of somatotropin was to increase growth in body size of normal as well as hypophysectomized rats. For this reason, the hormone is sometimes called ***growth hormone*** and abbreviated GH. The present chapter will discuss the endocrinology of GH secretion as well as the neurohormonal regulation and neuroendocrine integration of adenohypophysial release of GH. A final section will address some clinical disorders involving GH-related hormones in humans and the practical applications of GH biotechnology in agricultural animals.

The GH molecule is a single linear chain of 191 amino acids with two disulfide bridges between different parts of the chain creating two loops in the molecule. The molecular weight is 21,500 daltons (Da), and the molecule is not a dimer or glycosylated as are TSH, LH, and FSH. The three dimensional structure of GH is partly helical and is similar to that of PRL, with which it shares some sequence homology and immunological crossreaction (Sundaram and Sonenberg, 1969; Aloj and Edelhoch, 1970). The GH molecule of primate species differs considerably from GH in non-primate species. In fact, the human form of GH even possesses prolactin-like bioactivity and lactogenic binding properties that are lacking in the GH of ungulate species (Gertler et al., 1983, 1984). The biological and immunological similarities of human GH with PRL from human and other species caused some delay in the recognition of GH and prolactin as separate molecules in humans. They are now correctly recognized as separate molecules with different receptors and biological actions.

The adenohypophysial cells that synthesize and secrete GH are known as somatotrophs. As discussed in Chapter 2, there are monohormonal somatotrophs that secrete only GH as well as bihormonal cells (mammosomatotrophs) discovered in rats that contain and secrete both GH and PRL (Frawley et al., 1985; Nikitovitch-Winer et al., 1987). The relative numbers of mammosomatotrophs and somatotrophs vary widely among species, being very abundant in cattle and very rare in sheep (Kineman et al., 1991a; Thorpe et al., 1990). Moreover, the animal's endocrine state appears to influence the interconversion of somatotrophs and mammosomatotrophs. Surgical removal of the bovine testis increased the overall population of mammosomatotrophs from 9% to 22%, and the number of mammosomatotrophs was selectively increased during the early luteal phases of the bovine estrous cycle (Kineman et al., 1991a; 1991b). In both these endocrine-related changes, the increased numbers of mammosomatotrophs appear to result from induction of somatotrophs to secrete PRL, rather than mammotrophs being induced to secrete GH.

Cellular Actions of GH. Until recently, target tissues for GH were difficult to identify definitively. Specific binding of radiolabeled GH to homogenates of various tissues suggested a large number of target tissues. Chondrocytes in the epiphysial plates of the long bones as well as the liver are widely accepted as targets of GH action because of specific GH binding and also because GH-induced biological responses can be observed *in vitro* and *in vivo*. The structure of the GH receptor from liver tissue has recently been discovered, and the mRNA transcript that encodes the receptor can now be used to identify tissues that synthesize the GH receptor. The following additional organs have been identified in this way: kidney, mammary gland, and anterior pituitary gland (Hauser et al., 1990).

The major receptor-mediated action of GH appears to be stimulation of the secretion of a mediator molecule that was initially named ***somatomedin***. Multiple forms of somatomedin were subsequently discovered, and the amino acid sequences of some somatomedin forms were determined. After resolution of some differences in nomenclature, it is now generally accepted that there are least two forms of somatomedin bioactivity named ***insulin-like growth factor-I*** and ***insulin-like growth factor-II***. These molecules are abbreviated IGF-I and IGF-II, respectively. The molecule of IGF-I is identical to an entity known earlier as somatomedin-C. The molecule of IGF-II is very similar, if not identical, to entities known earlier as somatomedin-A, multiplication-stimulating activity, and non-suppressible insulin-like activity. These earlier names should be replaced with IGF-I and IGF-II when discussing the molecules that mediate the biological effects of GH. The term somatomedin is appropriately retained for generic use when discussing unidentified or unspecified compounds which mediate the biological effects of GH.

The term ***insulin-like growth factor*** was assigned to the identified somatomedin polypeptides because of structural homology with the beta chain of the insulin molecule (Rinderknect and Humbel, 1976). IGF-I is the primary somatomedin molecule secreted in response to GH binding to its receptor. The IGF-I secreted in response to GH may be released into the blood for an endocrine action, or it may act locally in the vicinity of its cellular source (either autocrine or paracrine action). The primary source of blood-borne IGF-I is the liver, but nutritional deficiencies can compromise the ability of blood-borne GH to stimulate the liver to secrete IGF-I into blood. Although IGF-II exists in the blood and tissues of adult mammals, it also represents the primary somatomedin molecule in mammalian fetuses. However, the role of adenohypophysial GH in the maintenance of IGF-II secretion in fetuses is not well understood (Daughaday, 1982; Hall and Sara, 1983). For example, hypophysectomy decreased blood-borne IGF-II in young pigs, but subsequent administration of GH failed to increase blood levels of IGF-II (Buonomo et al., 1988).

Both IGF-I and IGF-II should be considered as hormones and as local regulatory molecules. There are cellular receptors for both IGF-I and IGF-II, and there are also IGF-binding proteins which probably regulate the degradation and biopotency of the IGF molecules (Holly and Wass, 1989). Because the synthesis of IGF-I appears to depend upon blood-borne GH, it is possible to determine sites of GH action by looking for cells that transcribe the IGF-I gene. Liver, skeletal muscle and epiphysial plates in long bones were identified as sites of IGF-I synthesis, and exogenous GH increased the synthesis in all three tissues (Isgaard et al., 1988).

Transcription of the gene for IGF-II occurs in many tissues of the fetal rat, but after birth most tissues, with the exception of the CNS, lose the ability to transcribe the IGF-II gene (Lund et al., 1986). The precise role of IGF-II synthesis in the brain is not known, but it is probably not involved in feedback regulation of the neurohormones that regulate adenohypophysial GH. The feedback of blood-borne IGF-I will be discussed later in this chapter.

Patterns of GH Secretion. Release of GH into the blood by the adenohypophysis is highly episodic in all mammalian species investigated thus far. Such extreme fluctuations might appear inappropriate for a hormone that must sustain steady-state processes of cellular growth. However, it is the somatomedin molecules induced by the episodic GH that provide the relatively stable stimulation of cellular growth and other processes. Moreover, episodic administration of exogenous GH was more effective than continuous administration in stimulating IGF-I gene transcription in muscle and bone tissues of hypophysectomized rats (Isgaard et al., 1988).

Although GH secretion is episodic, the discharges of GH among individual animals or humans are usually not synchronized. For this reason, quantification of GH in sequential blood samples of a single animal will often reveal recurring discharges, but averaging the blood plasma concentration of GH across several individuals may obscure the individual profiles of a GH rhythm. Only if there is circadian ***entrainment*** of GH secretory rhythms among individuals will the mean profile from a group reveal a rhythm. Such entrainment seems to be relatively rare, and the best example of it occurs in male rats from which blood is sampled remotely and without disturbance. Figure 8-1 presents the average of plasma GH from 24 such rats sampled at various periods during the 24 h cycle

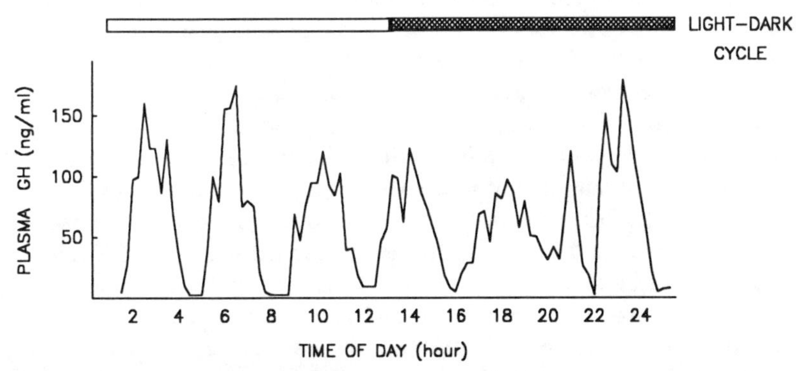

Figure 8-1. Synchronized ultradian rhythm of GH secretion in 24 male rats. Data redrawn from Tannenbaum and Martin (1976).

of alternating light and dark (Tannenbaum and Martin, 1976). This entrained rhythm of GH secretion is characterized by recurring discharges of GH about every 3.3 h. Because the period of the rhythm is less than 24 h, the rhythm is called ***ultradian***.

All rhythms of hormone secretion can be described in terms of either ***frequency*** (i.e., number of discharges per unit time) or ***period*** (i.e., time required for one complete cycle or the interval between successive discharges). It should be noted that frequency and period of a rhythm are reciprocals of each other. For example, a frequency of 8 pulses per 24 h equals an average period of 3 h (i.e., 24 h/8 pulses). Such rhythms can be entrained across individuals as in Figure 8-1, or not entrained as occurred when male rats were exposed to continuous light. Individual rats continued to discharge GH about every 3-4 h, but there was no synchrony of the discharges in constant light (Tannenbaum and Martin, 1976) or in rats in which the suprachiasmatic nucleus had been destroyed (Willoughby and Martin, 1978). Several experiments were conducted to discover what factors in addition to the light-dark cycle might be entraining the secretory rhythm of GH in male rats. There was no relationship between GH discharges and any of the following: feeding behavior, induced periods of hypoglycemia/hyperglycemia, or phases of sleep (Willoughby et al., 1976; Tannenbaum et al., 1976). A distinct difference was observed between the GH profiles of female and male rats with females having more frequent, lower amplitude discharges and less entrainment than males (Saunders et al., 1976; Clark et al., 1987). It was subsequently discovered that gonadectomy of male rats caused them to have GH secretory rhythms more like those of intact females (Painson et al., 1992).

Species other than rats also release GH in pulsatile discharges, but there is usually no group synchrony. Hormones of the male gonad appear to enhance the amplitude and decrease the frequency of GH discharges in male sheep (Davis et al., 1977). Secretory discharges of GH are generally greater in primate species during the period of sleep, and a large release of GH coincides with the first period of slow-wave sleep in humans and baboons but not monkeys (Parker et al., 1972; Weitzman et al., 1975; Quabbe et al., 1981). No such association between sleep and GH release was observed in goats (Tindal et al., 1978).

Stressful stimuli affect release of adenohypophysial GH in ways that vary widely among species. Stressors generally increase GH release in primate species, but in rodent species (e.g., rat, mouse, rabbit) stress-induced decreases are often observed, perhaps by interfering with the pulsatile discharges. By contrast, stressful stimuli do not seem to affect in any way the release of GH in domestic ungulates. Exercise that is too mild to be considered a stressor can provoke the release of GH in humans, and it is often used as a GH-releasing stimulus in clinical testing of children. Partial or total deprivation of food increases the release of GH in primates and domestic ungulates. This response to food deprivation probably results from the deprivation-induced decline in circulating IGF-I that decreases feedback on neural structures regulating adenohypophysial release of GH (Breier et al., 1986). Starvation in male rats, which may only reflect stress, decreases the release of GH concurrent with a decrease in transcription of the gene for GHRH, but no change in transcription of the gene for GHIH (Bruno et al., 1990).

Metabolic Signals Affecting GH Release. There are a number of blood-borne metabolic signals that provoke secretory discharges of adenohypophysial GH. These signals may not always be physiological regulators of endogenous release of GH, but they clearly have diagnostic value in determining the secretory capacity of GH in humans and animals. For this reason, a brief discussion of each type of signal is included here. A rapid, transient decrease in blood-borne glucose in primates and dogs will provoke an immediate increase of GH release. The usual clinical test is to inject humans with a small amount of insulin to induce hypoglycemia, which is normally quickly followed by GH release. In contrast to the established feedback relationship between blood glucose concentration and secretion of insulin, the absolute level of circulating glucose does not regulate the release of GH. Rather, it is the relative fall in blood-borne glucose that provokes GH release, even if that fall merely restores the patient from hyperglycemia to a normoglycemic state. Interference with the intracellular utilization of glucose by administration of 2-deoxyglucose can also stimulate release of GH.

Rapid changes in circulating concentrations of free fatty acids (FFA) also influence the release of GH. High levels of FFA tend to suppress GH release, and abruptly induced decreases in circulating FFA provoke GH discharges. One of the biological actions of GH is to mobilize body fat for use as energy, and as a result of this action GH elevates plasma levels of FFA. Therefore, an action of blood-borne FFA to suppress GH release is consistent with an inhibitory feedback loop, and the effect of GH released in response to a decrease of circulating FFA may act to stimulate lipolysis and increase FFA again.

Another metabolic change that can provoke release of GH is an abrupt increase in blood-borne amino acids. The provocative challenge used most often in clinical testing is an injection of arginine because this amino acid does not lead to an increase in blood glucose, which is known to antagonize the induced release of GH. It is not known whether arginine-induced stimulation of GH release represents a physiological mechanism that regulates the endogenous release of GH. There would never be a selective increase in only one amino acid following a large protein-rich meal. Moreover, some of the other amino acids that would also be elevated after consumption of such a meal are glucogenic and would lead to increased blood glucose. This may explain why GH release is not normally provoked by consumption of large quantities of protein.

Neurohormonal Regulation of GH Release

As stated previously in Chapter 4 (see Figure 4-6), adenohypophysial release of GH is regulated by GHRH stimulation and GHIH inhibition. As presented in Chapter 4, neuronal perikarya that synthesize GHRH are concentrated in the arcuate nucleus. In contrast, those neurons that synthesize GHIH and also project to the median eminence are located in the preoptic area and the periventricular parts of the anterior hypothalamus. There are also a few such GHIH perikarya in the PVN. Because of the dual control of GH release, it is difficult to determine whether a particular increase in GH secretion was caused by increased release of GHRH and/or decreased release of GHIH. One way to make such a determination is through selective immunoneutralization of individual blood-borne neurohormones. As an illustration of this approach, representative profiles of plasma GH are presented in Figure 8-2 for normal male rats (like those plotted in Figure 8-1) and for animals treated with antibodies that block the actions of either GHRH (anti-GHRH) or GHIH (anti-GHIH). Control animals are depicted as discharging two pulses of GH separated by an interval of about 3 h. When blood-borne GHRH was immunoneutralized with anti-GHRH, no GH discharges occurred. Therefore, the pulsatile discharges of GH must require release of GHRH in order to occur. When blood-borne GHIH was immunoneutralized with anti-GHIH in other rats, the concentration of plasma GH at the times of peak pulsatile discharge was the same as in controls, but plasma GH during the approximately 2-h period between the discharges occurring in controls was elevated due to immunoneutralization of GHIH. There may also have been pulsatile release of GH during this interval as illustrated in Figure 8-2, but this was not always observed. One additional observation, not shown in Figure 8-2, was that the magnitude of GH release that could be induced by

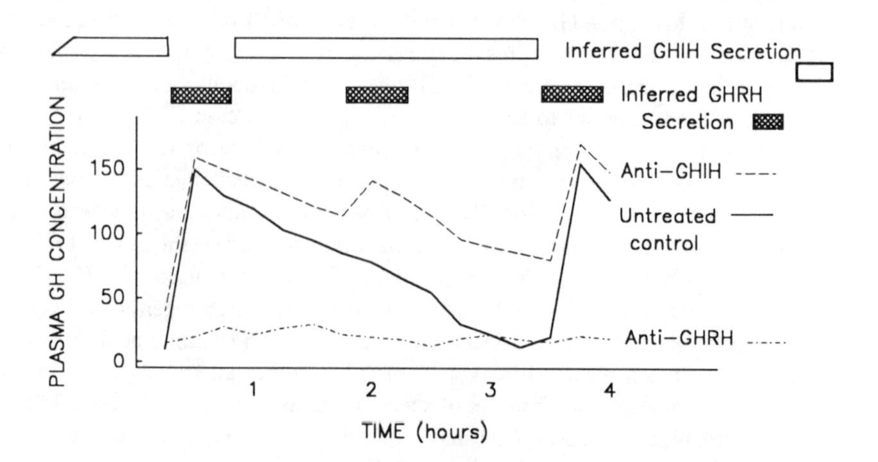

Figure 8-2. Plasma GH concentrations during immunoneutralization and inferred neurohormone secretion.

Although the data plotted in this figure are hypothetical, they reflect the data and conclusions of Tannenbaum and Ling (1984). The horizontal bars across the top represent the inferences drawn about spontaneous secretion of GHIH (open bars) and GHRH (filled bars) in male rats.

injection of exogenous GHRH varied depending on the time of injection. When GHRH injection occurred during a period of the light-dark cycle when entrained release of GH was minimal, the GHRH-induced release was smaller than when GHRH injection occurred at the time of an entrained GH discharge. Taken together, all these results formed the basis for the inferences illustrated by the bars at the top of Figure 8-2. The release of GHRH in male rats is probably pulsatile with recurring discharges occurring more frequently than the 3-h interval between GH discharges. The release of GHIH is probably intermittent with a prolonged period of GHIH release throughout the interval between spontaneous GH discharges, but this release of GHIH decreases about every 3 h coincident with some, but not all, discharges of GHRH to allow the observed rhythmic discharges of GH (Tannenbaum and Ling, 1984).

As stated earlier in this chapter, the spontaneous pattern of GH release differs between male and female rats. Female rats and castrated male rats display more frequent discharges of GH that have a lower amplitude than in intact males. The inferences drawn from immunoneutralization of GHRH and GHIH in female rats (Painson and Tannenbaum, 1991) are also somewhat different from those illustrated in Figure 8-2 for males. The inferred spontaneous release of GHIH in female rats is probably constant, not intermittent as it is in males. The inferred release of GHRH in female rats is less regular but occurs more often than in males. When transcription of the GHRH gene was quantified in male and female rats, transcription was greater in males and was also more sensitive to feedback suppression by exogenous GH in males (Maiter et al., 1991). When male rats were starved and spontaneous discharges of GH ceased to occur, transcription of the GHRH gene was reduced, whereas there was no change in transcription of the GHIH gene (Bruno et al., 1990). In summary, detailed investigations of GHRH and GHIH by immunoneutralization and by quantifying neurohormonal gene transcription illustrate the complexities of the antagonistic GH regulatory system. Moreover, it seems clear that each species, endocrine state, and provocative stimulus must be investigated before the relative contributions of GHRH and GHIH can be inferred for that situation because generalizations do not seem warranted.

The neurosecretory neurons that release GHRH and GHIH into hypophysial portal blood are influenced by a variety of synaptic inputs. Pharmacologic agents that affect aminergic neurotransmission have been observed to affect spontaneous as well as induced secretion of GH. Although it is not clear whether they release GH by stimulating GHRH or inhibiting GHIH, GH-stimulatory effects of dopamine, norepinephrine, serotonin, and acetylcholine have been observed. Among neuropeptides that may influence GH release, there is evidence from administration as well as immunoneutralization that endogenous NPY stimulates release of GHIH, which in turn inhibits release of GH (Rettori et al., 1990).

Neuroendocrine Integration of GH Release. Figure 8-3 presents a model that illustrates neuroendocrine integration of the release of adenohypophysial GH. Dual antagonistic control of GH release is depicted as one GHRH neuron and one GHIH neuron that are both influenced by a variety of stimuli discussed

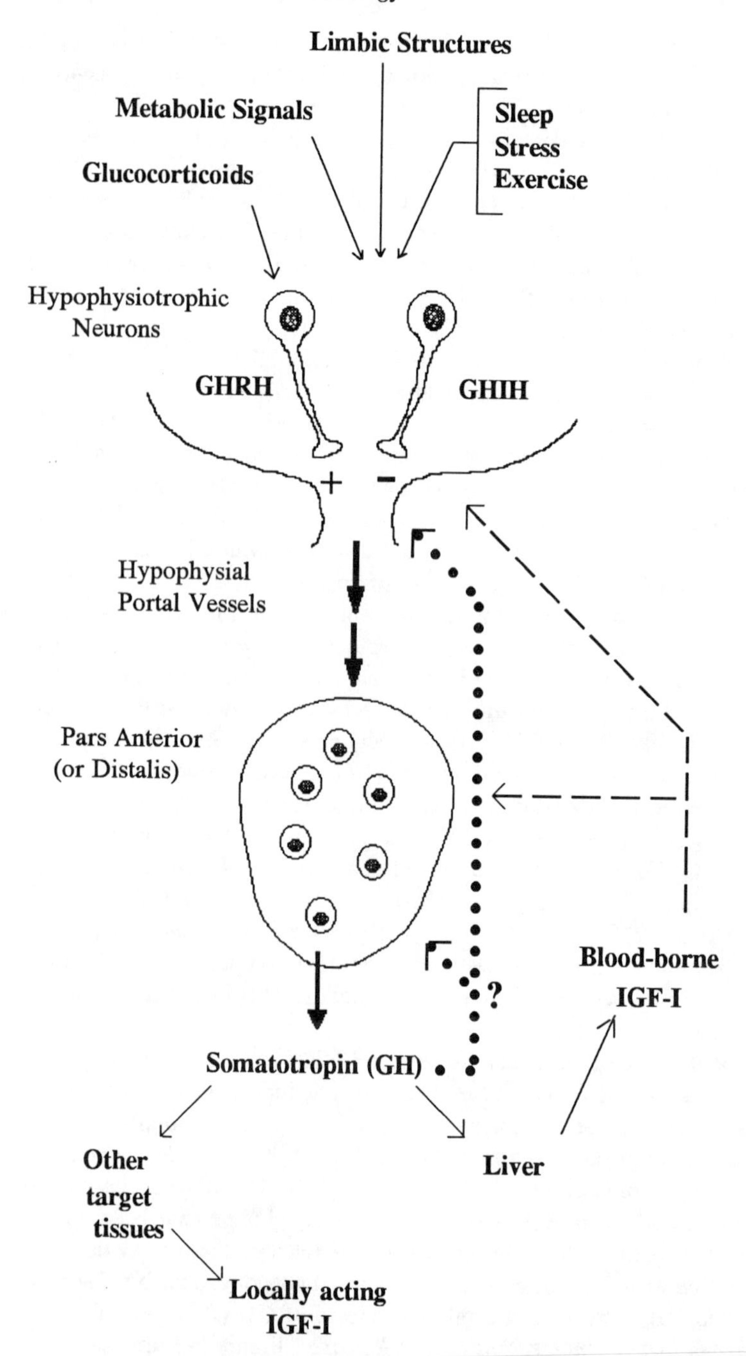

Figure 8-3. Neuroendocrine integration of GH release.
Illustration of mechanisms involved in the neuroendocrine integration of GH release.

previously (e.g., metabolic signals, sleep, stress, and exercise). Input from CNS regions comprising the limbic system is also included because it is known that lesion and stimulation of some limbic structures can affect the release of GH. An action of glucocorticoids from the adrenal cortex directly onto GHRH neurons is also depicted in Figure 8-3. The specificity of this action is based on the discovery of glucocorticoid receptors within GHRH neurons in the arcuate nucleus (Cintra et al., 1987). Glucocorticoids probably inhibit release of GHRH because elevated levels of exogenous or endogenous cortisol in humans clearly depress GH release and growth.

Adenohypophysial GH secreted into the blood acts upon the liver in well-nourished animals to cause IGF-I release into blood (Figure 8-3). The blood-borne IGF-I completes a feedback loop to decrease further release of GH in much the same manner that blood-borne hormones from thyroid gland, adrenal gland, and gonads feedback to inhibit their respective pituitary hormones. There is also the possibility that GH may directly participate in feedback inhibition of its own release without affecting circulating, liver-derived IGF-I. The adenohypophysis contains receptors for both GH and IGF-I (Hauser et al., 1990), whereas GHIH neurons contain GH receptors (Burton et al., 1992). Intracerebral administration of either GH or IGF-I inhibited episodic discharges of GH in normal rats (Tannenbaum, 1980; Abe et al., 1983). Although the exogenous GH used in those studies may have stimulated intracerebral synthesis of IGF-I (Wood et al., 1991), this author is reluctant to dismiss the possibility of direct GH feedback on the neurohormones that regulate GH release. Inhibitory feedback of IGF-I and GH in the hypothalamus may involve GHRH and/or GHIH. Hypophysectomy increased GHRH gene transcription without affecting that of GHIH (Wood et al., 1991). In addition, exogenous GH decreased hypothalamic content of GHRH without affecting the content of GHIH (Ganzetti et al., 1987), and exogenous IGF-I decreased GH release in fetal pigs even when GHIH was immunoneutralized (Spencer et al., 1991). However, GH may act directly on a subpopulation of GHIH neurons in the medial hypothalamus because they contain GH receptor mRNA (Burton et al., 1992). Another complexity, which is not illustrated in Figure 8-3, involves the potential interaction between GHRH and GHIH neurons. Each one may send axonal branches to the other type of neuron to influence their neurosecretory function. Intracerebral administration of GHRH suppressed the spontaneous release of GH presumably through activation of GHIH because immunoneutralization of GHIH prevented the induced suppression of GH release (Katakami et al., 1986).

Applied Aspects of GH Physiology

Chronic hypersecretion of adenohypophysial GH in humans leads to the well-recognized syndrome of acromegaly, and the syndrome of gigantism also results if the hypersecretion occurs in children before puberty-related fusion of the epiphysial plate. The hypersecretion of GH may be due to a defect within the adenohypophysis or to extrapituitary factors that overstimulate the release of GH (i.e., too much GHRH-like bioactivity or too little GHIH).

A deficiency of GH and/or somatomedin bioactivity will result in retardation of growth in humans and animals. A wide variety of causes have been discovered for the various types of human dwarfism (Van Wyk and Underwood, 1980). Hyposecretion of GH can result from a primary defect in the adenohypophysis or secondarily from dysfunction in the neurohormonal regulation of GH release (i.e., too little GHRH or too much GHIH). Both primary and secondary defects lead to deficient secretion of GH and low circulating levels of IGF-I, and such patients can benefit from therapy with exogenous GH or synthetic analogues of GHRH. There exists a third type of human dwarfism in which GH secretion is normal, or even elevated, but in which IGF-I secretion is deficient. Such target organ resistance to GH is characteristic of one particular genetic defect in humans known as Laron dwarfism and one environmentally induced syndrome (protein-calorie malnutrition). The hypersecretion of GH in these syndromes probably occurs because of reduced inhibitory feedback of blood-borne IGF-I. Patients with these syndromes can benefit from therapy with exogenous IGF-I but not GH. An animal counterpart to Laron dwarfism was recently discovered in cattle (Hammond et al., 1991). Blood levels of IGF-I in these miniature animals are quite low despite very high levels of GH in the blood.

Agricultural Applications of GH Biotechnology. The obvious role played by GH in modulation of growth and metabolism stimulated many efforts to manipulate the secretion or action of GH in agricultural animals. It was known for a very long time that administration of GH isolated from animal pituitaries increased milk production in lactating cows and modified tissue growth in ungulate species. There was relatively little interest in agricultural applications until recombinant gene technology had progressed to the point where GH could be synthesized by genetically modified bacteria. This ability to synthesize species-specific forms of GH for administration to agricultural animals has led to two major applications. In the first one, chronic administration of synthetic GH to cows that are already lactating causes them to produce more milk than untreated controls. One of the mechanisms by which milk yield is increased in GH-treated cows involves the redirection of metabolic energy away from body reserves and toward milk synthesis (Tyrrell et al., 1988). In the second application, young ungulates are administered synthetic GH during the period of rapid growth. The treated animals sometimes grow more rapidly, but they consistently have a greater efficiency of growth than controls as well as reduced fat deposition in the carcass. More research is needed to determine fully the optimum application of this technology as well as possible adverse side effects.

Another type of GH biotechnology involves the manipulation of *in vivo* release of GH in agricultural animals. One approach is to administer GHRH or its synthetic analogues. It appears that frequent, repetitive injections of GHRH will probably lead to increased release of GH and the same benefits observed following GH administration (Dahl et al., 1990). Immunoneutralization of neurohormones regulating the release of GH has also been investigated. Active immunization of animals so that they continuously produce antibodies against the neurohormone is the method of choice for such agricultural immunotherapy. Experimental immunoneutralization of GHRH in growing cattle resulted in decreased levels of GH and IGF-I as well as decreased growth and increased fat deposition (Trout and Schanbacher, 1990). However, immunoneutralization of GHIH by these same authors failed to increase the release of GH or to produce the expected benefits in terms of increased growth and/or reduced fat deposition.

Another potential application of GH biotechnology consists of inserting foreign genes into animals to cause overproduction of GH, GHRH, or IGF-I. Application of this biotechnology to laboratory mice has been quite successful. Overproduction of any one of these three molecules often leads to increased growth of one or more tissues. There are sometimes deleterious side effects in the transgenic mice that would need to be overcome before any practical application of this biotechnology. The situation in domestic ungulates seems quite different from that in mice. Transgenic pigs were created to overexpress GH or GHRH, but growth rates were not increased (Pursel et al., 1990). Transgenic sheep that overexpressed GH or GHRH also failed to grow at an increased rate (Rexroad et al., 1991). Both transgenic pigs and sheep displayed numerous deleterious side effects, and there were no compensatory benefits of increased growth as occurred in mice. Therefore, agricultural application of the transgenic technology involving GH-related events requires much more development.

REFERENCES

Abe, H., M.E. Molitch, J.J. Van Wyk, and L.E. Underwood. Human growth hormone and somatomedin C suppress the spontaneous release of growth hormone in unanesthetized rats. *Endocrinology* (1983) 113:1319-1324.

Aloj, S.M., and H. Edelhoch. Conformational similarity of ovine prolactin and bovine growth hormone. *Proc. Natl. Acad. Sci.* (1970) 66:830-836.

Breier, B.H., J.J. Bass, J.H. Butler, and P.D. Gluckman. The somatotrophic axis in young steers: influence of nutritional status on pulsatile release of growth hormone and circulating concentrations of insulin-like growth factor 1. *J. Endocrinol.* (1986) 111:209-215.

Bruno, J.F., D. Olchovsky, J.D. White, J.W. Leidy, J. Song, and M. Berelowitz. Influence of food deprivation in the rat on hypothalamic expression of growth hormone-releasing factor and somatostatin. *Endocrinology* (1990) 127:2111-2116.

Buonomo, F.C., D.L. Grohs, C.A. Baile, and D.R. Campion. Determination of circulating levels of insulin-like growth factor II (IGF-II) in swine. *Dom. Anim. Endocrinol.* (1988) 5:323-329.

Burton, K.A., E.B. Kabigting, D.K. Clifton, and R.A. Steiner. Growth hormone receptor messenger ribonucleic acid distribution in the adult male rat brain and its colocalization in hypothalamic somatostatin neurons. *Endocrinology* (1992) 131:958-963.

Cintra, A., K. Fuxe, A. Harfstrand, L.F. Agnati, A.C. Wikstrom, S. Okret, W. Vale, and J.A. Gustafsson. Presence of glucocorticoid receptor immunoreactivity in corticotropin releasing factor and in growth hormone releasing factor immunoreactive neurons of the rat di- and telencephalon. *Neurosci. Lett.* (1987) 77:25-30.

Clark, R.G., L.M.S. Carlsson, and I.C.A.F. Robinson. Growth hormone secretory profiles in conscious female rats. *J. Endocrinol.* (1987) 114:399-407.

Dahl, G.E., L.T. Chapin, S.A. Zinn, W.M. Moseley, T.R. Schwartz, and H.A. Tucker. Sixty-day infusions of somatotropin-releasing factor stimulate milk production in dairy cows. *J. Dairy Sci.* (1990) 73:2444-2452.

Daughaday, W.H. Divergence of binding sites, in vitro action, and secretory regulation of the somatomedin peptides, IGF-I and IGF-II. *Proc. Soc. Exp. Biol. Med.* (1982) 170:257-263.

Davis, S.L., D.L. Ohlson, J. Klindt, and M.S. Anfinson. Episodic growth hormone secretory patterns in sheep: relationship to gonadal steroid hormones. *Am. J. Physiol.* (1977) 233:E519-E523.

Frawley, L.S., F.R. Boockfor, and J.P. Hoeffler. Identification by plaque assays of a pituitary cell type that secretes both growth hormone and prolactin. *Endocrinology* (1985) 116:734-737.

Ganzetti, I., F. Petraglia, I. Capuano, F. Rosi, W.B. Wehrenberg, E.E. Muller, and D. Cocchi. Feed-back effect of growth hormone on hypothalamic opioid and somatocrinin neurons. *J. Endocrinol. Invest.* (1987) 10:241-245.

Gertler, A., A. Ashkenazi, and Z. Madar. Binding sites of human growth hormone and ovine and bovine prolactins in the mammary gland and liver of lactating dairy cow. *Mol. Cell. Endocrinol.* (1984) 34:51-57.

Gertler, A., N. Cohen, and A. Moaz. Human growth hormone but not ovine or bovine growth hormones exhibit galactopoietic prolactin-like activity in organ culture from bovine lactating mammary gland. *Mol. Cell. Endocrinol.* (1983) 33:169-182.

Hall, K., and V.R. Sara. Growth and somatomedins. *Vitam. Horm.* (1983) 40:175-233.

Hammond, A.C., T.H. Elsasser, and T.A. Olson. Endocrine characteristics of a miniature condition in Brahman cattle: circulating concentrations of some growth-related hormones. *Proc. Soc. Exp. Biol. Med.* (1991) 197:450-457.

Hauser, S.D., M.F. McGrath, R.J. Collier, and G.G. Krivi. Cloning and in vivo expression of bovine growth hormone receptor mRNA. *Mol. Cell. Endocrinol.* (1990) 72:187-200.

Holly, J.M.P., and J.A.H. Wass. Insulin-like growth factors; autocrine, paracrine or endocrine? New perspectives of the somatomedin hypothesis in the light of recent developments. *J. Endocrinol.* (1989) 122:611-618.

Isgaard, J., L. Carlsson, O.G.P. Isaksson, and J.O. Jansson. Pulsatile intravenous growth hormone (GH) infusion to hypophysectomized rats increases insulin-like growth factor I messenger ribonucleic acid in skeletal tissue more effectively than continuous GH infusion. *Endocrinology* (1988) 123:2605-2610.

Katakami, H., A. Arimura, and L.A. Frohman. Growth hormone (GH)-releasing factor stimulates hypothalamic somatostatin release: an inhibitory feed-back effect on GH secretion. *Endocrinology* (1986) 118:1872-1877.

Kineman, R.K., W.J. Faught, and L.S. Frawley. Mammosomatotrophs are abundant in bovine pituitaries: influence of gonadal status. *Endocrinology* (1991a) 128:2229-2233.

Kineman, R.D., D.M. Hendricks, W.J. Faught, and L.S. Frawley. Fluctuations in the proportions of growth hormone and prolactin-secreting cells during the bovine estrous cycle. *Endocrinology* (1991b) 129:1221-1225.

Lund, P.K., B.M. Moats-Staats, M.A. Hynes, J.G. Simmons, M. Jansen, A.J. D'Ercole, and J.J. Van Wyk. Somatomedin-C/insulin-like growth factor-I and insulin-like growth factor-II mRNAs in rat fetal and adults tissues. *J. Biol. Chem.* (1986) 261:14539-14544.

Maiter, D., J.I. Koenig, and L.M. Kaplan. Sexually dimorphic expression of the growth hormone-releasing hormone gene is not mediated by circulating gonadal hormones in the adult rat. *Endocrinology* (1991) 128:1709-1716.

Nikitovitch-Winer, M.B., J. Atkin, and B.E. Maley. Colocalization of prolactin and growth hormone within specific adenohypophysial cells in male, female and lactating female rats. *Endocrinology* (1987) 121:625-630.

Painson, J.C., and G.S. Tannenbaum. Sexual dimorphism of somatostatin and growth hormone-releasing factor signaling in the control of pulsatile growth hormone in the rat. *Endocrinology* (1991) 128:2858-2866.

Painson, J.C., M.O. Thorner, R.J. Krieg, and G.S. Tannenbaum. Short term adult exposure to estradiol feminizes the male pattern of spontaneous and growth hormone-releasing factor stimulated growth hormone secretion in the rat. *Endocrinology* (1992) 130:511-519.

Parker, D.C., M. Morishima, D.J. Koerker, C.C. Gale, and C.J. Goodner. Pilot study of growth hormone release in sleep of the chair-adapted baboon: potential as model of human sleep release. *Endocrinology* (1972) 91:1462-1467.

Pursel, V.G., R.E. Hammer, D.J. Bolt, R.D. Palmiter, and R.L. Brinster. Integration, expression and germ-line transmission of growth-related genes in pigs. *J. Reprod. Fertil.* (1990) Suppl 41:77-87.

Quabbe, H.J., M. Gregor, C. Bumke-Vogt, A. Eckhof, and I. Witt. Twenty-four hour-pattern of growth hormone secretion in the rhesus monkey: studies including alterations of the sleep/wake and sleep stage cycles. *Endocrinology* (1981) 109:513-522.

Rettori, V., L. Milenkovic, M.C. Aguila, and S.M. McCann. Physiologically significant effect of neuropeptide Y to suppress growth hormone release by stimulating somatostatin discharge. *Endocrinology* (1990) 126:2296-2301.

Rexroad, C.E., K. Mayo, D.J. Bolt, T.H. Elsasser, K.F. Miller, R.R. Behringer, R.D. Palmiter, and R.L. Brinster. Transferrin- and albumin-directed expression of growth-related peptides in transgenic sheep. *J. Anim. Sci.* (1991) 69:2995-3004.

Rinderknecht, E., and R.E. Humbel. Amino-terminal sequences of two polypeptides from human serum with nonsuppressible insulin-like and cell-growth-promoting activities: evidence for structural homology with insulin B chain. *Proc. Natl. Acad. Sci.* (1976) 73:4379-4381.

Saunders, A., L.C. Terry, J. Audet, P. Brazeau, and J.B. Martin. Dynamic studies of growth hormone and prolactin secretion in the female rat. *Neuroendocrinology* (1976) 21:193-203.

Spencer, G.S.C., A.A. MacDonald, S.S. Carlyle, and L.G. Moore. Decreased circulating growth hormone levels following centrally administered insulin-like growth factor-I is not mediated by somatostatin in the pig fetus. *Reprod. Nutr. Devel.* (1991) 31:585-590.

Sundaram, K., and M. Sonenberg. Immunological studies of human growth hormone, ovine prolactin, bovine growth hormone and a tryptic digest of bovine growth hormone. *J. Endocrinol.* (1969) 44:517-522.

Tannenbaum, G.S. Evidence for autoregulation of growth hormone secretion via the central nervous system. *Endocrinology* (1980) 107:2117-2120.

Tannenbaum, G.S., and N. Ling. The interrelationship of growth hormone (GH)-releasing factor and somatostatin in generation of the ultradian rhythm of GH secretion. *Endocrinology* (1984) 115:1952-1957.

Tannenbaum, G.S., and J.B. Martin. Evidence for an endogenous ultradian rhythm governing growth hormone secretion in the rat. *Endocrinology* (1976) 98:562-570.

Tannenbaum, G.S., J.B. Martin, and E. Colle. Ultradian growth hormone rhythm in the rat: effects of feeding, hyperglycemia, and insulin-induced hypoglycemia. *Endocrinology* (1976) 99:720-727.

Thorpe, J.R., K.P. Ray, and M. Wallis. Occurrence of rare somatomammotrophs in ovine anterior pituitary tissue studied by immunogold labelling and electron microscopy. *J. Endocrinol.* (1990) 124:67-73.

Tindal, J.S., G.S. Knaggs, I.C. Hart, and L.A. Blake. Release of growth hormone in lactating and non-lactating goats in relation to behavior, stages of sleep, and electroencephalograms, environmental stimuli and levels of prolactin, insulin, glucose and free fatty acids in the circulation. *J. Endocrinol.* (1978) 76:333-346.

Trout, W.E., and B.D. Schanbacher. Growth hormone and insulin-like growth factor-I responses in steers actively immunized against somatostatin or growth hormone-releasing factor. *J. Endocrinol.* (1990) 125:123-129.

Tyrrell, H.F., A.C.G. Brown, P.J. Reynolds, G.L. Haaland, D.E. Bauman, C.J. Peel, and W.D. Steinhour. Effect of bovine somatotropin on metabolism of lactating dairy cows: energy and nitrogen utilization as determined by respiration calorimetry. *J. Nutr.* (1988) 118:1024-1030.

Van Wyk, J.J., and L.E. Underwood. Growth hormone, somatomedins and growth failure. In: Krieger, D.T., and J.C. Hughes (eds). *Neuroendocrinology* (Sinauer Associates Inc., Sunderland, MA, 1980), pg 299-309.

Weitzman, E.D., R.M. Boyar, S. Kapen, and L. Hellman. The relationship of sleep and sleep stages to neuroendocrine secretion and biological rhythms in man. *Rec. Prog. Horm. Res.* (1975) 31:399-441.

Willoughby, J.O., and J.B. Martin. The suprachaismatic nucleus synchronizes growth hormone secretory rhythms with the light-dark cycle. *Brain Res.* (1978) 151:413-417.

Willoughby, J.O., J.B. Martin, L.P. Renaud, and P. Brazeau. Pulsatile growth hormone release in the rat: failure to demonstrate a correlation with sleep phases. *Endocrinology* (1976) 98:991-996.

Wood, T.L., M. Berelowitz, M.C. Gelato, C.T. Roberts, D. LeRoith, W.J. Millard, and J.F. McKelvy. Hormonal regulation of rat hypothalamic neuropeptide mRNAs: Effect of hypophysectomy and hormone replacement on growth hormone-releasing factor, somatostatin and the insulin-like growth factors. *Neuroendocrinology* (1991) 53:298-305.

Chapter 9

PROLACTIN

A population of acidophilic cells in the adenohypophysis (pars anterior or pars distalis), that are known as mammotrophs, synthesize and secrete ***prolactin*** (abbreviated PRL). These mammotrophs constitute the largest population of any cell type in the adenohypophysis. Alternative names for prolactin found in the older literature include ***mammotropin***, ***lactogen***, and ***luteotropin***. Although mammotropin and lactogen remain valid names, the term luteotropin should not be used as an alternative for prolactin. One biological action of early prolactin-rich extracts of adenohypophysis was to exert a luteotrophic action on the corpora lutea of rats and mice. However, it is now known that this biological action is not consistent enough across species to warrant general use of the term luteotropin.

The PRL molecule is a single linear chain of approximately 200 amino acids with a molecular weight of about 23,000 daltons (Da). There are three disulfide bridges linking different parts of the linear chain, which creates loops not unlike those in GH to which PRL has an evolutionary relationship. In contrast to GH, the monomer molecule of PRL is sometimes glycosylated as are the dimeric molecules of TSH, LH, and FSH (see Figure 7-1). This glycosylation occurs at residue 31 (asparagine) in PRL from sheep, pigs, mice, and humans, but not from cattle which have an aspartate residue at position 31 (Strickland and Pierce, 1985). Within the pars anterior tissue, only a fraction of the total PRL is glycosylated (Sinha, 1992), and this proportion varies greatly among species (e.g., 8-50%). Glycosylated PRL also occurs in blood plasma, but its biological significance is not fully understood. The glycosylated molecules have bioactivity, but it is usually less than the major nonglycosylated form (Sinha, 1992). In addition to the major 23,000 Da form of PRL, there are several other variants of the molecule that also possess bioactivity, but additional research is needed to understand the physiological relevance of the PRL variants.

Biological Actions of Prolactin

Adenohypophysial PRL is a hormone that occurs across all vertebrate species, and it is involved in many different functions. Unlike other hormones of the adenohypophysis, PRL was apparently not committed early in evolution to a particular target organ or biological process. Therefore, the biological actions of PRL are diversified and highly adaptive in nature. A major function of PRL in female mammals is stimulation of milk production by the mammary gland for support of neonatal offspring. This nurturing role for PRL first appears in pigeons where the PRL-stimulated crop sac produces the functional equivalent of mammalian milk for the offspring. Other actions of PRL in vertebrates modulate the processes of reproduction, osmoregulation, growth/metabolism, as well as maternal and migratory behaviors.

Mammotrophic Actions of PRL. After mammary tissue has been suitably prepared by the maternal hormones of pregnancy, stimulation by PRL is required in *all* mammals to initiate lactogenesis (defined as copious secretion of milk). This mammary requirement for PRL has been demonstrated *in vitro* as well as *in vivo* using hypophysectomized females or those in which the secretion of PRL has been inhibited by pharmacological agents. In both types of *in vivo* models, it is essential to demonstrate that injections of PRL will correct the mammary deficits resulting from hypophysectomy or PRL suppression. When this result occurs, as it usually does, one can conclude that the mammary deficits probably resulted from a deficiency of adenohypophysial PRL. In one example, exogenous PRL restored milk secretion and mammary cytological features in periparturient cows that had been treated with a dopamine agonist drug to suppress the release of endogenous PRL (Akers et al., 1981a, b).

Although PRL is required in all mammals for initiation of milk secretion, only some species also require adenohypophysial PRL to sustain the secretion of milk once it is initiated. While rodents, primates, carnivores, and pigs appear to require the action of PRL for continued milk secretion, lactating females of the ruminant ungulates continue to secrete milk normally in the relative absence of blood-borne PRL (Forsyth, 1986).

The biochemical actions of PRL to initiate milk secretion in appropriately prepared mammary epithelial cells have been studied extensively. The hormone interacts initially with PRL receptors located in the plasma membrane of the cells of the mammary alveoli. The molecular structure of these PRL receptors was recently determined for rat liver tissue and the same receptors were also present in rat mammary tissue (Jahn et al., 1991). These authors also found that progesterone antagonized the usual periparturient increase in mammary receptors for PRL. This action may partly explain the observation that progesterone inhibits PRL-dependent lactogenesis in many species. As a result of PRL binding to its receptor in mammary cells, intracellular levels of mRNA encoding casein and α-lactalbumin proteins are increased. Both of these proteins are constituents of milk, but α-lactalbumin also comprises a part of the enzyme, lactose synthetase, that catalyzes the production of lactose for secretion into milk. As stated above, the liver has many PRL receptors but their functional role has not been fully determined. However, there is some evidence that PRL stimulates the liver to produce an unidentified factor which, together with PRL, acts on the mammary tissue to promote lactogenesis (English et al., 1990).

Luteotrophic Actions of PRL. As indicated above, the name luteotropin was initially assigned to PRL because the hormone serves an important luteotrophic function in rats. It is now known that luteotrophic effects (i.e., functional maintenance of the corpus luteum) require a complex of hormones and that the essential components of this ***luteotrophic complex*** do vary among species. In species such as the rat, mouse, hamster, dog, and mink, adenohypophysial PRL is a primary element of the luteotrophic complex. In other mammals such as the primate, rabbit, cow, and sheep, PRL plays either no role or a very minor role in the luteotrophic complex, with LH being the primary luteotrophic factor. In other species such as the goat and ferret, PRL acts together with LH to maintain the secretory function of corpora lutea (Buttle, 1983; Agu et al., 1986). In pigs, it is only older corpora lutea late in their life span that seem to require PRL for maintenance (Li et al., 1989). In summary, progesterone secretion by the corpus luteum is required for at least part of pregnancy in all mammals, perhaps as an adaptation for viviparity. Throughout the evolutionary development of mammals, most of them appear to utilize adenohypophysial PRL, together with other pituitary and placental hormones, to stimulate the luteal production of progesterone. However, the precise role of PRL in the luteotrophic complex varies widely among mammals. Additional details about the luteotrophic complex for each species will be presented in Chapter 12. In addition, corpus luteum-containing ovaries of rats were found to express multiple forms of the PRL receptor isolated originally from liver and also found in mammary tissue (Shirota et al., 1990; Hu and Dufau, 1991).

Other PRL Effects. All the biological actions of PRL discussed so far occur only in female mammals. The hormone is also present in and secreted from the adenohypophysis of males. Although perhaps not of major significance, other biological effects of PRL have been noted in both males and females, and one additional effect has been observed only in males. The male-specific effect of PRL involved stimulation of the seminal vesicles of rats, mice, sheep, and guinea pigs. Rarely did PRL stimulate the seminal vesicles by itself, but the hormone appeared instead to synergize with exogenous or endogenous testosterone to increase tissue growth.

Another action of PRL appears to be exerted on the adrenal gland in certain situations. Together with ACTH, exogenous PRL stimulated the bovine and human adrenal glands to produce androgenic steroids (Higuchi et al., 1984, 1985). In human females who secrete very high levels of PRL (a condition known as hyperprolactinemia), pharmacological inhibition of blood-borne PRL selectively decreased circulating concentrations of the adrenal androgen, dehydroepiandrosterone (Carter et al., 1977).

In submammalian species, PRL plays a role in osmoregulation, and in mammals the most consistently observed osmotic effect is that PRL increases renal retention of sodium ions (Nicoll, 1974). Administration of PRL also promotes absorption of sodium and water from the lumen of the intestine (Mainoya, 1975). However, pharmacological inhibition of endogenous PRL failed to alter renal processing of sodium in either humans or sheep (DelPozo and Ohnhaus, 1976; Bell et al., 1991). Although osmoregulation in adult mammals may not be regulated by PRL, the hormone may be important for osmoregulation in mammalian fetuses. Amniotic fluid contains very high concentrations of PRL, and it was recently shown that decidual cells of the human placenta transcribe the gene for adenohypophysial PRL (Wu et al., 1991). Therefore, the very large quantities of placental PRL are probably derived from decidual cells as well as from the adenohypophysis, and PRL may serve important osmoregulatory functions in the fetoplacental unit.

Another effect of PRL that appears to have been lost during the evolutionary transition between birds and mammals is the ability to promote migratory behavior. The hormone promotes premigratory increases in body reserves as well as in locomotor activity in birds, perhaps culminating in migratory behavior (Ensor, 1978). Another behavioral action of PRL in female rats is related to the previously described effects on the mammary gland to benefit the survival of offspring. Intrahypothalamic administration of PRL in steroid-primed female rats induced maternal nursing-related behaviors (Bridges et al., 1990a). Suppression of endogenous PRL decreased and replacement with exogenous PRL rapidly restored these maternal behaviors in similarly treated female rats (Bridges et al., 1990b).

The influence of PRL on specific behaviors suggests a direct CNS effect of the hormone. One established CNS-related action of PRL in rats and cattle is to activate dopaminergic neurons of the tuberoinfundibular tract that release dopamine into the hypophysial portal vessels (Gudelsky and Porter, 1980; Zinn et al., 1990). This action is part of the short-feedback regulation of PRL release because the PRL-activated dopamine delivered to the adenohypophysis inhibits further release of PRL. However, the action of PRL on the CNS may also represent a separate biological effect of the hormone, namely suppression of the hypothalamic-pituitary-gonadal axis. Chronic hyperprolactinemia is often associated with suppressed levels of adenohypophysial gonadotropins, primarily LH. In some situations, this association does not appear to be cause and effect because several factors (e.g., suckling and stressors) independently stimulate PRL and inhibit LH. However, there is experimental evidence in rats and humans that hyperprolactinemia can suppress adenohypophysial LH secretion, and that this suppression is mediated by inhibition of LHRH release. In male and female rats, specific experimental conditions are necessary to demonstrate that hyperprolactinemia suppresses LH secretion. However, spontaneously occurring hyperprolactinemia in human females often coexists with an anovulatory state. Pharmacological suppression of the elevated PRL secretion in these women with a dopamine agonist drug is often followed by a resumption of ovulatory menstrual cycles (Frantz and Kleinberg, 1978). It should be noted that these human clinical results, while therapeutically important, do not prove that PRL was directly suppressing the LHRH neurosecretory systems. The dopamine agonist drug may have exerted direct CNS effects unrelated to its inhibitory action on adenohypophysial mammotrophs. Moreover, endogenous opioid peptides (see Chapter 5) may be elevated in hyperprolactinemic women, and these opioid neuropeptides may be responsible for inhibiting the neurosecretion of LHRH. Because opioid neuropeptides are regulated by dopaminergic neurotransmission, it is possible that pharmacological manipulation of dopaminergic neurons may decrease opioidergic suppression of LHRH and allow ovulatory menstrual cycles to resume.

Prolactin is unusual among adenohypophysial hormones because it is present in milk at concentrations equivalent to those in maternal blood (Malven, 1983). In fact, colostrum contains even higher concentrations of PRL than maternal blood. This situation results in the neonatal offspring orally consuming a substantial quantity of maternal PRL. In contrast to maternal thyroid hormones ingested in milk (see Chapter 7), the biological significance of maternal PRL in the neonatal gut or in the blood of the neonate following intestinal absorption is not fully understood. In those species such as rats in which the gut can absorb intact proteins for several weeks after birth, PRL is absorbed across the intestinal wall and into blood (Whitworth and Grosvenor, 1978; Gonnella et al., 1989). In those species in which the offspring are born in a more mature state, it is likely that maternal PRL from orally consumed milk would not be absorbed into neonatal blood (Malven et al., 1976). However, direct intralumenal effects within the gut could still occur. More research will be required to determine the physiological role of maternal PRL delivered to the offspring in milk.

Patterns of Prolactin Secretion

The release of PRL from the adenohypophysial mammotrophs occurs as pulsatile discharges superimposed on steady-state secretion. The one stimulus or situation that causes pulsatile release of PRL in all species is suckling and/or milk removal from the mammary gland. Figure 9-1 depicts the results of one such study conducted in women who were nursing their infants (Noel et al., 1974). In this experimental protocol, the mothers were allowed to cuddle and play with their infants for an initial 30-min period during which no suckling was permitted. After about 5 min of this activity and 25 min before suckling was initiated, milk-ejection was observed in all mothers. Although oxytocin was not quantified in the blood of these mothers, it is valid to assume that a discharge of oxytocin preceded the observed milk-ejection (see Chapter 3). The PRL data in Figure 9-1 show very clearly that PRL release did not begin to increase until after the infants began suckling the breast. Consequently, these data illustrate that discharges of PRL and oxytocin can be dissociated even though they usually coincide. In other subjects, mechanical breast pumps were used to remove milk, and in these cases release of PRL was also stimulated. These results suggest that oxytocin discharges can be triggered by psychological and/or conditioned factors while PRL discharges require neural input from tactile receptors in the nipple. Of course, the neuroendocrine reflex occurring in most natural situations (animals and humans) will discharge both oxytocin and PRL because the tactile stimulation will coincide very closely with the onset of oxytocin-induced ejection of milk. Moreover, it has been demonstrated in cows and rats that electrical stimulation of the nipple or the mammary nerve triggered the release of PRL even though no milk was removed from the mammary gland (Reinhardt and Schams, 1975; Mena et al., 1980).

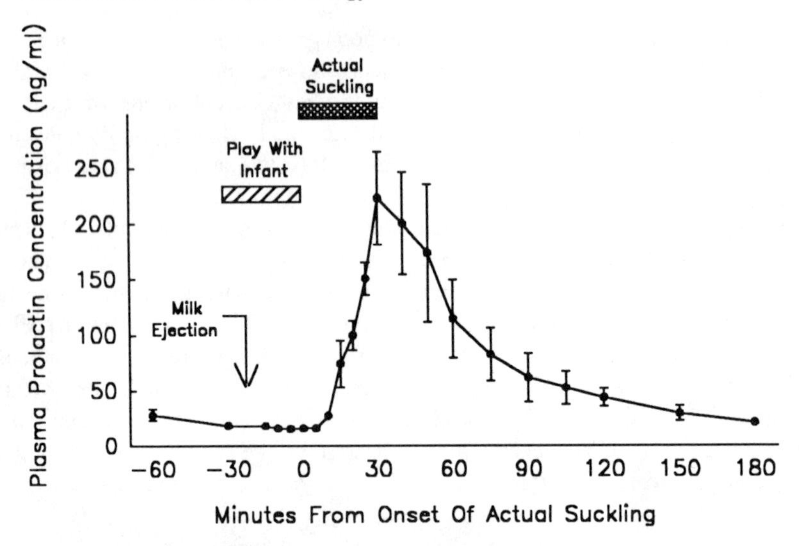

Figure 9-1. Suckling-induced release of human PRL.
Average plasma PRL concentration in three human females during a 30-min period of play with their infants followed by 30-min period of suckling. Data were averaged and redrawn from Noel et al. (1974). Milk ejection occurred in all women about 25 min before onset of actual suckling and is denoted by an arrow. Standard errors are plotted when they were larger than the symbols representing the average.

Neuroendocrine events in the reproductive cycle of female rats are highly circadian. For example, discharges of adenohypophysial hormones (notably LH and PRL) occur only at specific times of the day. As will be discussed later in Chapter 12, the preovulatory surge of LH release can only occur in the late afternoon hours. Ovulation and formation of new corpora lutea occur during the nocturnal hours between the days designated, based on vaginal cytology, as proestrus and estrus of the 4-5 day estrous cycle of unmated female rats. The female is sexually receptive during this same nocturnal period. When copulation occurs during this period of female receptivity, the newly formed corpora lutea are activated to secrete more progesterone and to have a functional life span of approximately 12 days during which proestrus/estrus changes are blocked. This period of mating-induced diestrus is called ***pseudopregnancy***. Although pseudopregnancy does not depend on conception, in cases of fertile matings it allows development of the fetoplacental unit to the point where it can provide luteotrophic support of the corpora lutea for maintenance of pregnancy to term at 22 days. As discussed earlier in this chapter, PRL exerts an important luteotrophic function in the rat. In fact, it is the only hormone required for functional maintenance of rat corpora lutea during the early part of pseudopregnancy (Smith et al., 1975). The secretory profiles of adenohypophysial PRL in non-mated and mated female rats are illustrated in Figure 9-2. Non-mated females have a surge of PRL release during the afternoon of proestrus approximately coincident with the preovulatory surge of LH release denoted in Figure 9-2. However, it is possible with experimental means to dissociate the PRL and LH surges on the afternoon of proestrus (Alexander et al., 1989).

When a sexually receptive female copulates with a male during the night between proestrus and estrus, a unique secretory profile of PRL discharges is initiated as shown in Figure 9-2. Beginning on the afternoon of the next day (i.e., estrus) and continuing on each subsequent afternoon (between 1500 and 2100 h), there occurs a discharge of PRL known as the daily ***diurnal surge***. Beginning during the nocturnal period between the days of estrus and diestrus-1, there occurs a daily ***nocturnal surge*** of PRL every night between 0100 and 0700 h. This pattern of twice-daily surges of PRL (diurnal and nocturnal) also occurs after a female rat, even one that is not sexually receptive, receives manual or electrical stimulation of the vagina and cervix. When such induced pseudopregnancies were artificially terminated, there occurred one estrous cycle followed by another pseudopregnancy, even though no mating or vaginal stimula-

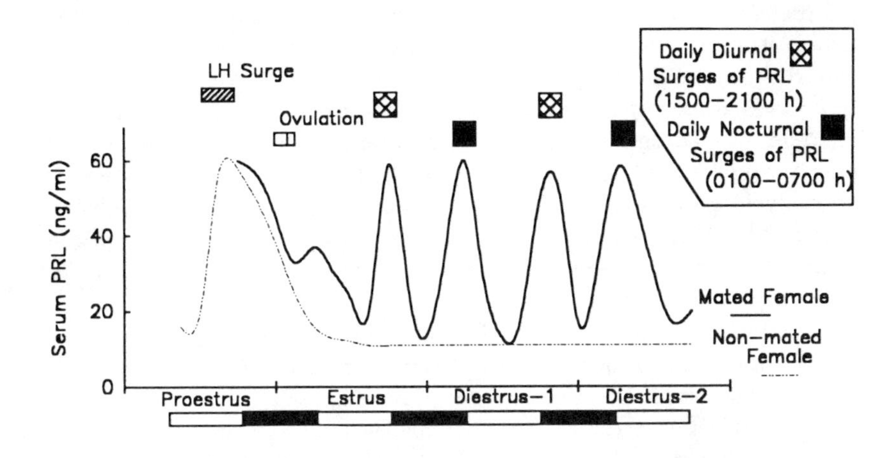

Figure 9-2. Effect of mating on serum profiles of PRL in female rats.

Representative profiles of serum PRL in one nonmated cyclic rat and in one female that was mated during the nocturnal period on the day of proestrus (alternating 12 h periods of light and dark denoted by horizontal bars of light and dark below the x-axis). Times of the LH surge and ovulation are also denoted. Twice-daily surges of PRL (diurnal and nocturnal) occur in mated females and contribute to activation of functional corpora lutea during the day known as diestrus-2. Hormone profiles were derived from the data of Smith et al. (1975).

tion occurred around the time that the second pseudopregnancy was initiated (Everett, 1968). This situation is called ***delayed pseudopregnancy***, and the neuroendocrine stimuli that cause the twice-daily surges of PRL necessary to activate the corpora lutea are apparently "*remembered*" during the intervening estrous cycle when these PRL surges do not occur (DeGreef and Zeilmaker, 1976).

Initiation and continued maintenance of the diurnal and nocturnal surges of PRL during pseudopregnancy have been investigated in great detail. Although the time of cervical stimulation does not affect timing of the surges, destruction of the suprachiasmatic nuclei (i.e., circadian timekeeper) abolishes the PRL surges in cervically stimulated rats (Bethea and Neill, 1980). The continuation of PRL surges during pseudopregnancy depends, in part, on continued ovarian secretion (Freeman et al., 1974), and their termination during the middle of pregnancy appears related to fetoplacental development and perhaps production of PRL-like hormones (Yogev and Terkel, 1978).

Other Secretory Profiles. As stated earlier in this chapter, initiation of lactogenesis at the end of pregnancy is the most consistent biological effect of PRL in mammals. Although many species secrete a placental hormone with PRL-like actions (i.e., placental lactogen), secretion of adenohypophysial PRL is almost always increased just before parturition. The lower panel of Figure 9-3 presents mean PRL data for dairy cows, and it illustrates a surge of PRL release approximately 12 h before parturition (Chew et al., 1979). Similar peri-parturient increases of adenohypophysial PRL have been observed in many species. Pharmacological suppression of this discharge of PRL profoundly interferes with lactogenesis in cattle (Akers et al., 1981a, b) and in goats bearing only a single fetus (Forsyth et al., 1985). However, lactogenesis occurred normally in goats bearing twins despite selective suppression of adenohypophysial PRL (Forsyth et al., 1985). Apparently, the greater quantity of placental lactogen secreted by twin fetuses was able to compensate for the induced deficiency of adenohypophysial PRL.

Concentrations of PRL in the blood of many mammals vary greatly between seasons of the year. Blood levels of PRL in both males and females are highest in the summer and lowest in the winter (Curlewis, 1992). The greatest seasonal differentials are observed in domestic ungulates (cow, sheep, goat, horse). In those species where it has been investigated, some type of seasonal differential in PRL persisted even after removal or denervation of the pineal gland. Seasonal variation of PRL can even be observed in fetal blood of sheep (Serone-Ferre et al., 1989).

The upper panel of Figure 9-3 illustrates this seasonal difference for cows around the time of parturition. Concentrations of plasma PRL for all three periparturient periods were greater in summer than in winter. Similar seasonal differentials in PRL secretion have also been reported in deer and mink (Schulte et al., 1980; Martinet et al., 1982). Domesticated pigs appear to have lost most of the seasonal differences in PRL secretion, but these differences can be still observed in wild pigs (Ravault et al. 1982; Trudeau et al., 1988).

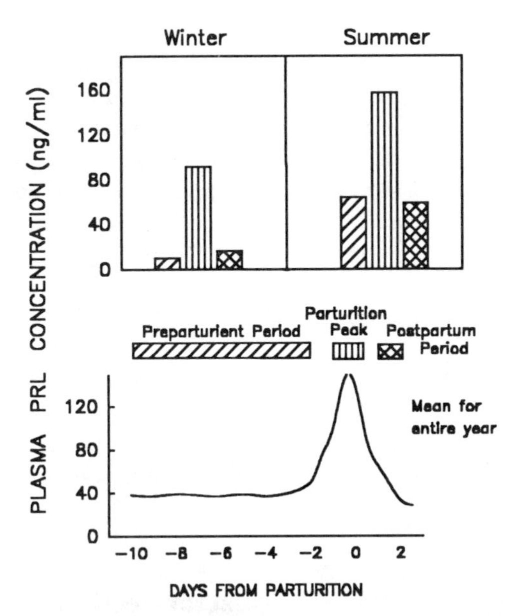

Figure 9-3. Plasma concentrations of PRL in dairy cattle around parturition and the effect of season.

Data redrawn from Chew et al. (1979). Winter values represent the month of December while summer values represent the month of June. All values in the upper panel are plotted with vertical bars that correspond to the time-based horizonal bars in the lower panel.

Seasonal differences in PRL depend primarily upon environmental lighting. Exposure of animals to controlled photoperiods (hours of light per day) can produce changes in PRL secretion similar to those that occur in natural environments. Therefore, the long photoperiods of summer are responsible for the high secretion of PRL, and the short photoperiods of winter are responsible for the depressed secretion of PRL. When photoperiods are held experimentally constant and ambient temperatures are manipulated, release of PRL in cattle can also be influenced, with higher temperatures being associated with increased release of PRL (Tucker et al., 1991). Although statistical associations between seasonal profiles of PRL and *either* photoperiod *or* temperature are quite strong (Chew et al., 1979), ambient temperature is still thought to be secondary to photoperiod as a causative factor in the natural release of PRL because seasonal profiles can be so easily duplicated by exposure to controlled photoperiods. When ewes were maintained for three years in an equatorial photoperiod (i.e., 12 light : 12 dark), they retained some annual periodicity of PRL secretion which may have been entrained by ambient temperatures or by an endogenous circannual rhythm (Jackson and Jansen, 1991).

Release of adenohypophysial PRL is also provoked by stressful stimuli in many species. Stressors stimulate the release of PRL most consistently in rats and other rodent species (Gala, 1990). Exercise-related stress of humans also stimulates release of PRL (Moretti et al., 1983). In domestic ungulates, intense stressors usually provoke discharges of PRL (Lamming et al., 1974), whereas milder stressors do not consistently increase the release of PRL (Lefcourt et al., 1986; Thomas et al., 1988). In domestic pigs, severe and prolonged restraint stress increased PRL release (Klemcke et al., 1987), whereas more brief, but still severe, restraint did not increase release of PRL (Hoagland et al., 1981; Smith and Wagner, 1985).

As discussed earlier in this chapter, a primitive biological effect of PRL in mammals is to promote the retention of sodium. Experimental induction of hyperosmotic states sometimes increased the release of PRL in humans and rats (Buckman and Peake, 1973; Relkin, 1974). However, more physiological alterations of plasma osmolarity in rats, taking special care to avoid experimental stressors, failed to confirm that osmotic stimuli influence the spontaneous release of PRL (Mattheij, 1977).

Neurohormonal Control of Prolactin Secretion

Prolactin is the only adenohypophysial hormone that can be secreted in substantial quantities without significant hypophysiotrophic stimulation from the hypothalamus. As discussed in Chapter 4, the primary hypothalamic compound with PRL-inhibiting hormone (PIH) activity is the catecholamine dopamine (DA). Acceptance of DA as an endogenous PIH was based upon the following evidence: (1) high concentrations of DA in hypophysial portal blood that are inversely related to PRL release in most situations, (2) presence of DA receptors on mammotrophs and other cells of the pars anterior, and (3) PRL-inhibiting bioactivity resulting from physiological concentrations of DA. The perikarya that synthesize and release DA to act on the adenohypophysis are located in the arcuate nucleus and periventricular hypothalamus, and they contribute axons to either the tuberoinfundibular tract (to median eminence) or the tuberohypophysial tract (to pars nervosa). These two dopaminergic systems appear to be regulated independently because of differential responses to various experimental manipulations (Ben-Jonathan et al., 1989). The tuberoinfundibular axons that discharge DA into the long portal vessels are probably more important in the tonic regulation of PRL. The biochemical steps mediating the inhibition of PRL by DA are not fully understood. However, there is evidence for multiple actions including (1) antagonism of intracellular second messengers, (2) activation of ion channels, and (3) reduction of intracellular calcium to interfere with exocytosis of PRL granules.

Although basal release of PRL is under inhibitory DA control, other mechanisms are probably involved in generating transient discharges of PRL as occur after suckling, mating, or stress. Of course, a rapid decline in delivery of

DA to the lactotrophs may also contribute to PRL discharges. Several hypothalamic compounds have PRL-releasing hormone (PRH) bioactivity. Two such PRH candidates discussed earlier (Chapters 4 and 5) are thyrotropin-releasing hormone (TRH) and vasoactive intestinal peptide (VIP). Both TRH and VIP have satisfied the criteria for an endogenous PRH. Hypophysial portal blood contains enriched concentrations of both peptides capable of stimulating PRL release *in vitro*, and immunoneutralization of each peptide interferes with some types of PRL discharges. Angiotensin II (AII) is another PRH candidate that is widely distributed in the mediobasal hypothalamus, pars nervosa, and pars anterior. Administration of AII in very low dosages is able to stimulate the release of PRL, but AII has not yet satisfied the criterion of its immunoneutralization compromising the release of PRL.

There are also PRH candidates that are concentrated in the neurointermediate lobe (NIL) of the hypophysis. One of these candidates is an unidentified peptide, less than 5000 Da molecular weight and different from any known PRL-releasing compound (Ben-Jonathan et al., 1989). Of the known NIL compounds, oxytocin constitutes the one most likely to also be an endogenous PRH. A few oxytocinergic axons that terminate in the median eminence discharge their contents into long portal vessels, perhaps creating an enriched concentration for delivery to the pars anterior. In addition, the release of oxytocin into the vasculature of the pars nervosa makes it likely that the short portal vessels carry blood with very high concentrations of oxytocin. Administration of antisera against oxytocin or antagonists of oxytocin receptors only sometimes interferes with suckling- or mating-induced discharges of PRL in rats. Moreover, relatively large quantities of oxytocin must be delivered to the lactotrophs in order to stimulate PRL release. This result and others suggest that oxytocin may also promote the release of PRL by acting on those neural systems that regulate the release of adenohypophysial PRL.

The complicated neurohormonal regulation of adenohypophysial PRL secretion is illustrated in Figure 9-4. A single DA neuron is depicted as the only source of PIH in hypophysial portal blood. Various authors have suggested that gamma-aminobutyric acid may also possess PIH activity, but definitive evidence is lacking. The cleaved portion of the precursor of LHRH, called GnRH-associated peptide, also possesses PIH bioactivity, but evidence for a physiological role of GnRH-associated peptide as an endogenous PIH is relatively weak (Yu et al., 1989).

Figure 9-4 also depicts two types of PRH neurons distinguished by the location of their axon terminals (i.e., hypothalamus and NIL of hypophysis). Hypothalamic PRH candidates are TRH and VIP, with both peptides meeting the criteria for a hypophysiotrophic hormone. Neurointermediate PRH candidates are oxytocin and an unidentified peptide with PRH bioactivity, which was mentioned above. As stated earlier, neither of the PRH candidates in the NIL has yet satisfied all the criteria. Indicated across the top of Figure 9-4 are the major factors that modulate the neurohormonal regulation of PRL release. The precise roles of DA and the various PRH candidates in each of these physiological situations will be discussed later in this chapter.

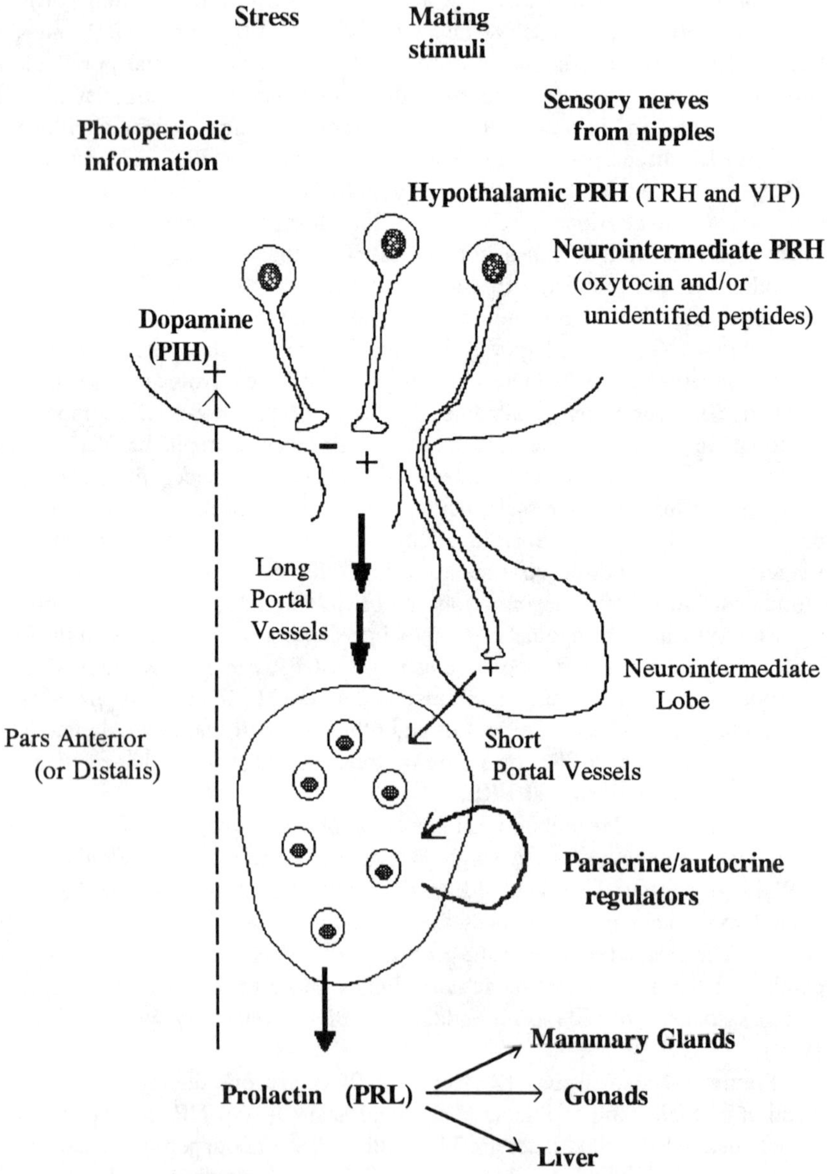

Stress
Mating stimuli
Sensory nerves from nipples
Photoperiodic information
Hypothalamic PRH (TRH and VIP)
Neurointermediate PRH (oxytocin and/or unidentified peptides)
Dopamine (PIH)
\+
\-
\+
Long Portal Vessels
\+
Neurointermediate Lobe
Pars Anterior (or Distalis)
Short Portal Vessels
Paracrine/autocrine regulators
Mammary Glands
Prolactin (PRL)
Gonads
Liver

Figure 9-4. Neuroendocrine integration of PRL release.

Illustration of mechanisms involved in the neuroendocrine integration of PRL release. This diagram depicts three types of neurosecretory neurons that modulate the release of PRL.

Mammotroph cells of the pars anterior are subject to autocrine/paracrine regulation in addition to neurohormonal regulation by PIH and PRH compounds (Figure 9-4). Autocrine/paracrine regulation could involve the short feedback inhibitory action of PRL (Kadowski et al., 1984), or perhaps an action of galanin which is colocalized with PRL in mammotrophs and may be released during exocytosis of PRL granules (Hyde et al., 1991). Intrapituitary VIP may also modulate release of PRL by autocrine/paracrine mechanisms (O'Halloran et al., 1991; Carrillo and Phelps, 1992). One candidate specifically for paracrine regulation of mammotrophs is AII, which is produced by gonadotrophs to exert PRH bioactivity on adjacent mammotrophs possessing AII receptors (Ganong, 1989).

The neuroendocrine integration depicted in Figure 9-4 also illustrates short feedback of blood-borne PRL onto hypothalamic structures that regulate the release of PIH and PRH compounds for delivery to the adenohypophysis. Evidence for PRL-induced increases in DA neurosecretion by tuberoinfundibular neurons is very strong. Such an increase in DA delivered to the pars anterior would inhibit release of PRL and constitute an inhibitory feedback loop. An additional action of blood-borne PRL on the hypothalamic and neurointermediate PRH systems remains a possibility.

Neurohormonal Regulation in Specific Situations. The relative roles of DA and the various putative PRH compounds must be determined individually for each type of PRL secretion. Suckling of female rats leads to depressed concentrations of DA in portal blood as well as to increased portal concentrations of TRH (DeGreef and Visser, 1981). The results of other investigations confirm that PRL release in suckled rats requires a decrease in DA release as well as an increase in TRH release (Plotsky and Neill, 1982). The results of immunoneutralization studies during suckling-induced release of PRL are complicated. Experimental immunoneutralizations of TRH, VIP, and oxytocin sometimes partly antagonized the PRL discharge in suckled rats. In a recent report, immunoneutralization of vasopressin also attenuated suckling-induced release of PRL (Nagy et al., 1991). In specially prepared sheep from which portal blood was collected without anesthesia, the correlation between suckling-induced discharges of PRL and observed increases of TRH in portal blood was relatively weak (Thomas et al., 1988).

Stress-induced discharges of PRL appear to involve decreases in tuberoinfundibular DA secretion probably caused by increased opioidergic inhibition of DA neurons. Antagonism of opioid receptors or immunoneutralization of specific opioid neuropeptides consistently decreases the magnitude of PRL release following various stressors. In contrast, the twice-daily surges of PRL release in mated pseudopregnant rats (Figure 9-2) appear to be timed by complicated mechanisms involving several PRH compounds. Neuronal activation of oxytocin release coincides with both the nocturnal and diurnal surges of PRL in mated rats (Arey and Freeman, 1992). In addition, VIP release appears to participate in the nocturnal PRL surge but not in the diurnal surge (Arey and Freeman, 1989).

Mechanisms that mediate the enhanced secretion of PRL by ungulates maintained in long photoperiods are not fully understood. Secretion of melatonin by the pineal gland appears to be involved in photoperiodic regulation of PRL release in sheep but not cattle (Kennaway et al., 1982; Stanisiewski et al., 1988a, b). However, daily administration of melatonin to cattle during the 16-h light period, in order to simulate a reduced ambient photoperiod, decreased the secretion of PRL (Sanchez-Barcelo et al., 1991). Photoperiodic regimens that altered PRL secretion in cattle failed to affect the steady-state turnover of DA in the mediobasal hypothalamus (Zinn et al., 1991). Concentrations of TRH quantified by push-pull perfusion of the median eminence in sheep were not correlated with photoperiod-induced changes in PRL release (Leshin and Jackson, 1987). In summary, neuroendocrine mediation of photoperiodic influences on release of PRL in ungulates is poorly understood. However, transient increases and decreases of ambient temperature to increase and decrease PRL release, respectively, altered DA turnover in the infundibulum of cattle in ways that were consistent with a cause-effect relationship (Tucker et al., 1991).

Effects of Estrogen. Secretion of PRL is enhanced by estrogenic hormones acting on both the adenohypophysis and the hypothalamus. Direct estrogenic stimulation of mammotrophs is readily demonstrated *in vitro*, but estrogen also appears to promote PRL secretion by decreasing PIH and/or increasing PRH neurosecretion. The relative magnitude of estrogenic stimulation of PRL secretion is much greater in rats and than in sheep (Shupnik et al., 1979). Chronic administration of estradiol to female rats altered the proportion of dispersed pars anterior cells which released PRL *in vitro* as well as transcription of the PRL gene and quantity of PRL released per cell (Wise et al., 1992). In addition to estrogenic effects on mammotrophs and on PIH or PRH neurons, it has also been suggested that estrogen stimulates pars intermedia cells to secrete α-MSH into short portal vessels for delivery to pars anterior where it recruits more mammotrophs to secrete PRL (Ellerkmann et al., 1992).

Applied Aspects of Prolactin Physiology

Most tumors of the adenohypophysis contain many mammotroph cells that secrete excessive quantities of PRL. These tumors have been studied mainly in rats and humans because their incidence in non-rodent animal species is low. Vascular defects that interfere with delivery of hypothalamic PIH to the adenohypophysis represent one probable cause of these tumors. In addition, estrogenic stimulation of mammotrophs may also contribute to the development of tumors. In rats, there are large differences between strains in the ability of exogenous estradiol to promote tumor development, and these differences are related to vascularity of the adenohypophysis (Elias and Weiner, 1984). In humans with PRL-secreting tumors, administration of a dopamine agonist drug inhibits PRL release and often decreases tumor size (Frantz and Kleinberg, 1978).

The possible role of PRL in carcinogenic development of mammary tissue has been studied extensively. Through its lactogenic actions, PRL promotes cell division in rat mammary tissue, thereby increasing the probability that other agents (e.g., carcinogens) will initiate tumorigenesis. Although this contributory role of PRL has been demonstrated for experimental tumorigenesis in rodents (Kim and Furth, 1976), there is no definitive evidence that hypersecretion of PRL plays a role in spontaneous development of mammary tumors as occurs in women.

As stated earlier in this chapter, secretion of PRL is required for lactogenesis in all mammals and for maintenance of lactation in non-ungulate species. Because the quantity of milk secreted has economic importance in agricultural species, research has been conducted on the possible beneficial effects of providing additional endogenous or exogenous PRL to ungulate species. When lactations have been initiated by parturition with the associated increase of endogenous PRL, there have not been any lactational benefits as a result of providing additional PRL (Plaut et al., 1987). Exposure of dairy cattle to long photoperiods results in small but consistent increases in milk production as well as major increases in PRL secretion (Tucker, 1985). However, these photoperiodic treatments have effects other than those involving PRL (e.g., feed intake), and it is unlikely that the photoperiod-induced increase of PRL caused the increase in milk production.

Lactation can be induced to occur in nonpregnant ungulates by administration of estrogen and progesterone. Milk yield during these so-called ***induced lactations*** is usually much less than that occurring in natural lactations, and individual lactational responses are highly variable. Secretion of endogenous PRL at the onset of these induced lactations is lower than that associated with parturition, but the regular removal of milk stimulates regular discharges of endogenous PRL (Erb et al., 1976). Concentrations of circulating PRL in cows during the period preceding the onset of induced lactation were positively correlated with subsequent milk yield (Chakriyarat et al., 1978). Compared to induction of lactation during the winter, when PRL secretion is low, treatments initiated in the spring resulted in higher yields of milk (Kensinger et al., 1979). In summary, it appears that endogenous secretion of PRL is fully adequate to initiate maximum lactational performance following normal parturition. Supplemental PRL can always improve milk production when blood-borne PRL has been inhibited by dopaminergic drugs. When lactation is initiated in nonpregnant ungulates by exogenous hormones, the spontaneous secretion of PRL necessary to initiate lactogenesis may be either marginal or minimally adequate.

REFERENCES

Agu, G.O., K. Rajkumar, and B.D. Murphy. Evidence for dopaminergic regulation of prolactin and a luteotropic complex in the ferret. *Biol. Reprod.* (1986) 35:508-515.

Akers, R.M., D.E. Bauman, A.V. Capuco, G.T. Goodman, and H.A. Tucker. Prolactin regulation of milk secretion and biochemical differentiation of mammary epithelial cells in periparturient cows. *Endocrinology* (1981a) 109:23-30.

Akers, R.M., D.E. Bauman, G.T. Goodman, A.V. Capuco, and H.A. Tucker. Prolactin regulation of cytological differentiation of mammary epithelial cells in periparturient cows. *Endocrinology* (1981b) 109:31-41.

Alexander, M.J., P.D. Mahoney, C.F. Ferris, R.E. Carraway, and S.E. Leeman. Evidence that neurotensin participates in the central regulation of the preovulatory surge of luteinizing hormone in the rat. *Endocrinology* (1989) 124:783-788.

Arey, B.J., and M.E. Freeman. Hypothalamic factors involved in the endogenous stimulatory rhythm regulating prolactin secretion. *Endocrinology* (1989) 124:878-883.

Arey, B.J., and M.E. Freeman. Activity of oxytocinergic neurons in the paraventricular nucleus mirrors the periodicity of the endogenous stimulatory rhythm regulating prolactin secretion. *Endocrinology* (1992) 130:126-132.

Bell, F.R., S.L. Lightman, and A. Simmonds. Dopamine modulation of prolactin and vasopressin but not behavior on satiation of sheep. *Am. J. Physiol.* (1991) 260:R1194-R1199.

Ben-Jonathan, N., L.A. Arbogast, and J.F. Hyde. Neuroendocrine regulation of prolactin release. *Prog. Neurobiol.* (1989) 33:399-477.

Bethea, C.L., and J.D. Neill. Lesions of the suprachiasmatic nuclei abolish the cervically stimulated prolactin surges in the rat. *Endocrinology* (1980) 107:1-5.

Bridges, R.S., M. Numan, P.M. Ronsheim, P.E. Mann, and C.E. Lupini. Central prolactin infusions stimulate maternal behavior in steroid-treated nulliparous female rats. *Proc. Natl. Acad. Sci.* (1990a) 87:8003-8007.

Bridges, R.S., and P.M. Ronsheim. Prolactin (PRL) regulation of maternal behavior in rats: bromocriptine treatment delays and PRL promotes the rapid onset of behavior. *Endocrinology* (1990b) 126:837-848.

Buckman, M.T., and G.T. Peake. Osmolar control of prolactin secretion in man. *Science* (1973) 181:755-757.

Buttle, H.L. The luteotrophic complex in hysterectomzied and pregnant goats. *J. Physiol.* (1983) 342:399-402.

Carrillo, A.J., and C.J. Phelps. Quantification of vasoactive intestinal peptide immunoreactivity in the anterior pituitary glands of intact male and female, ovariectomized, and estradiol benzoate-treated rats. *Endocrinology* (1992) 131:964-969.

Carter, J.N., J.E. Tyson, G.L. Warne, A.S. McNeilly, C. Faiman, and H.G. Friesen. Adrenocortical function in hyperprolactinemic women. *J. Clin. Endocrinol. Metab.* (1977) 45:973-980.

Chakriyarat, S., H.H. Head, W. W. Thatcher, F.C. Neal, and C.J. Wilcox. Induction of lactation: lactational, physiological, and hormonal responses in the bovine. *J. Dairy Sci.* (1978) 61:1715-1724.

Chew, B.P., P.V. Malven, R.E. Erb, C.N. Zamet, M.F. D'Amico, and V.F. Colenbrander. Variables associated with peripartum traits in dairy cows. IV. Seasonal relationships among temperature, photoperiod, and blood plasma prolactin. *J. Dairy Sci.* (1979) 62:1394-1398.

Curlewis, J.D. Seasonal prolactin secretion and its role in seasonal reproduction. A review. *Reprod. Fertil. Devel.* (1992) 4:1-23.

De Greef, W.J., and T.J. Visser. Evidence for involvement of hypothalamic dopamine and thyrotropin-releasing hormone in suckling-induced release of prolactin. *J. Endocrinol.* (1981) 91:213-223.

De Greef, W.J., and G.H. Zeilmaker. Prolactin and delayed pseudopregnancy in the rat. *Endocrinology* (1976) 98:305-310.

DelPozo, E., and E.E. Ohnhaus. Lack of effect of acute prolactin supression on renal water, sodium and potassium excretion during sleep. *Horm. Res.* (1976) 7:11-15.

English, D.E., S.M. Russell, L.S. Katz, and C.S. Nicoll. Evidence for a role of the liver in the mammotropic action of prolactin. *Endocrinology* (1990) 126:2252-2256.

Elias, K.A., and R.I. Weiner. Direct arterial vascularization of estrogen-induced prolactin-secreting anterior pituitary tumors. *Proc. Natl. Acad. Sci.* (1984) 81:4549-4553.

Ellerkmann, E., G.M. Nagy, and L.S. Frawley. α-Melanocyte-stimulating hormone is a mammotrophic factor released by neurointermediate lobe cells after estrogen treatment. *Endocrinology* (1992) 130:133-138.

Ensor, D.M. *Comparative Endocrinology of Prolactin.* Chapman & Hall (London, 1978), pp 309.

Erb, R.E., P.V. Malven, E.L. Monk, T.A. Mollett, K.L. Smith, F.L. Schanbacher, and L.B. Willett. Hormone induced lactation in the cow. IV. Relationships between lactational performance and hormone concentrations in blood plasma. *J. Dairy Sci.* (1976) 59:1420-1428.

Everett, J.W. "Delayed pseudopregnancy" in the rat, a tool for the study of central neural mechanisms in reproduction. In: Diamond, M. (ed). *Reproduction and Sexual Behavior* (Indiana Univ. Press, Bloomington, 1968), pg 25-31.

Forsyth, I.A. Variation among species in the endocrine control of mammary growth and function: the roles of prolactin, growth hormone and placental lactogen. *J. Dairy Sci.* (1986) 69:886-903.

Forsyth, I.A., J.C. Byatt, and S. Iley. Hormone concentrations, mammary development and milk yield in goats given long-term bromocriptine treatment in pregnancy. *J. Endocrinol.* (1985) 104:77-85.

Frantz, A.G., and D.L. Kleinberg. The pathophysiology of hyperprolactinemic states and the role of newer ergot compounds in their treatment. *Fed. Proc.* (1978) 37:2192-2197.

Freeman, M.E., M.S. Smith, S.J. Nazian, and J.D. Neill. Ovarian and hypothalamic control of the daily surges of prolactin secretion during pseudopregnancy in the rat. *Endocrinology* (1974) 94:875-882.

Gala, R.R. The physiology and mechanisms of the stress-induced changes in prolactin secretion in the rat. *Life Sci.* (1990) 46:1407-1420.

Ganong, W.F. Angiotensin II in the brain and pituitary: contrasting roles in the regulation of adenohypophyseal secretion. *Horm. Res.* (1989) 31:24-31.

Gonnella, P.A., P. Harmatz, and W.A. Walker. Prolactin is transported across the epithelium of the jejunum and ileum of the suckling rat. *J. Cell. Physiol.* (1989) 140:138-149.

Gudelsky, G.A., and J.C. Porter. Release of dopamine from tuberoinfundibular neurons into pituitary stalk blood after prolactin or haloperiodol administration. *Endocrinology* (1980) 106:526-529.

Higuchi, K., H. Nawata, K. Kato, and H. Ibayashi. Ovine prolactin potentiates the action of adrenocorticotropic hormone on the secretion of dehydroepiandrosterone sulfate and dehydroepiandrosterone from cultured bovine adrenal cells. *Horm. Metab. Res.* (1985) 17:451-453.

Higuchi, K. H. Nawaka, T. Maki, M. Higashizima, K.I. Kato, and H. Ibayashi. Prolactin has a direct effect on adrenal adrogen secretion. *J. Clin. Endocrinol. Metab.* (1984) 59:714-718.

Hoagland, T.A., M.A. Diekman, and P.V. Malven. Failure of stress and supplemental lighting in swine to affect release of prolactin in swine. *J. Anim. Sci.* (1981) 53:467-472.

Hyde, J.F., M.G. Engle, and B.E. Maley. Colocalization of galanin and prolactin within secretory granules of anterior pituitary cells in estrogen-treated Fischer 344 rats. *Endocrinology* (1991) 129:270-276.

Hu, Z.Z., and M.L. Dufau. Multiple and differential regulation of ovarian prolactin receptor messenger RNAs and their expression. *Biochem. Biophys. Res. Commun.* (1991) 181:219-225.

Jackson, G.L., and H.T. Jansen. Persistence of a circannual rhythm of plasma prolactin concentrations in ewes exposed to a constant equatorial photoperiod. *Biol. Reprod.* (1991) 44:469-475.

Jahn, G.A., M. Edery, L. Belair, P.A. Kelly, and J. Djiane. Prolactin receptor gene expression in rat mammary gland and liver during pregnancy and lactation. *Endocrinology* (1991) 128:2976-2984.

Kadowski, J., N. Ku, W.S. Oetting, and A.M. Walker. Mammotroph autoregulation: uptake of secreted prolactin and inhibition of secretion. *Endocrinology* (1984) 114:2060-2067.

Kennaway, D.J., T.A. Gilmore, and R.F. Seamark. Effect of melatonin feeding on serum prolactin and gonadotropin levels and the onset of seasonal estrous cyclicity in sheep. *Endocrinology* (1982) 110:1766-1772.

Kensinger, R.S., D.E. Bauman, and R.J. Collier. Season and treatment effects on serum prolactin and milk yield during induced lactation. *J. Dairy Sci.* (1979) 62:1880-1888.

Kim, U., and J. Furth. The role of prolactin in carcinogenesis. *Vitam. Horm.* (1976) 34:107-136.

Klemcke, H.G., J.A. Nienaber, and G.L. Hahn. Stressor-associated alterations in porcine plasma prolactin. *Proc. Soc. Exp. Biol. Med.* (1987) 186:333-343.

Lamming, G.E., S.R. Moseley, and J.R. McNeilly. Prolactin release in the sheep. *J. Reprod. Fertil.* (1974) 40:151-168.

Lefcourt, A.M., S. Kahl, and R.M. Akers. Correlation of indices of stress with intensity of electrical shock for cows. *J. Dairy Sci.* (1986) 69:833-842.

Leshin, L.S., and G.L. Jackson. Effects of photoperiod and morphine on plasma prolactin concentrations and thyrotropin-releasing hormone secretion in the ewe. *Neuroendocrinology* (1987) 46:461-467.

Li, Y., J.R. Molina, J. Klindt, D.J. Bolt, and L.L. Anderson. Prolactin maintains relaxin and progesterone secretion by aging corpora lutea after hypophysial stalk transection or hypophysectomy in the pig. *Endocrinology* (1989) 124:1294-1304.

Mainoya, J.R. Analaysis of the role of endogenous prolactin on fluid and sodium chloride absorption by the rat jejunum. *J. Endocrinol.* (1975) 67:343-349.

Malven, P.V. Transfer of prolactin from plasma into milk and associated physiological benefits to mammary cells. *Endocrinol. Exp.* (1983) 17:283-299.

Malven, P.V., A.M. Hollister, and J.E. Morningstar. Failure to demonstrate transfer of milk prolactin into blood of milk-fed rats and calves. *J. Dairy Sci.* (1976) 59:889-893.

Martinet, L., J.P. Ravault, and M. Meunier. Seasonal variations in mink (*Mustela vison*) plasma prolactin measured by heterologous radioimmunoassay. *Gen. Comp. Endocrinol.* (1982) 48:71-75.

Mattheij, J.A.M. Evidence against a role for prolactin in osmoregulation in the rat: water balance studies. *Endocr. Res. Commun.* (1977) 4:1-9.

Mena, F., P. Pacheco, and C.E. Grosvenor. Effect of electrical stimulation of mammary nerve upon pituitary and plasma prolactin concentrations in anesthetized lactating rats. *Endocrinology* (1980) 106:458-462.

Moretti, C., A. Fabbri, L. Gnessi, M. Cappa, A. Calzolari, F. Fraioli, A. Grossman, and G.M. Besser. Naloxone inhibits exercise-induced release of PRL and GH in athletes. *Clin. Endocrinol.* (1983) 18:135-138.

Nagy, G.M., T.J. Gorcs, and B. Halasz. Attenuation of the suckling-induced prolactin release and the high afternoon oscillations of plasma prolactin secretion of lactating rats by antiserum to vasopressin. *Neuroendocrinology* (1991) 54:566-570.

Nicoll, C.S. Physiological actions of prolactin. In: Greep, R.O., and E.B. Astwood (eds). *Handbook of Physiology. Sect. 7. Vol. IV. Part 2.* (Amer. Physiol. Soc., Washington, DC, 1974), pg 253-292.

Noel, G.L., H.K. Suh, and A.G. Frantz. Prolactin release during nursing and breast stimulation in postpartum and nonpostpartum subjects. *J. Clin. Endocrinol. Metab.* (1974) 38:413-423.

O'Halloran, D.J., P.M. Jones, M.A. Ghatei, and S.R. Bloom. Rat anterior pituitary neuropeptides following chronic prolactin manipulation: a combined radioimmunoassay and mRNA study. *J. Endocrinol.* (1991) 131:411-419.

Plaut, K., D.E. Bauman, N. Agergaard, and R.M. Akers. Effect of exogenous prolactin administration on lactational performance of dairy cows. *Dom. Anim. Endocrinol.* (1987) 4:279-290.

Plotsky, P.M., and J.D. Neill. Interactions of dopamine and thyrotropin-releasing hormone in the regulation of prolactin release in lactating rats. *Endocrinology* (1982) 111:168-173.

Ravault, J.P., F. Martinat-Botte, R. Mauget, N. Martinat, A. Locatelli, and F. Bariteau. Influence of the duration of daylight on prolactin secretion in the pig: hourly rhythm in ovariectomized females, monthly variation in domestic (male and female) and wild strains during the year. *Biol. Reprod.* (1982) 27:1084-1089.

Reinhardt, V., and D. Schams. Prolactin release in dairy heifers relative to duration of teat stimulation. *Neuroendocrinology* (1975) 17:54-61.

Relkin, R. Effects of alterations in serum osmolality on pituitary and plasma prolactin levels in the rat. *Neuroendocrinology* (1974) 14:61-64.

Sanchez-Barcelo, E.J., M.D. Mediavilla, S.A. Zinn, B.A. Buchanan, L.T. Chapin, and H.A. Tucker. Melatonin suppression of mammary growth in heifers. *Biol. Reprod.* (1991) 44:875-879.

Schulte, B.A., J.A. Parsons, U.S. Seal, E.D. Plotka, L.J. Verme, and J.J. Ozoga. Heterologous radioimmunoassay for deer prolactin. *Gen. Comp. Endocrinol.* (1980) 40:59-68.

Serone-Ferre, M., M. Vergara, V.H. Parraguez, R. Riquelme, and A.J. Llanos. The circadian variation of prolactin in fetal sheep is affected by the seasons. *Endocrinology* (1989) 125:1613-1616.

Shirota, M., D. Banville, S. Ali, C. Jolicoeur, J.M. Boutin, M. Edery, J. Djiane, and P.A. Kelly. Expression of two forms of prolactin receptor in rat ovary and liver. *Mol. Endocrinol.* (1990) 4:1136-1142.

Shupnik, M.A., L.A. Baxter, L.R. French, and J. Gorski. In vivo effects of estrogen on ovine pituitaries: prolactin and growth hormone biosynthesis and messenger ribonucleic acid translation. *Endocrinology* (1979) 104:729-735.

Sinha, Y.N. Prolactin variants. *Trends Endocrinol. Metab.* (1992) 3:100-106.

Smith, B.B., and W.C. Wagner. Effect of dopamine agonists or antagonists, TRH, stress and piglet removal on plasma prolactin concentrations in lactating gilts. *Theriogenology* (1985) 23:283-296.

Smith, M.S., M.E. Freeman, and J.D. Neill. The control of progesterone secretion during the estrous cycle and early pseudopregnancy in the rat: prolactin, gonadotropin and steroid levels associated with rescue of the corpus luteum of pseudopregnancy. *Endocrinology* (1975) 96:219-226.

Stanisiewski, E.P., N.K. Ames, L.T. Chapin, C.A. Blaze, and H.A. Tucker. Effect of pinealectomy on prolactin, testosterone and luteinizing hormone concentrations in plasma of bull calves exposed to 8 or 16 hours of light per day. *J. Anim. Sci.* (1988a) 66:464-469.

Stanisiewski, E.P., L.T. Chapin, N.K. Ames, S.A. Zinn, and H.A. Tucker. Melatonin and prolactin concentrations in blood of cattle exposed to 8, 16 or 24 hours of daily light. *J. Anim. Sci.* (1988b) 66:727-734.

Strickland, T.W., and J.G. Pierce. Glycosylation of ovine prolactin during cell-free biosynthesis. *Endocrinology* (1985) 116:1295-1298.

Thomas, G.B., J.T. Cummins, B. Yao, K. Gordon, and I.J. Clarke. Release of prolactin is independent of the secretion of thyrotropin-releasing hormone into hypophysial portal blood of sheep. *J. Endocrinol.* (1988) 117:115-122.

Trudeau, V.L., D.F.M. van de Wiel, J. Erkens, and L.M. Sanford. Influence of season and social environment on basal and thyrotropin releasing hormone-induced prolactin secretion in the adult domestic boar. *Acta Endocrinol.* (1988) 118:277-282.

Tucker, H.A. Photoperiodic influences on milk production in dairy cows. In: Haresign, W., and D.J.A. Cole (eds). *Recent Advances in Animal Nutrition - 1985.* (Butterworths, London, 1985), pg 211-221.

Tucker, H.A., L.T. Chapin, K.J. Lookingland, K.E. Moore, G.E. Dahl, and J.M. Evers. Temperature effects on serum prolaction concentrations and activity of dopaminergic neurons in the infundibulum/pituitary stalk of calves. *Proc. Soc. Exp. Biol. Med.* (1991) 197:74-76.

Whitworth, N.S., and C.E. Grosvenor. Transfer of milk prolactin to the plasma of neonatal rats by intestinal absorption. *J. Endocrinol.* (1978) 79:191-199.

Wise, P.M., G.H. Larson, K. Scarbrough, S. Chiu, N.G. Weiland, J.M. Lloyd, D.A. Hinkle, and A. Cai. Simultaneous monitoring of pituitary hormone secretion and gene expression within individual cells. *Biol. Reprod.* (1992) 46:178-185.

Wu, W.X., J. Brooks, M.R. Millar, W.L. Ledger, P.T.K. Saunders, A.F. Glasier, and A.S. McNeilly. Localization of the sites of synthesis and action of prolactin by immunocytochemistry and in-situ hybridization within the human utero-placental unit. *J. Mol. Endocrinol.* (1991) 7:241-247.

Yogev, L., and J. Terkel. The temporal relationship between implantation and termination of nocturnal prolactin surges in pregnant lactating rats. *Endocrinology* (1978) 102:160-165.

Yu, W.H., M. Arisawa, R.P. Millar, and S.M. McCann. Effects of gonadotropin-releasing hormone associated peptide (GAP) on the release of luteinizing hormone (LH), follicle stimulating hormone (FSH) and prolactin (PRL) in vivo. *Peptides* (1989) 10:1133-1138.

Zinn, S.A., L.T. Chapin, K.J. Lookingland, K.E. Moore, and H.A. Tucker. Effects of photoperiod on lactotrophs and on dopaminergic and 5-hydroxytryptaminergic neurones in bull calves. *J. Endocrinol.* (1991) 129:141-148.

Zinn, S.A., K.J. Lookingland, L.T. Chapin, K.E. Moore, and H.A. Tucker. Prolactin regulation of dopaminergic neurons in the infundibulum pituitary stalk of bull calves. *Proc. Soc. Exp. Biol. Med.* (1990) 193:98-103.

Chapter 10

PINEAL GLAND

The pineal gland is a small diverticulum of ependymal cells dorsal to the third ventricle and adjacent to the epithalamus. The name derives from the Latin term ***pinea*** (pine cone), the shape of which it resembles. Because the pineal gland is not bilaterally represented as are most CNS structures, ancient scholars attributed important functions to it. The terms pineal ***gland*** and pineal ***organ*** are both used, and each name is correct. Pineal gland, the name used in this book, is an appropriate term because the cells secrete a hormone (melatonin) into the blood. Pineal organ is also an appropriate name because the structure is one of several circumventricular organs (see Chapter 4). The cells of the pineal gland are called ***pinealocytes***, and they function as neuroendocrine transducers as described in Chapter 1 (Figure 1-1). This transduction consists of converting action potentials (AP) of postganglionic sympathetic neurons into melatonin secretion by the pinealocytes.

Changes in the pineal gland during evolution leading up to mammals reflect a progressive loss of photoreception by the pineal structure. In amphibians and lower species, the pineal gland appears to function as a ***third eye***, having photoreception and electrical activities consistent with visual function. Although visual functions were lost during evolutionary development, the pineal gland of birds and mammals retains the capacity to be influenced by ambient light. The pineal gland of birds appears, at least in part, to be directly sensitive to ambient light passing through the cranium. In mammals, the pinealocytes are modulated by ambient light, but this effect is mediated by retinal photoreception occurring in the eye.

The gross morphology of the pineal gland ranges from a spherical shape in ruminant ungulates to a cylindrical structure containing deep and superficial portions in rodent species (Rollag et al., 1985). The pineal tissue consists of non-neuronal parenchymal cells, the pinealocytes, that share some antigenic characteristics with retinal cells but not neuroglia (Korf et al., 1986). The pineal gland contains a few neuroglial cells but lacks neuronal perikarya. Although there is a tissue connection between the pineal gland and adjacent epithalamus, there do not appear to be any axons passing through this connecting neural tissue. However, there are numerous axonal terminals in the pineal gland; they are derived from postganglionic axons that originate outside the CNS in the superior cervical ganglion (SCG). The SCG is the rostral-most ganglion of the sympathetic chain, and it contains the perikarya of postganglionic sympathetic neurons. The perikarya in the SCG are activated by preganglionic axons originating in the CNS. Postganglionic axons enter the pineal gland from outside the CNS and innervate pinealocytes. Norepinephrine (NE) is the neurotransmitter released at these secretomotor terminals (see Figure 1-1). The ultrastructure of these terminals is similar to synaptic contacts between neurons.

Pineal Biochemistry

The primary secretory product of the pineal gland is melatonin and Figure 10-1 summarizes melatonin biosynthesis. Pinealocytes take up blood-borne tryptophan and convert it into 5-hydroxytryptophan using the enzyme tryptophan hydroxylase. The 5-hydroxytryptophan is then decarboxylated to yield the indolamine compound serotonin (also known as 5-hydroxytryptamine or 5-HT). Neither enzymatic conversion leading to the formation of serotonin is thought to be rate-limiting to biosynthesis of melatonin. However, the conversion of serotonin into N-acetylserotonin by the enzyme N-acetyltransferase (NAT) is a highly regulated and very rate-limiting reaction in melatonin synthesis. The final biosynthetic step involves the conversion of N-acetylserotonin to melatonin (also known as 5-methyoxy N-acetylserotonin) by the enzyme hydroxyindole-O-methyltransferase (abbreviated HIOMT). Most of the melatonin synthesized in pinealocytes is secreted into blood. A small portion is secreted into the cerebrospinal fluid (CSF) of the third ventricle, and some may be metabolized within the pinealocytes to biologically weaker compounds. Although the concentrations of melatonin in blood and in CSF are approximately equivalent, quantitative analysis of melatonin secretion by the sheep pineal gland revealed that far greater amounts were secreted into blood than into CSF, reflecting the much larger volume of blood compared to CSF (Rollag et al., 1978).

In addition to production of melatonin, the pineal gland also produces various polypeptides. One such biologically active peptide reported to be present in the pineal gland is ***arginine vasotocin*** (Reiter and Vaughan, 1977). This neuropeptide is closely related to both oxytocin and vasopressin, and in submammalian species this molecule appears to have many of the functions of these two neurohypophysial hormones (see Chapter 3). Further investigation of pineal arginine vasotocin using specialized chromatography and highly specific immunoassays suggests that the vasotocin-like compound in rat and bovine pineal glands is not identical to the compound isolated from the neurohypophyses of submammalian species (Pevet et al., 1981; Fisher and Fernstrom, 1981). In summary, the pineal gland of some mammalian species may contain arginine vasotocin or a closely related polypeptide, but the biological significance of this pineal compound is not known (Nieuwenhuis, 1984).

Influence of Ambient Light on the Pineal Gland

As stated above, pinealocytes are influenced in mammals by ambient light acting on the retina. Figure 10-2 diagrams the pathways between the retina and pinealocyte and also summarizes the effects of light on various elements of the pathway. Light-induced information travels from the retina and optic nerve down to the brain stem by two pathways as shown in Figure 10-2. The inferior accessory optic tract conveys impulses directly to the brain stem, whereas the alternative pathway involves the suprachiasmatic nucleus (SCN) of the

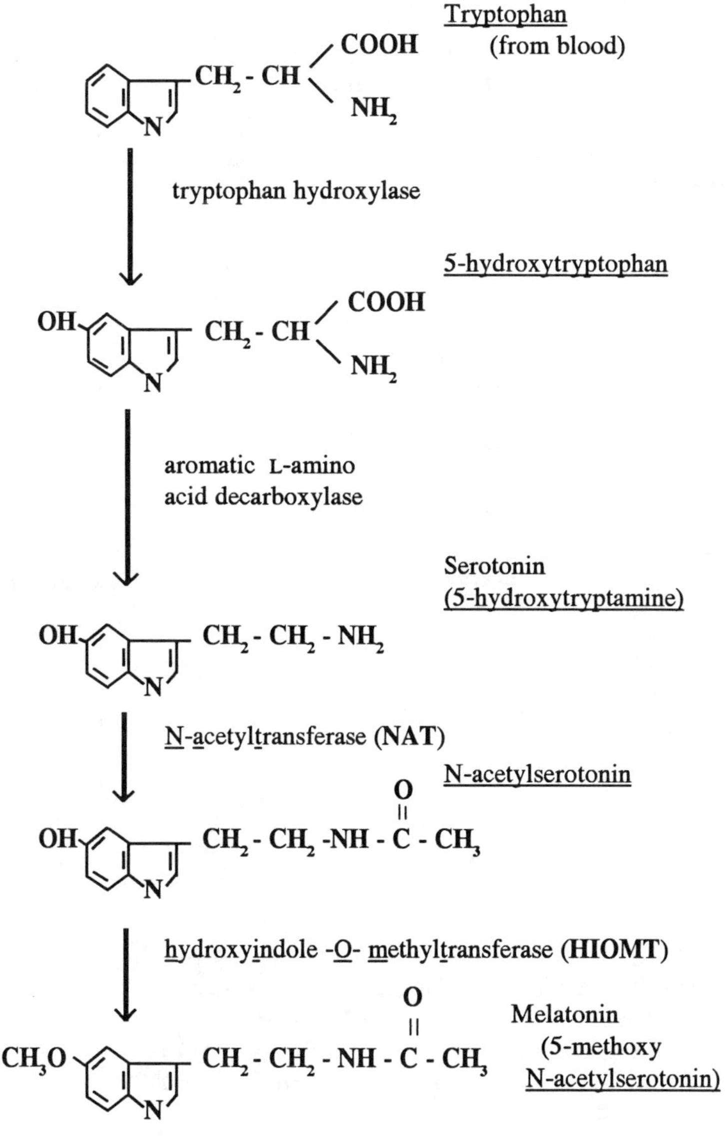

Tryptophan
(from blood)

$CH_2 - CH(COOH)(NH_2)$

tryptophan hydroxylase

5-hydroxytryptophan

OH $CH_2 - CH(COOH)(NH_2)$

aromatic L-amino
acid decarboxylase

Serotonin
(5-hydroxytryptamine)

OH $CH_2 - CH_2 - NH_2$

N-acetyltransferase (**NAT**)

N-acetylserotonin

OH $CH_2 - CH_2 - NH - C(=O) - CH_3$

hydroxyindole -O- methyltransferase (**HIOMT**)

Melatonin
(5-methoxy
N-acetylserotonin)

CH_3O $CH_2 - CH_2 - NH - C(=O) - CH_3$

Figure 10-1. Biosynthesis of melatonin in pinealocytes.

This figure contains chemical names and structures of melatonin and its precursors as well as the names of enzymes involved in each step of the biosynthesis.

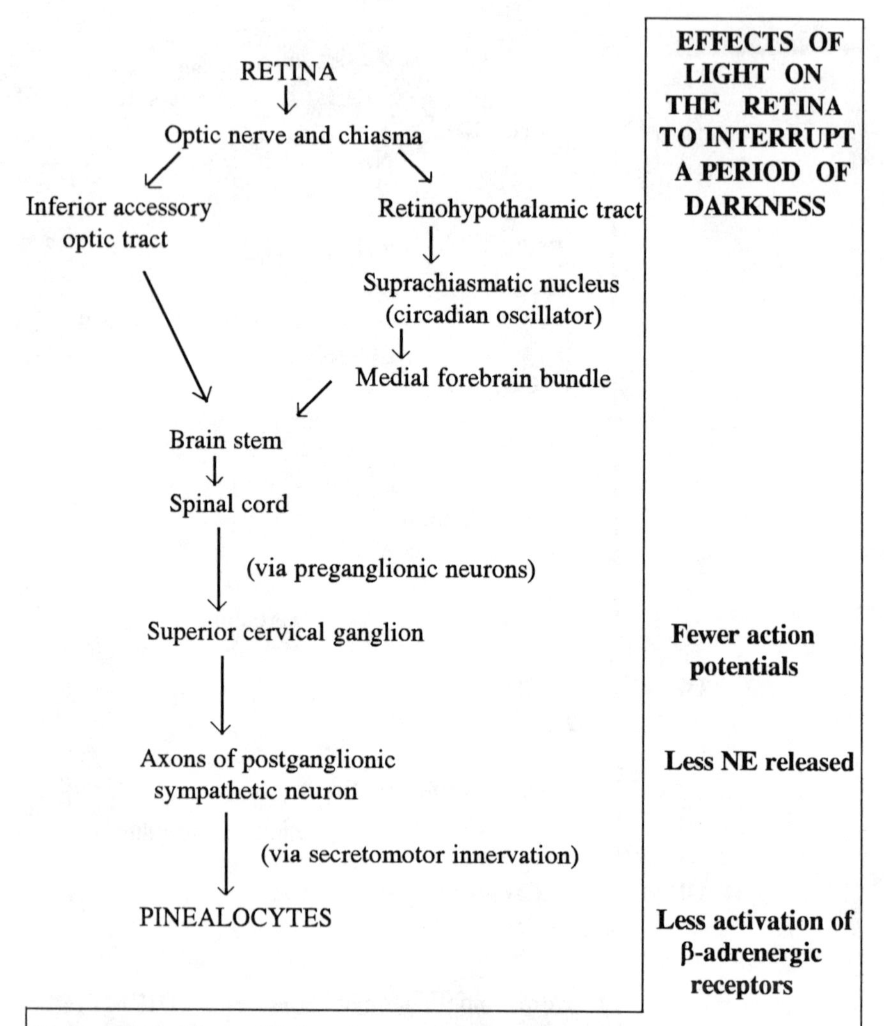

Light-induced intracellular changes in pinealocytes:

1. **Decreased production of cyclic AMP.**
2. **Large decrease in activity of the enzyme NAT.**
3. **Small decrease in activity of the enzyme HIOMT.**

Figure 10-2. Pathways between retina and pinealocytes: effect of light.
This diagram presents (**a**) structures of the pathways between retina and pinealocytes through which light-induced information passes and (**b**) effects of transient exposure of the retina to ambient light on various structures in the pathway. All effects of ambient light are enclosed within the lines of the polygon along the right and bottom of the figure.

hypothalamus. As discussed in Chapter 6, the SCN contains the mechanisms that govern the endogenous, free-running circadian rhythm. Light-induced information passes from the optic tract to the SCN via the retinohypothalamic tract, and it leaves the SCN via the medial forebrain bundle to the brain stem. After light information arrives in the brain stem, whether via the inferior accessory optic tract or the medial forebrain bundle, the subsequent pathways to the pineal gland are shared and involve the sympathetic nervous system. The perikarya of the preganglionic sympathetic neurons are located in the spinal cord, and they send their axons to synapse in the SCG with the perikarya of postganglionic sympathetic neurons.

The effects of light acting on the retina of mammalian eyes are also indicated in Figure 10-2. Through mechanisms which are only partly understood, exposure of the eye to light results in reduced firing of neurons in the SCG. This light-induced inhibition of the SCG neurons probably involves neuronal inhibition somewhere in the pathway because light activates most axons in the optic nerve. As a consequence of light-induced inhibition of action potentials in SCG neurons, there is less NE release at their secretomotor terminals with the pinealocytes. Because of reduced release of NE, there is less activation of β-adrenergic receptors in the pinealocytes. The intracellular changes in pinealocytes as a consequence of ambient light include decreased production of the second messenger compound cyclic AMP and a dramatic decrease in the activity of the enzyme NAT. There may also be a smaller light-induced decrease in the activity of the enzyme HIOMT.

When rats are maintained in an environment of alternating light and dark, there are profound time-dependent changes in pineal metabolism. Figure 10-3 presents the circadian rhythms for various pineal compounds and pineal enzymes in rats maintained on a daily lighting regimen of 14 light (L):10 dark (D). The onset of the dark period at 1900 h is followed by a rapid decrease in pineal content of serotonin, and this compound serves as a precursor to N-acetylserotonin and melatonin, both of which increase during the dark period. The enzymatic change that most closely parallels and explains these dark-induced changes in indole content is a dark-related increase in the activity of the enzyme NAT. There is a much smaller dark-related increase in activity of the enzyme HIOMT, but this increased activity may only reflect the much greater nocturnal quantities of N-acetylserotonin to be converted into melatonin. The circadian rhythms of indole compounds and enzyme activity in the pineal gland that are illustrated in Figure 10-3 were obtained in rats, but they are probably representative of most mammalian and avian species. Among birds, one notable difference occurred in the chicken pineal gland because no dark-induced depletion of serotonin could be observed. However, other changes illustrated in Figure 10-3 did occur in chickens. Moreover, these changes could be induced by ambient light and dark in chickens that had been surgically blinded (Binkley et al., 1975). It was subsequently discovered that *in vitro* oscillations of melatonin production by dissociated cells of the chicken pineal gland could be directly

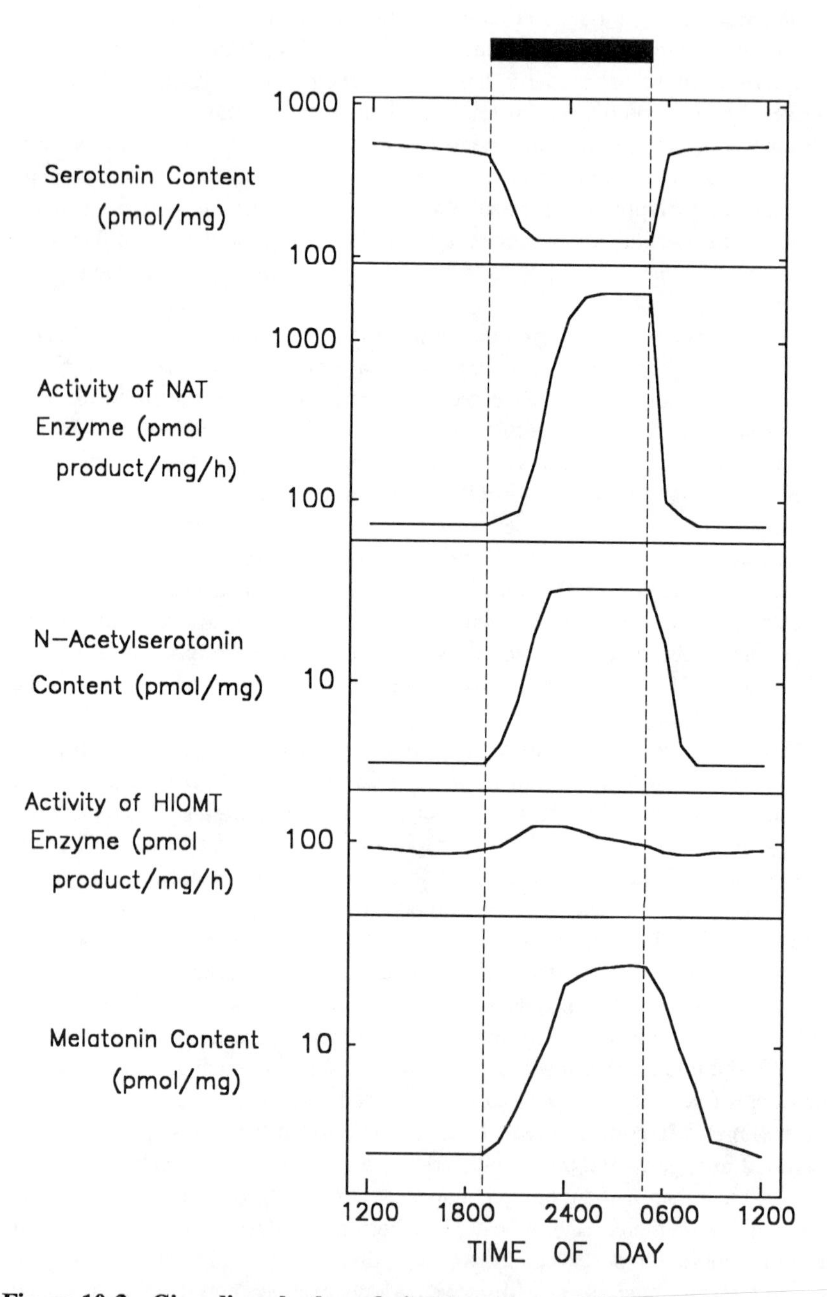

Figure 10-3. Circadian rhythm of pineal enzymes and pineal indoles. These data were redrawn from Klein (1974), and they were derived from rats maintained under conditions of darkness between 1900 h and 0500 h each day (black bar at top of graph.

altered by ambient light (Robertson and Takahashi, 1988). Moreover, light is able to penetrate into the cranial caivity in some species (Ganong et al., 1963).

To supplement information derived from measurement of enzymes and indoles in the pineal gland, circulating concentrations of blood-borne melatonin can also be quantified, and this approach is especially valuable in larger species. Figure 10-4 contains such data obtained from sheep, and these results clearly demonstrate the major effect that ambient light exerts on secretion of melatonin from the pineal gland. The middle panel depicts the circadian rhythm of serum

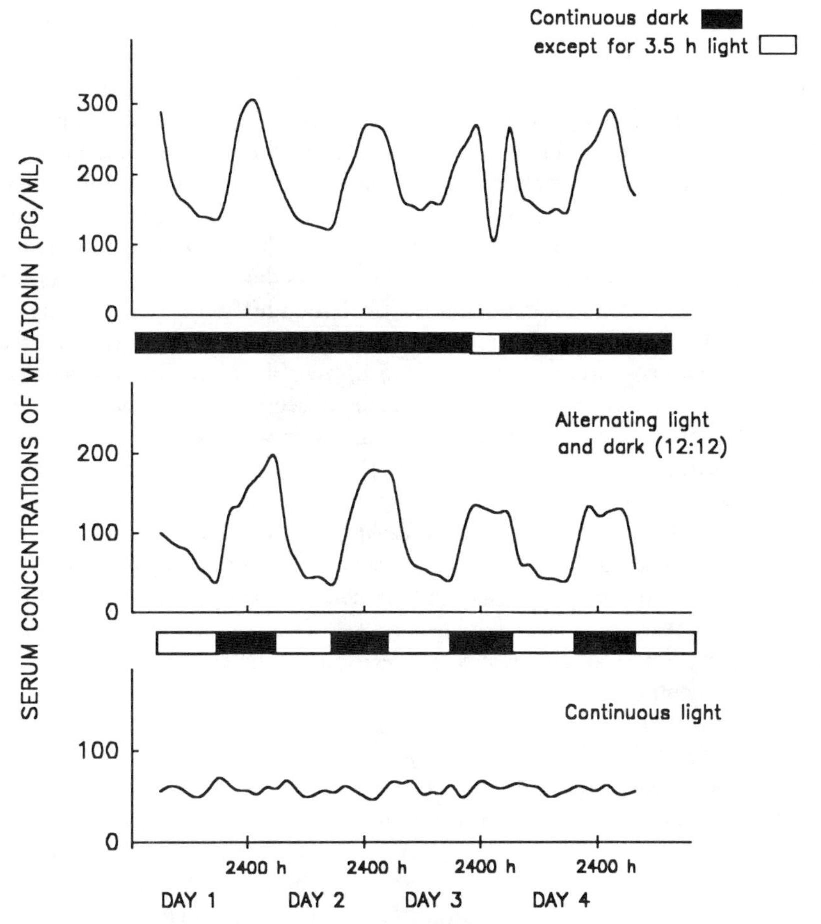

Figure 10-4. Serum profiles of melatonin in sheep housed under different lighting schedules.

Data estimated from those of Rollag and Niswender (1976) for ewes housed in continuous light (lowest panel), alternating 12 h light and 12 h dark (middle panel) and continuous dark interrupted by 3.5 h light around midnight on Day 3 (upper panel).

melatonin that was observed when sheep were maintained under a lighting regimen of 12L:12D. There was a nocturnal elevation in blood-borne melatonin during each dark period, and the duration of elevated melatonin corresponded approximately to the duration of the dark period. When these sheep were placed in continuous light (lowest panel, Figure 10-4), serum concentrations of melatonin remained relatively stable at the low concentrations that had been observed during the alternating periods of light. As suggested by Figures 10-2 and 10-3, the suppression of serum melatonin in the lowest panel of Figure 10-4 probably reflects reduced NE activation of β-adrenergic receptors in the pinealocytes and a decrease in the activity of the NAT enzyme.

The top panel of Figure 10-4 depicts the circadian profile of serum melatonin that usually occurs when mammalian species are maintained in continuous darkness. The recurring periods of elevated blood-borne melatonin during Days 1 and 2 corresponded in time to the former periods of darkness in these animals. Because these periods of increased release of melatonin occurred in the absence of light-dark cues, they represent an endogenous free-running circadian rhythm. As indicated in Figure 10-2, such a free-running rhythm is probably entrained by the circadian oscillator resident in the SCN. It should be noted that pineal gland expression of this circadian oscillation occurs during continuous darkness but not during continuous light (lowest panel, Figure 10-4). The top panel of Figure 10-4 also illustrates an experiment in which sheep being maintained in continuous darkness were exposed for 3.5 h to ambient light around midnight on Day 3. Serum concentrations of melatonin declined immediately and remained low during light exposure. When darkness was restored, serum melatonin increased immediately. This profile demonstrates, as does the lowest panel, that ambient light overrides the pineal expression of the endogenous circadian oscillator, but that a return to darkness allows pineal expression of the oscillator. It should also be noted that exposure to 3.5 h of light did not alter in any major way the timing of the circadian rhythm, as reflected in the melatonin increase around midnight of Day 4. Other research on endogenous circadian rhythms confirms that a single exposure to altered lighting is not sufficient to entrain the endogenous rhythm to a different schedule.

Sensitivity to Light. There are large differences between species in pineal sensitivity to ambient light. Rodents are extremely sensitive to light-induced suppression of melatonin secretion, and moderate intensities of light between 50 and 500 lux can inhibit melatonin secretion in sheep and cattle (Kennaway et al., 1983; Stanisiewski et al., 1988b). In contrast, humans require light intensities of over 2000 lux to modulate pineal secretion of melatonin (Lewy et al., 1980).

Although light sensitivity may differ among mammalian species, light-induced suppression of pineal secretion of melatonin during daylight hours leads to consistent and large nocturnal increases in blood-borne melatonin in almost all species. However, the existence of nocturnal increases of melatonin in domesticated pigs is controversial. In one early study, blood-borne melatonin was inhibited during the lighted period only when pigs were maintained on a balanced 12L:12D lighting regimen (McConnell and Ellendorff, 1987). In other work with pigs studied in many environments, no consistent differences were observed in melatonin concentrations during dark versus light periods (Diekman et al., 1992). However, research from another laboratory concluded that pigs had nocturnal rises in blood-borne melatonin in a variety of ambient photo-periods (Paterson et al., 1992). Because of these contradictory results, the influence of ambient light on melatonin secretion in pigs cannot be stated with any certainty.

Endogenous Free-Running Circadian Rhythms

The rhythm of serum melatonin in sheep maintained in continuous darkness (upper panel Figure 10-4) is but one of several manifestations of an endogenous free-running rhythm. The description free-running means that the rhythm occurs in the absence of time-related signals from the environment. Voluntary locomotor activity in rodent species is readily quantified in activity-monitoring cages, and this parameter is often used in the study of endogenous rhythms. Nocturnal mammals such as hamsters and rats initiate a period of intense locomotor activity at the onset of each daily period of darkness. Figure 10-5 describes the effects of several types of lighting changes on the daily profiles of voluntary locomotion. When hamsters were changed from an alternating 14L:10D daily regimen to continuous darkness, this voluntary locomotor activity became free-running. Moreover, its onset gradually became less intense, and the interval between successive onsets of locomotion became about 24.5 h instead of 24 h as occurred when entrained by 14L:10D lighting. As shown in the lower part of Figure 10-5, the circadian rhythm of locomotion remained fully entrained to an interval of exactly 24 h when the hamsters were changed from 14L:10D to an environment of continuous darkness that was interrupted by two 1-sec pulses of light at the beginning and end of the former 14-h period of light. As illustrated, this experimental regimen created a 14-h period of darkness followed by a 10-h period of darkness, with each period separated by the pulses of light. Based on the occurrence of intense locomotion at 1400 h each day, the 10-h period between the two light pulses was apparently perceived as the nocturnal period. When the 1-sec pulse of light occurring at 1400 h was shifted 3 h earlier to 1100 h so that it occurred at the end of an 11-h period of darkness, the newly created 13-h period of darkness was still perceived as the nocturnal period, and the onset of intense locomotion shifted to occur at 1100 h. Altering the time of occurrence of such an endogenously regulated parameter as voluntary locomotion by altering an environmental factor is called ***phase-shifting***, and it demonstrates entrainment of the endogenous rhythm by the environmental factor.

Destruction of the SCN abolishes the ability of lighting transitions such as those described in Figure 10-5 to entrain locomotor activity. Moreover, endogenous free-running rhythms of locomotor activity are disrupted by experimental lesions of the SCN. These various results established that the SCN was either (1) involved as a center or essential pathway in endogenous timekeeping or

(2) essential for the expression of the effects of a separate timekeeping mechanism. A series of other studies has recently established that the SCN is indeed a center for endogenous timekeeping in hamsters (Ralph et al., 1990). As described previously in Chapter 6, these workers established a mutant strain of hamsters in which the free-running rhythm of voluntary locomotion was 20 h rather than 24 h. By selective lesioning and transplantation of SCN tissue from the opposite strain, these authors demonstrated that the period of the free-running rhythm was a characteristic of transplanted SCN tissue rather than the host brain. The hypothalamic SCN is also an important element in lighting-induced entrainment of circadian rhythms in hypophysial secretion of ACTH and GH (see Chapters 6 and 8).

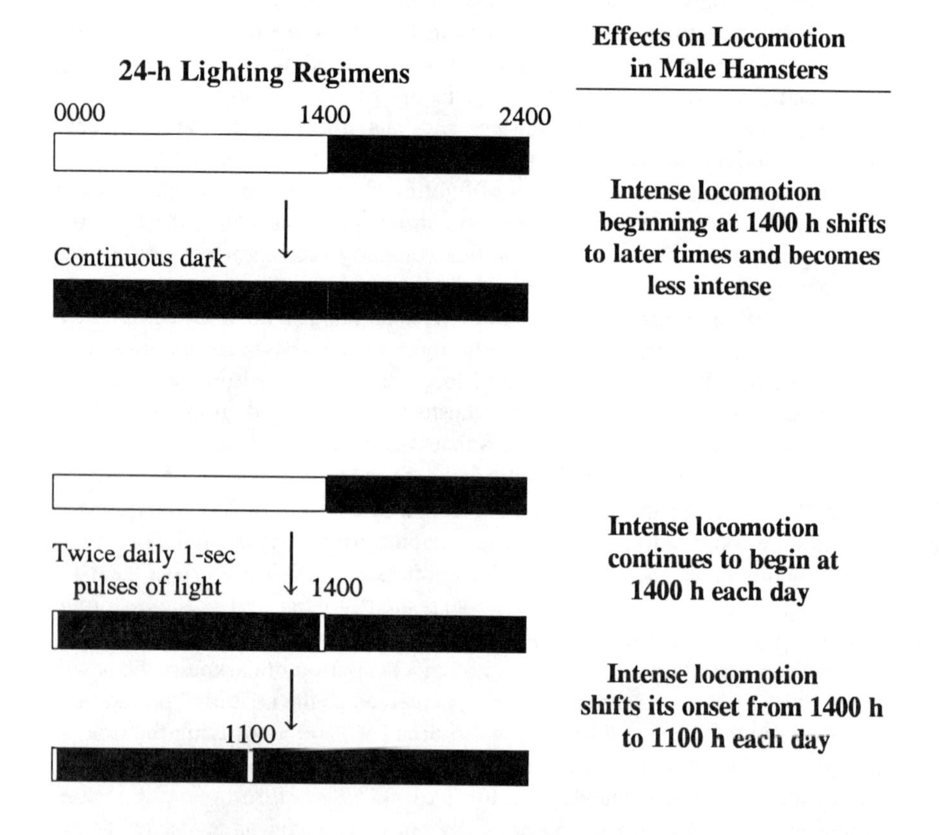

Figure 10-5. Effect of ambient lighting on profiles of voluntary locomotion in male hamsters.

This figure presents on the left the lighting regimens used by Earnest and Turek (1983), and to the right of each transition between lighting regimens, a description of the long-term effects of that transition on daily profiles of voluntary locomotor activity.

Possible Free-Running Circannual Rhythms. Various processes of reproduction, migration, hibernation, antler growth/shedding, hair-coat changes, and hormone secretions display season-related changes. The yearly recurrence of any such process constitutes an ***annual rhythm*** that is also sometimes called a circannual (approximately annual) rhythm using terminology similar to that used for circadian (approximately daily) rhythms. This author prefers to limit the name ***circannual rhythm*** to those free-running rhythms which can be demonstrated to recur at approximately yearly intervals when seasonal factors that might entrain the rhythm have been held constant. With such a strict definition for circannual rhythms, they become very difficult to demonstrate. The environment of the experimental animal must be held completely constant for many years or at least not be influenced by the natural changes in climate. The term annual rhythm should be used to describe those seasonal changes that recur at annual intervals under natural conditions in which one or more environmental signals probably govern the timing of the seasonal change.

The strongest evidence in support of endogenous free-running circannual rhythms involves the molting activity of certain species of birds (Gwinner, 1986). During maintenance of such birds in a constant 10L:14D photoperiod for 7 years, there occurred nine complete cycles of pre-nuptial and post-nuptial molting. This molting activity recurred on the average every 9-10 months. Because the molting did not occur at a consistent time of the year, it could not have been entrained by unregulated seasonal elements.

Although free-running circannual rhythms have been demonstrated for some parameters in mammals (Farner, 1985), their existence for reproductive hormones has only been suggested by recent experiments. For example, ovariectomized sheep treated continuously with estradiol display annual increases in hypophysial LH secretion which approximate the onset and duration of the photoperiod-regulated breeding season. When such animals were maintained for 5 years in a constant photoperiod of 8L:16D, there occurred periodic increases in LH secretion that were not synchronous among individual animals and did not recur at intervals of exactly 12 months (Karsch et al., 1989). When sheep in another experiment were housed under a 12L:12D photoperiod for 3 years, there were some significant fluctuations of average prolactin secretion across all animals (Jackson and Jansen, 1991). However, in neither of these studies with sheep was it possible to control ambient temperatures that could have been entraining annual rhythms of LH and prolactin secretion. When woodchucks (*Marmoto monax*) were maintained for over 4 years without variation in either photoperiod (12L:12D) or ambient temperature, both males and females displayed recurrent regular increases in reproductive hormones suggesting the existence of a free-running circannual rhythm with a period of 10-11 months (Concannon et. al., 1992).

Pineal Gland Influences on Hormone Secretion

Hypophysial secretion of prolactin and LH varies among seasons, and in most mammals the seasonal patterns of hormone secretion are regulated by the environmental photoperiod. Because the daily duration of darkness usually determines the daily duration of elevated melatonin secretion, endogenous melatonin may mediate the effects of annual variation in photoperiod on secretion of both PRL and LH.

Prolactin. As discussed in Chapter 9, long summer-like photoperiods increase the secretion of PRL in ruminant ungulates. Mediation of this PRL effect by the pineal gland through its secretion of melatonin is suggested by some studies, but the extent of pineal mediation may differ among species. Elimination of pineal gland function by surgical removal or denervation partly antagonized seasonal or photoperiod-induced changes in PRL secretion of goats in one study (Maeda et al., 1988), but not in another study (Buttle, 1977). Pinealectomy in sheep disrupted the seasonal changes in PRL secretion (Brown and Forbes, 1980; Munro et al., 1980; Kennaway et al., 1981). However, pinealectomy in cattle did not interfere with photoperiod-induced increases of PRL release (Stanisiewski et al., 1988a).

Administration of melatonin to intact or pineal-denervated ungulates has, with only a few exceptions, produced more consistent results. Animals exposed to short photoperiods would be expected to secrete more melatonin during the increased hours of darkness each day. Therefore, the predicted effect of exogenous melatonin would be to decrease the release of PRL if endogenous melatonin is responsible for seasonal and/or photoperiod-induced changes in PRL secretion. This effect of exogenous melatonin on PRL release was observed in pineal-denervated goats (Maeda et al., 1988), intact female sheep (Kennaway et al., 1982), and fetal sheep (Bassett et al., 1989), but it was not observed in cattle (Stanisiewski et al., 1988b). It should be noted that sheep and goats are seasonal breeding species, and that photoperiod-induced changes in melatonin influence the pituitary-gonadal axis in these species (see next section on LH), but not in cattle.

Luteinizing Hormone. In mammals that reproduce only during specific seasons, it appears that photoperiodic signals acting through the pineal gland control the onset of the annual breeding season (Hastings et al., 1985). This pineal transduction of photoperiod occurs in both males and females, and it occurs in those species that breed only in the spring (i.e., increasing photoperiods) or only in the fall (i.e., decreasing photoperiods).

Hamsters represent a species in which the annual breeding season occurs during long photoperiods (denoted as long-day breeders). The pineal gland is clearly involved in regulating the reproductive hormones that mediate this effect of long days. Such photoperiod effects have been extensively studied in male hamsters because testes size can be readily and repeatedly monitored without surgery. When reproductively active males with large testes are exposed to short (<12 h) photoperiods or exogenous melatonin regimens which simulate short days, the testes undergo a rapid pineal-mediated atrophy (Turek and Campbell, 1979). It is possible to prevent this induced atrophy of the testes by experimental disruption of any specific link in the pathway between the hypothalamic SCN and increased secretion of melatonin by pinealocytes (Figure 10-2). However, when male hamsters are experimentally blinded by transection of the optic nerve, the testes atrophy irrespective of the environmental photoperiod. Presumably, blinded animals produce excessive melatonin, and the testes atrophy as though the animals are being maintained in short days. Testicular atrophy in short days or darkness is due to deficient secretion of LH (Turek and Campbell, 1979). A major part of this deficient LH secretion results from an enhanced inhibitory feedback of testicular androgens, but there is also a steroid-independent component to the LH deficiency (Ellis and Turek, 1980). Another contributing factor in testicular atrophy may be decreases in testicular receptors for LH due to short-day induced hypoprolactinemia (Bex et al., 1978).

The regrowth of atrophied testes (termed recrudescence) can occur by two different means. Daily environmental photoperiod can be increased to 13 h or longer, and testicular recrudescence will be induced. Alternatively, male hamsters maintained continuously on short days for prolonged periods will become refractory to these short days, and their testes will undergo a spontaneous or endogenous recrudescence. In both situations, hypophysial secretion of LH increases and initiates testicular growth and androgen secretion. Spontaneous recrudescence of the testes is not dependent on the pineal gland, but long-day induced recrudescence is pineal-mediated. One evolutionary benefit associated with spontaneous testicular recrudescence would be that males that spend the winter in darkened burrows can enter the external environment ready for mating in the spring.

Internal perception of the duration of an ambient photoperiod does not depend upon the total duration of light per day, but rather on when the light occurs during the day and how that light entrains the endogenous circadian rhythm (Elliott and Goldman, 1981). As illustrated in Figure 10-5, a second brief (1 sec) pulse of light occurring 14 h after a previous brief pulse (that was apparently perceived as dawn in the circadian oscillator) could entrain the circadian profile of locomotion, and it also retarded the darkness-induced atrophy of testes in some animals (Earnest and Turek, 1983). A similar brief pulse of light occurring 8 h after the end of a once-daily 6-h period of light completely prevented testicular atrophy suggesting that the light of 6 h duration together with an additional 1-sec pulse at 14 h after the perceived dawn could maintain testes as well as a full 14 h of daily light (Earnest and Turek, 1983).

Reproductive function of female hamsters is also affected by photoperiod, with short days causing previously cyclic females to become anestrus. Female hamsters can be induced to become anestrus even in long photoperiods by administration of melatonin each afternoon. Apparently, the combination of exogenous melatonin each afternoon and endogenous melatonin secreted from the pineal gland during the dark period was perceived by the reproductive axis as a short day, and estrous cycles were suppressed probably due to defects in LH secretory mechanisms (Goldman and Darrow, 1983).

Domestic sheep represent the most thoroughly studied species that are short-day breeders. Because of economic advantages associated with causing female sheep to breed at times other than the autumn, extensive research has been conducted in these females. However, male sheep (rams) also display annual cycles of testicular activity and sperm production. Lengthening photoperiods that occur in the spring and summer are associated with decreased testicular function. Denervation of the pineal gland by surgical removal of the superior cervical ganglion in rams from one primitive breed of sheep (Soay) blocked the gonadal atrophy induced by long photoperiods (Lincoln, 1979). However, the atrophied testes of Soay rams maintained on long days eventually underwent spontaneous recrudescence (Almeida and Lincoln, 1982) similar to the testes of male hamsters maintained continuously in short days (see above). If fertility of female sheep can be induced during any season, it is equally desirable to maintain rams in a highly fertile state throughout the year. One experimental lighting regimen that achieved this goal in domestic rams involved rapidly changing environmental photoperiods (Almeida and Pelletier, 1988). The lighting environment of these rams alternated between 16L:8D and 8L:16D at such a rapid rate that one complete annual cycle, that took 12 months in nature, was completed in this experiment every 2 months. Serum concentrations of LH and testosterone were maintained at levels that were intermediate between the peaks and nadirs observed in rams housed in naturally varying photoperiods. The authors concluded that inhibitory feedback of androgen in the experimental rams was insufficient to suppress the secretion of LH below that needed to maintain normal testicular function, and the rams remained continuously fertile. It should also be noted that melatonin secretion in these photoperiod-regulated rams always increased during the daily period of darkness irrespective of the phase of the alternating experimental photoperiods (Pelletier and Ravault, 1988).

Female sheep (ewes) initiate estrous cycles in the autumn and deliver offspring about 5 months later in the spring. Ewes of most breeds then enter a period of anestrus during the remainder of the spring and the early part of the summer. Ovulation does not occur during this period of anestrus, and it has been well established that the secretion of LH is effectively suppressed during anestrus by the low levels of circulating estrogens derived from the otherwise quiescent ovaries (Karsch et al., 1984). This period of anestrus ends with the onset of the breeding season because the inhibitory feedback of estrogen is no longer able to prevent increases in LH secretion. This change in feedback sensitivity to estrogen is synchronized among individual ewes by perceived changes in the duration of daily elevations of melatonin secretion during the dark period (Wayne et al., 1988). The reduced sensitivity to LH-inhibitory feedback of estrogen can also occur spontaneously, but in such cases it is not especially synchronous among individuals (Woodfill et al., 1991). Such spontaneous changes in feedback sensitivity to estrogen appear to result from development of a photorefractory state in which exposure to long days is no longer able to maintain high sensitivity to estrogen suppression of LH. This photorefractory state may have a role in initiating the breeding season under natural conditions because onset of ovulation often occurs when photoperiods have hardly begun to decrease (Thimonier and Mauleon, 1969). Pineal secretion of melatonin was not disrupted in ewes that had become photorefractory since increased serum concentrations of melatonin still faithfully reflected each period of darkness (Karsch et al., 1986). Therefore, unknown components of the hypothalamic-hypophysial system for LH secretion change as a result of photorefractoriness to the melatonin signal of long days (i.e., short duration of elevated nocturnal melatonin), and the resulting increase of LH secretion initiates ovarian follicular development leading to the resumption of ovulatory estrous cycles. In ewes that are not mated, these estrous cycles continue for several months and usually end early in the winter. The termination of ovulatory estrous cycles is caused by an increased sensitivity to estrogen suppression of LH secretion. This resumption of enhanced estrogen feedback appears to be caused by development of photorefractoriness to short days and the associated nocturnal elevations of melatonin secretion (Malpaux et al., 1988).

Biotechnological Alteration of Breeding Seasons. In those species reared for commercial production, there are often economic advantages to be gained from altering the breeding season. Alteration of ambient photoperiods represents the simplest approach, and this technique works best in long-day breeders because extra light can be provided in the evening to stimulate early onset of breeding in the spring. As discussed above for male hamsters, internal perception of the duration of environmental photoperiod depends on light occurring during a specific period of the internal circadian rhythm. When environmental light occurs during this so-called ***photoinducible period***, the animal responds as though the light occurred continuously since dawn of that day. For example, premature onset of ovulatory estrous cycles could be induced in horses by providing a 2-h pulse of light in the evening of a short day (Malinowski et al., 1985).

Because there are economic advantages to be gained from inducing ewes to breed out of season, much research has been conducted in this short-day breeder. Administration of gonadotrophic hormones combined with progesterone-like compounds has been used with success to induce breeding in anestrous ewes for many years (Brunner et al., 1964). Exogenous melatonin has more recently been used alone or in combination with the hormones used previously. There are two primary methods by which melatonin is administered. It can be administered as a bolus (oral or injected) at a specific time of the afternoon to cause the ewe to perceive the combination of exogenous and endogenous melatonin as a short day (Nett and Niswender, 1982). Melatonin can also be administered via a continuous release implant that can cause premature onset of estrous cycles as well as prolongation of the breeding season in unmated ewes. Continuous release of exogenous melatonin from these implants creates a profile of constant melatonin in the blood that obviously never occurs spontaneously in nature. Even ewes maintained in continuous darkness would not secrete melatonin continuously because of the endogenous circadian oscillator in the hypothalamic SCN (see Figure 10-4). Nevertheless, the profile of constant exogenous melatonin together with the nocturnal elevations of endogenous melatonin appears to be perceived by the neuroendocrine mechanisms governing LH secretion as short days rather than complete darkness or functional pinealectomy (O'Callaghan et al., 1991).

Maternal Influences on Pineal Rhythms in Fetuses and Neonates. Fetal mammals display circadian rhythms of melatonin secretion that coincide with environmental lighting as well as with the maternal rhythm of pineal melatonin. The pineal rhythm occurring *in utero* persists after birth and appears to influence both PRL secretion and reproductive development in the neonates of some species.

Through various experimental manipulations of environmental lighting and pinealectomy of pregnant ewes, it was demonstrated that the fetal rhythm of melatonin depended upon the rhythm of blood-borne melatonin in the mother (Yellon and Longo, 1988). Moreover, radiolabelled melatonin in the maternal circulation of ewes was readily transferred into the fetal circulation (Zemdegs et al., 1988). Similar transfer of photoperiodic information occurs between mother and fetus in hamsters, and postnatal reproduction in this species is influenced by this information (Stetson et al., 1989). Fetal hamsters also possess a photoinducible period because daily exposure to maternal melatonin during each afternoon resulted in postnatal development of the neonates as though their mothers had been exposed to short days (Horton et al., 1989). Once the fetuses have been born, the neonates are directly influenced by environmental photoperiods, and neonatal sheep display nocturnal elevations of pineal melatonin within a few weeks after birth, even when born to pinealectomized mothers (Yellon and Longo, 1988).

REFERENCES

Almeida, G, and J. Pelletier. Abolition of seasonal testis change in the Ile-de-France ram by short light cycles: relationship to luteinizing hormone and testosterone release. *Theriogenology* (1988) 29:681-691.

Almeida, O.F.X., and G.A. Lincoln. Photoperiodic regulation of reproductive activity in the ram: evidence for the involvement of circadian rhythms in melatonin and prolactin secretion. *Biol. Reprod.* (1982) 27:1062-1075.

Bassett, J.M., N. Curtis, C. Hanson, and C.M. Weeding. Effect of altered photoperiod or maternal melatonin administration on plasma prolactin concentrations in fetal lambs. *J. Endocrinol.* (1989) 122:633-643.

Bex, F., A. Bartke, B.D. Goldman, and S. Dalterio. Prolactin, growth hormone, luteinizing hormone receptors, and seasonal changes in testicular activity in the golden hamster. *Endocrinology* (1978) 103:2069-2080.

Binkley, S., S.E. MacBride, D.C. Klein, and C.L. Ralph. Regulation of pineal rhythms in chickens: refractory period and nonvisual light perception. *Endocrinology* (1975) 96:848-853.

Brown, W.B., and J.M. Forbes. Diurnal variation of plasma prolactin in growing sheep under two lighting regimes and the effect of pinealectomy. *J. Endocrinol.* (1980) 84:91-99.

Brunner, M.A., W. Hansel, and D.E. Hogue. Use of 6-methyl-17-acetoxyprogesterone and pregnant mare serum to induce and synchronize estrus in ewes. *J. Anim. Sci.* (1964) 23:32-36.

Buttle, H.L. The effect of anterior cervical ganglionectomy on the seasonal variation in prolactin concentration in goats. *Neuroendocrinology* (1977) 23:121-128.

Concannon, P.W., J.E. Parks, P.J. Roberts, and B.C. Tennant. Persistent free-running circannual reproductive cycles during prolonged exposure to a constant 12L:12D photoperiod in laboratory woodchucks (*Marmota monax*). *Lab. Anim. Sci.* (1992) 42:382-391.

Diekman, M.A., K.E. Brandt, M.L. Green, J.A. Clapper, and J.L. Malayer. Lack of nocturnal rise of serum melatonin in prepubertal gilts. *Dom. Anim. Endocrinol.* (1992) 9:161-167.

Earnest, D.J., and F.W. Turek. Effect of one-second light pulses on testicular function and locomotor activity in the golden hamster. *Biol. Reprod.* (1983) 28:557-565.

Elliott, J.A., and B.D. Goldman. Seasonal reproduction. Photoperiodism and biological clocks. In: Adler, N.T. (ed). *Neuroendocrinology of Reproduction* (Plenum Press, New York, 1981), pg 377-423.

Ellis, G.B., and F.W. Turek. Photoperiodic regulation of serum luteinizing hormone and follicle-stimulating hormone in castrated and castrated-adrenalectomized male hamsters. *Endocrinology* (1980) 106:1338-1344.

Farner, D.S. Annual rhythms. *Ann. Rev. Physiol.* (1985) 47:65-82.

Fisher, L.A., and J.D. Fernstrom. Measurement of nonapeptides in pineal and pituitary using reversed-phase, ion-pair chromatography with post-column detection by radioimmunoassay. *Life Sci.* (1981) 28:1471-1481.

Ganong, W.F., M.D. Shepherd, J.R. Wall, E.E. VanBrunt, and M.T. Clegg. Penetration of light into the brain of mammals. *Endocrinology* (1963) 72:962-963.

Goldman, B.D., and J.M. Darrow. The pineal gland and mammalian photoperiodism. *Neuroendocrinology* (1983) 37:386-396.

Gwinner, E. Circannual rhythms in the control of avian migration. In: Rosenblatt, J.S., C. Beer, M.C. Busnel, and P.J.B. Slater (eds). *Advances in the Study of Behavior, Vol. 16,* (Academic Press, Orlando, FL, 1986), pg 191-228.

Hastings, M.H., J. Herbert, N.D. Martensz, and A.C. Roberts. Annual reproductive rhythms in mammals: mechanisms of light synchronization. In: Wurtman, R.J., M.J. Baum, and J.T. Potts (eds). *The Medical and Biological Effects of Light.* (*Ann. N.Y. Acad. Sci.* Vol. 453, 1985) pg 182-204.

Horton, T.H., S.L. Ray, and M.H. Stetson. Maternal transfer of photoperiodic information in Siberian hamsters. III. Melatonin injections program postnatal reproductive development expressed in constant light. *Biol. Reprod.* (1989) 41:34-39.

Jackson, G.L., and H.T. Jansen. Persistence of a circannual rhythm of plasma prolactin concentrations in ewes exposed to a constant equatorial photoperiod. *Biol. Reprod.* (1991) 44:469-475.

Karsch, F.J., E.L. Bittman, D.L. Foster, R.L. Goodman, S.J. Legan, and J.E. Robinson. Neuroendocrine basis of seasonal reproduction. *Rec. Prog. Horm. Res.* (1984) 40:185-232.

Karsch, F.J., E.L. Bittman, J.E. Robinson, S.M. Yellon, N.L. Wayne, D.H. Olster, and A.H. Kaynard. Melatonin and photorefractoriness: Loss of response to the melatonin signal leads to seasonal reproductive transitions in the ewe. *Biol. Reprod.* (1986) 34:265-274.

Karsch, F.J., J.E. Robinson, C.J.I. Woodfill, and M.B. Brown. Circannual cycles of luteinizing hormone and prolactin secretion in ewes during prolonged exposure to a fixed photoperiod: evidence for an endogenous reproductive rhythm. *Biol. Reprod.* (1989) 41:1034-1046.

Kennaway, D.J., T.A. Gilmore, and R.F. Seamark. Effect of melatonin feeding on serum prolactin and gonadotropin levels and the onset of seasonal estrous cyclicity in sheep. *Endocrinology* (1982) 110:1766-1772.

Kennaway, D.J., J.M. Obst, E.A. Dunstan, and H.G. Friesen. Ultradian and seasonal rhythms in plasma gonadotropins, prolactin, cortisol, and testosterone in pinealectomized rams. *Endocrinology* (1981) 108:639-646.

Kennaway, D.J., L.M. Sanford, B. Godfrey, and H.G. Friesen. Patterns of progesterone, melatonin and prolactin secretion in ewes maintained in four different photoperiods. *J. Endocrinol.* (1983) 97:229-242.

Klein, D.C. Circadian rhythms in indole metabolism in the rat pineal gland. In: Schmitt, F.O., and F.G. Worden (eds). *The Neuroesciences, Third Study Program* (MIT Press, Cambridge, MA, 1974), pg 509-515.

Korf, H.W., A. Oksche, A., P. Ekstrom, I. Gery, J.S. Zigler, and D.C. Klein. Pinealocyte projections into mammalian brain revealed with S-antigen antiserum. *Science* (1986) 231:735-737.

Lewy, A.J., T.A. Wehr, F.K. Goodwin, D.A. Newsome, and S.P. Markey. Light suppresses melatonin in humans. *Science* (1980) 210:1267-1269.

Lincoln, G.A. Photoperiodic control of seasonal breeding in the ram: participation of the cranial sympathetic nervous system. *J. Endocrinol.* (1979) 82:135-147.

Maeda, K.I., Y. Mori, and Y. Kano. Involvement of melatonin in the seasonal changes of the gonadal function and prolactin secretion in female goats. *Reprod. Nutr. Devel.* (1988) 28:487-497.

Malinowski, K., A.L. Johnson, and C.G. Scanes. Effects of interrupted photoperiods on the induction of ovulation in anestrous mares. *J. Anim. Sci.* (1985) 61:951-955.

Malpaux, B., J.E. Robinson, M.B. Brown, and F.J. Karsch. Importance of changing photoperiod and melatonin secretory pattern in determining the length of the breeding season in the Suffolk ewe. *J. Reprod. Fertil.* (1988) 83:461-470.

McConnell, S.J., and F. Ellendorff. Absence of nocturnal plasma melatonin surge under long and short artificial photoperiods in the domestic sow. *J. Pineal Res.* (1987) 5:201-210.

Munro, C.J., K.P. McNatty, and L. Renshaw. Circa-annual rhythms of prolactin secretion in ewes and the effect of pinealectomy. *J. Endocrinol.* (1980) 84:83-89.

Nett, T.M., and G.D. Niswender. Influence of exogenous melatonin on seasonality of reproduction in sheep. *Theriogenology* (1982) 17:645-653.

Nieuwenhuis, J.J. Arginine vasotocin (AVT), an alleged hormone of the mammalian pineal organ. *Life Sci.* (1984) 35:1713-1724.

O'Callaghan, D., F.J. Karsch, M.P. Boland, and J.F. Roche. What photoperiodic signal is provided by continuous-release melatonin implant? *Biol. Reprod.* (1991) 45:927-933.

Paterson, A.M., G.B. Martin, A. Foldes, C.A. Maxwell, and G.P. Pearce. Concentrations of plasma melatonin and luteinizing hormone in domestic gilts reared under artificial long and short days. *J. Reprod. Fertil.* (1992) 94:85-95.

Pelletier, J., and J.P. Ravault. Daylength measurement in rams made permanent breeders by short light cycles. *Neuroendocrinol. Lett.* (1988) 10:329-336.

Pevet, P., C. Neacsu, F.C. Holder, A. Reinharz, J. Dogterom, R.M. Buijs, J.M. Guerne, and B. Vivien-Roels. The vasotocin-like biological activity present in the bovine pineal is due to a compound different from vasotocin. *J. Neural Trans.* (1981) 51:295-302.

Ralph, M.R., R.G. Foster, F.C. Davis, and M. Menaker. Transplanted suprachiasmatic nucleus determines circadian period. *Science* (1990) 247:975-978.

Reiter, R.J., and M.K. Vaughan. Pineal antigonadotrophic substances: polypeptides and indoles. *Life Sci.* (1977) 21:159-172.

Robertson, L.M., and J.S. Takahashi. Circadian clock in cell culture: II. In vitro photic entrainment of melatonin oscillation from dissociated chick pineal cells. *J. Neurosci.* (1988) 8:22-30.

Rollag, M.D., R.J. Morgan, and G.D. Niswender. Route of melatonin secretion in sheep. *Endocrinology* (1978) 102:1-8.

Rollag, M.D., and G.D. Niswender. Radioimmunoassay of serum concentrations of melatonin in sheep exposed to different lighting regimens. *Endocrinology* (1976) 98:482-489.

Rollag, M.D., M.N. Sheridan, and M.G. Welsh. Melatonin contents within the deep pineal following superficial pinealectomy. In: Brown, G.M., and S.D. Wainwright (eds). *The Pineal Gland: Endocrine Aspects* (Pergamon Press, Oxford, UK, 1985), pg 191-196.

Stanisiewski, E.P., N.K. Ames, L.T. Chapin, C.A. Blaze, and H.A. Tucker. Effect of pinealectomy on prolactin, testosterone and luteinizing hormone concentrations in plasma of bull calves exposed to 8 or 16 hours of light per day. *J. Anim. Sci.* (1988a) 66:464-469.

Stanisiewski, E.P., L.T. Chapin, N.K. Ames, S.A. Zinn, and H.A. Tucker. Melatonin and prolactin concentrations in blood of cattle exposed to 8, 16 or 24 hours of daily light. *J. Anim. Sci.* (1988b) 66:727-734.

Stetson, M.H., S.L. Ray, N. Creyaufmiller, and T.H. Horton. Maternal transfer of photoperiodic information in Siberian hamsters. II. The nature of the maternal signal, time of signal transfer, and the effect of the maternal signal on peripubertal reproductive development in the absence of photoperiodic input. *Biol. Reprod.* (1989) 40:458-465.

Thimonier, J., and P. Mauleon. Variations saisonnieres du comportement d'oestrus et des activites ovarienne et hypophysaire chez les ovins. *Ann. Biol. Anim. Bioch. Biophys.* (1969) 9:233-250.

Turek, F.W., and C.S. Campbell. Photoperiodic regulation of neuroendocrine-gonadal activity. *Biol. Reprod.* (1979) 20:32-50.

Wayne, N.L., B. Malpaux, and F.J. Karsch. How does melatonin code for day length in the ewe: duration of nocturnal melatonin release or coincidence of melatonin with a light-entrained sensitive period. *Biol. Reprod.* (1988) 39:66-75.

Woodfill, C.J.I., J.E. Robinson, B. Malpaux, and F.J. Karsch. Synchronization of the circannual reproductive rhythm of the ewe by discrete photoperiodic signals. *Biol. Reprod.* (1991) 45:110-121.

Yellon, S.M., and L.D. Longo. Effect of maternal pinealectomy and reverse photoperiod on the circadian melatonin rhythm in the sheep and fetus during the last trimester of pregnancy. *Biol. Reprod.* (1988) 39:1093-1099.

Zemdegs, I.Z., I.C. McMillen, D.W. Walker, G.D. Thorburn, and R. Nowak. Diurnal rhythms in plasma melatonin concentrations in the fetal sheep and pregnant ewe during late gestation. *Endocrinology* (1988) 123:284-289.

Chapter 11

GONADOTROPINS IN THE MALE

This chapter will cover both luteinizing hormone (LH) and follicle-stimulating hormone (FSH) because these two gonadotropins have many similarities, including a common hypophysiotrophic neurohormone (***luteinizing hormone-releasing hormone***; LHRH) and sometimes a common cell of origin in the adenohypophysis. Because neuroendocrine regulation and biological effects of the gonadotropins differ so much between males and females, the discussion will be divided between the present chapter on the male and Chapter 12 on the female. Because the molecules of LH and FSH do not appear to differ between the sexes (Courte et al., 1972), their molecular structure for both sexes will be discussed in the present chapter.

Each molecule of LH and of FSH is a heterodimer composed of two glycosylated subunits (alpha and beta), as described earlier for the TSH molecule (see Figure 7-1). The alpha subunits of LH and FSH are identical, but the beta subunits differ, thus conferring biological specificity on the dimeric molecule. The molecular weight of both LH and FSH is approximately 29,000 daltons (Da). There are multiple isoforms of each hormone existing in the tissue of the adenohypophysis, and some of these isoforms may be secreted into the blood. Because of variation in the biopotency of various isoforms of the gonadotropins, differential secretion of these isoforms could be physiologically important (Pierce, 1988).

Hormones that are related to the pituitary gonadotropins include human chorionic gonadotropin (hCG) and pregnant mare serum gonadotropin (PMSG). The hCG molecule is similar in structure and bioactivity to LH, and the secretion of blood-borne hCG by the human chorion occurs very early in fetoplacental development. The molecule of PMSG is produced by specializations of the uterine endometrium of pregnant horses, and the bioactivity of blood-borne PMSG resembles that of adenohypophysial FSH. Both hCG and PMSG act on the ovary of the pregnant female in ways that promote the continuance of pregnancy.

In both sexes, a population of basophilic cells in the pars anterior (or pars distalis) and a portion of the cells in the pars tuberalis contain and secrete either LH or FSH, but in some species many adenohypophysial cells contain *both* LH and FSH. Such bihormonal (LH + FSH) gonadotrophic cells are very common in the human, pig, and rat (Phifer et al., 1973; Batten and Hopkins, 1978; Dada et al., 1983), but they could not be demonstrated in the bovine adenohypophysis (Bastings et al., 1991). In addition to species differences in the incidence of bihormonal gonadotrophs, the intensity of gonadotropin secretion may influence the proportion of bihormonal cells. Administration of LHRH to a fraction of dispersed monohormonal gonadotrophs from the rat hypophysis converted some of them into bihormonal gonadotrophs (Lloyd and Childs, 1988).

Biological Actions of Gonadotropins in the Male

Circulating LH secreted into blood by the gonadotroph cells of the adenohypophysis binds to cell-surface receptors on the interstitial cells of the testis. These cells (also known as Leydig cells) are stimulated by LH to synthesize and secrete testosterone. Not only is testosterone secreted into blood, but it diffuses into the adjacent seminiferous tubules where it acts to promote spermatogenesis. Blood-borne testosterone acts on a variety of tissues to produce the sexually dimorphic features unique to males. In addition, testosterone acts upon the brain to promote libido and to exert inhibitory feedback regulation on the neurohormone LHRH.

Blood-borne FSH binds to cell-surface receptors on the Sertoli cells located within the seminiferous tubules. As part of a general stimulation of Sertoli cells, FSH increases the secretion of various proteins including ***androgen-binding protein*** (ABP) into the fluid of the seminiferous tubule. The ABP molecule binds testosterone, and thereby promotes diffusion of the hormone into the lumen of the seminiferous tubules and epididymides. The FSH-stimulated Sertoli cells also secrete the hormone ***inhibin***, a dimeric molecule consisting of two dissimilar subunits. It circulates in blood and acts on the gonadotroph cells of the adenohypophysis to suppress the release of FSH in a selective closed-loop inhibitory feedback system.

Patterns of LH and FSH Secretion in the Male

Adult males release LH in a pulsatile pattern that is reflected by transient changes in circulating concentrations, and each LH discharge is usually followed by an increase in circulating testosterone. On the other hand, the release of FSH, as detected in the circulation, is much more continuous. Although degradation of blood-borne FSH is slower than blood-borne LH, it is unlikely that the differential patterns of each gonadotropin in blood could be due completely to different rates of degradation. After removal of the testes to eliminate feedback by both testosterone and inhibin, circulating concentrations of LH become even more episodic, but those of FSH remain relatively steady.

Detailed investigations of the episodic profiles of LH in adult males provide definitive evidence for pulsatile release that only sometimes displays an ultradian rhythm (i.e., regular recurrence with period less than 24 h). There is very little evidence for circadian (day-night) variation in the incidence of LH discharges except for human males around the time of puberty. In such peripubertal boys, episodic secretion of LH is greater during the nocturnal period of sleep (Boyar et al., 1974).

Secretory profiles of LH secretion in adult males reflect the factors of environment and perhaps sexual excitement (see next paragraph) interacting within a closed-loop feedback system. However, normal males do not secrete the large surge of LH that characterizes the preovulatory surge of normal females.

Effects of Sexual Stimuli on LH Release. There is some disagreement about whether sexual stimuli and/or ejaculation increases the secretion of LH in adult males. A transient increase in LH secretion followed copulation in male pigs (Ellendorf et al., 1975) and bulls (Katongole et al., 1971). However, other workers failed to observe increased LH release in bulls after ejaculation (Convey et al., 1971; Gombe et al., 1973). In a more recent study of young bulls in natural mating situations (Lunstra et al., 1989), there was no association between ejaculation and LH release, but there was increased LH release after ejaculation in those bulls that displayed multiple ***flehmen*** responses of the upper lip (perhaps indicative of greater sexual excitement). It was also demonstrated in adult male rats that secretion of LH and testosterone could be conditioned to occur when rats anticipated sexual activity (Graham and Desjardins, 1980). In summary, spontaneous discharges of hypophysial LH occur regularly in adult males with or without sexual activity, but intense sexual excitement may sometimes facilitate their occurrence.

Secretion of FSH. As stated above, blood-borne FSH is maintained at a rather stable concentration in adult males. Concentrations may be elevated during early stages of prepubertal development in some species, and presumably this situation is due to minimal secretion of inhibin by the immature testes. As the Sertoli cells mature and increase their secretion of inhibin, FSH secretion is brought under effective inhibitory feedback. Experimental situations that compromise the ability of the Sertoli cells to secrete inhibin are likely to result in greater secretion of FSH. One such situation is ***cryptorchidism*** occurring naturally or involving a surgical technique that causes the testes to be carried close to the abdominal wall and exposed to higher than normal temperatures. Although circulating testosterone was normal in cryptorchid bulls and rams, there were increased concentrations of FSH as well as LH (Schanbacher and Ford, 1977; Schanbacher, 1979). It is reasonable to hypothesize that the higher temperature of the cryptorchid testis decreased the testicular production of inhibin (Ishii et al., 1990) and reduced the inhibitory feedback on secretion of both gonadotropins.

Sexual Differentiation

In addition to sexual differentiation of reproductive tract morphology and reproductive behaviors, neuroendocrine mechanisms regulating the secretion of LH, but not FSH, appear to undergo sexual differentiation (Breedlove, 1992). Moreover, these LH-regulatory mechanisms appear to be neural rather than adenohypophysial. Because undifferentiated mechanisms are primarily female, the process of sexual differentiation actually involves masculinization of LH-regulating mechanisms. The time during fetal development when this masculinization occurs varies greatly among mammalian species. In rats and other rodents that are born in a very immature state, masculinization of LH mechanisms occurs after birth, usually by Day 3 of age (Harlan et al., 1979). By contrast, sexual differentiation in fetal sheep occurs during the first half of gestation (Wilson and Tarttelin, 1978).

Figure 11-1 summarizes sexual differentiation of LH secretion as it occurs in rats. The undifferentiated female state is masculinized under the organizing influence of an intracerebral estrogen that is formed by conversion of blood-borne androgen into neurally active estradiol by a process known as aromatization (Naftolin et al., 1975). However, a direct action of testosterone or its 5-alpha reduced product, dihydrotestosterone, on the neonatal brain to produce some aspects of masculinization remains a possibility (Dohler, 1986). Figure 11-1 lists examples of both the non-masculinized (female) and masculinized states. Removal of the testes from male rats on or before Day 3 prevents the process of

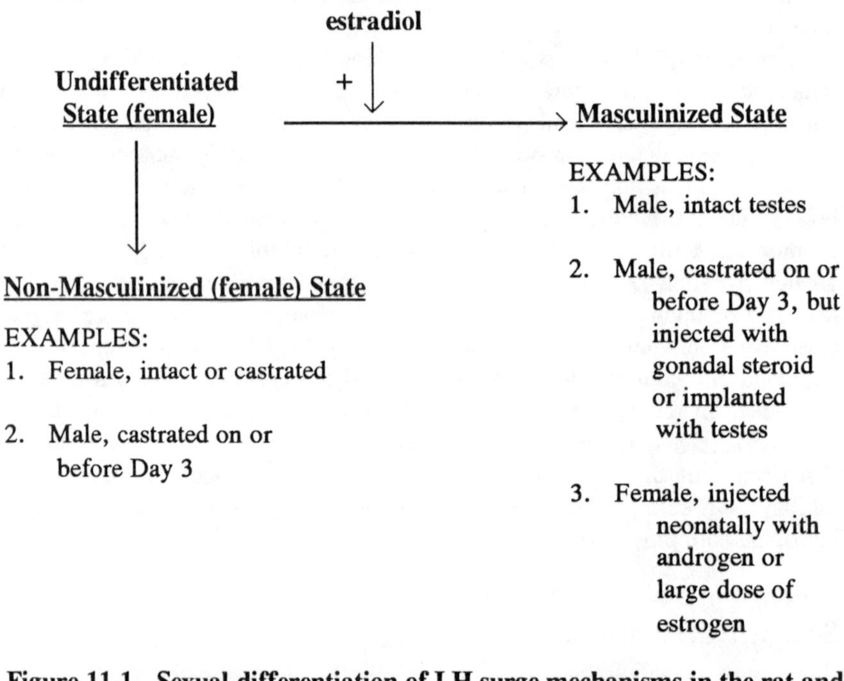

Figure 11-1. Sexual differentiation of LH surge mechanisms in the rat and examples of both states.

The top of this figure shows the conversion from an undifferentiated state to a masculinized state. Examples of non-masculinized and masculinized states in adult rats are listed under the heading for each state.

masculinization by removing the source of blood-borne androgen that is converted intracerebrally into estradiol. Blood-borne estradiol in normal female rats is rendered partly inactive by binding to blood-borne α-fetoprotein secreted by the liver (Greenstein, 1992). The small quantities of estradiol that ultimately reach the brain are insufficient to masculinize the LH-regulatory mechanisms. Thus, intact female rats do not undergo masculinization because intracerebral levels of estrogen are too low. Examples of masculinized states in Figure 11-1 include genetic males either with intact testis or gonadectomized and provided replacement therapy (e.g., exogenous testosterone, transplanted testes, or large doses of estrogen). Genetic females injected neonatally with testosterone or large dosages of estradiol are also masculinized. Such pharmacological masculinization sometimes results from injection as late as Day 10 of age, but the adult neuroendocrine defects after such delayed injections are slightly different from those of natural masculinization. Some of these androgenized females display normal female LH-regulatory mechanisms for a few estrous cycles before the mechanisms become fully masculinized (Van der Werff ten Bosch et al., 1971).

Differences between male and female mechanisms for regulation of LH release have been studied extensively, and species seem to fall into one of two categories. In rodents and some ungulates (sheep, cattle, horses), those males that are castrated as adults and injected with a bolus of estrogen fail to secrete a surge of LH as occurs when castrated females are treated in the same way. On the other hand, similar castration and administration of estrogen to males of primate species, pigs, and goats will induce a surge of LH release similar to that observed in normal or castrated females of these species (Karsch et al., 1973; Dial et al., 1985). In summary, masculinization of LH-surge generating mechanisms, as outlined in Figure 11-1, may only be valid for certain species. It should be noted that normal males of these other species (primate, pig, and goat) would never experience in nature the estrogen signal necessary to induce the type of LH surges observed experimentally. Nevertheless, the masculinized brains of these species do retain LH-surge generating mechanisms that are responsive to exogenous estrogen.

Embryonic and neonatal exposure to androgens also determines sexual differentiation of neural mechanisms that mediate reproductive behavior, and this subject will be covered later in Chapter 14. Sexual differentiation of the genitalia and sex accessory organs results from embryonic exposure to androgens either secreted by the testes (testosterone) or derived from blood-borne testosterone (dihydrotestosterone) in the androgen-sensitive tissues. This peripheral conversion of testosterone into dyhydrotestosterone is mediated by the enzyme known as ***5-alpha reductase***, and a unique deficiency of this enzyme occurs in some human males. Boys, who are afflicted with a genetic deficiency of 5-alpha reductase, undergo masculine development of some sex accessory organs, but the prostate gland, external genitalia, and general body morphology tend to develop in a feminine manner until puberty is reached. After puberty in these afflicted patients, there occurs a testes-dependent masculinization of the formerly feminine characteristics (Wilson et al., 1983). Research in these human males who experience a deficiency of dihydrotestosterone has provided valuable information about the role of this compound in sexual differentiation in primates, but similar models for other species are not available (Dohler, 1986).

Central Nervous System Correlates of Sexual Differentiation. A variety of morphological differences have been observed in the CNS of males and females. However, demonstration of such a difference cannot, by itself, establish that the morphological difference reflects sexual differentiation of LH-secretory mechanisms. For example, male-female differences in the CNS may be due to either (1) post-differentiation effects of sex-specific gonadal hormones, or (2) genetically determined sex differences unrelated to sexual differentiation. To prove that a particular CNS difference between males and females is associated with sexual differentiation of LH secretory mechanisms, it is necessary to experimentally modify sexual differentiation and observe that the CNS change is consistently associated with the potential (or inability) to secrete a surge of adenohypophysial LH in response to estrogen administration. Because there are species differences in the effect of masculinization on this parameter, this criterion can only be satisfied in certain species.

Morphological correlates of sexual differentiation in the rat occur in the preoptic area (POA) of the hypothalamus. In non-masculinized (female) rats, axons arriving in the POA from CNS regions other than the amygdala tend to synapse more on dendritic processes known as *spines* than on other parts of the dendrites, and this synaptic pattern is reversed in masculinized rats (Raisman and Field, 1973). In addition, the number and volume of neuronal perikarya in one portion of the POA were greater in masculinized rats (Jacobson et al., 1981). These specific POA neurons in the rat hypothalamus have been named the ***sexually dimorphic nucleus*** (SDN). The distribution of opioid receptors in the SDN, but not other areas, was also influenced by masculinization (Hammer, 1985).

Other sexual differences that have not yet been linked to sexual differentiation of LH-surge mechanisms include axonal projections from the vomernasal organ to certain limbic structures being more extensive in male rats (Simerly, 1990). Two small interstitial nuclei in the anterior hypothalamus of the adult human were also reported to be larger in males than females (Allen et al., 1989). Also, the number and size of oxytocin- and vasopressin-containing neurons in a unique hypothalamic location in pigs were shown to be greater in postpubertal females as compared to males (Van Eerdenburg and Swaab, 1991). Perikarya and dendrites of LHRH neurons received less synaptic innervation in male rats (Chen et al., 1990). Also, male rats had fewer LHRH perikarya that also contained the neuropeptide galanin than did female rats (Merchenthaler et al., 1991).

Post-differentiation administration of estrogen can also influence CNS morphology but, as stated above, this does not represent sexual differentiation because it can occur in adults. Exogenous estrogen stimulated dendritic modifications of perikarya in the ventromedial nucleus (VMN) of the hypothalamus (Segarra and McEwen, 1991). Similar treatment of female rats with estrogen also stimulated synthesis of the receptor for progesterone in a portion of the VMN (Lauber et al., 1991). Elimination of ovarian hormones in female monkeys decreased the degree of post-synaptic specializations in LHRH perikarya (Witkin et al., 1991). A sexual difference in the amount and distribution of enkephalin in the rat POA was shown to depend on circulating estrogen rather than on genetic sex or sexual differentiation (Watson et al., 1986).

Neurohormonal Control of LH and FSH in the Male

As discussed in Chapter 4, the decapeptide neurohormone known as LHRH stimulates the synthesis and release of both LH and FSH from the adenohypophysis. Various efforts have been made to isolate another hypothalamic releasing factor specific for FSH release. However, evidence for a separate endogenously acting FSH-releasing neurohormone is weak despite the fact that experimental inhibition of the action of endogenous LHRH produces only modest decreases in FSH secretion. There have also been some efforts toward demonstration of a hypothalamic compound that inhibits LH, but again physiological evidence for an endogenous LH-inhibiting neurohormone is not strong. Therefore, neuroendocrine integration of LH and FSH release in the male consists primarily of (1) neural input via LHRH neurons and (2) inhibitory feedback of gonadal hormones (Figure 11-2).

As discussed earlier in this chapter, the release of LH can occur from cells that contain only LH or those that contain both LH and FSH. Release of FSH can similarly occur from monohormonal cells (only FSH) as well as from bihormonal cells. Irrespective of the cell type, all gonadotrophs have receptors for LHRH and activation of these receptors stimulates release of those gonadotropins that are present within the cell. There is even some evidence that both LH and FSH may be present within individual secretory vesicles (Batten and Hopkins, 1978), and in such cases independent release would be highly unlikely. Experimental immunoneutralization of blood-borne LHRH antagonized the pulsatile LH discharges that occur in castrated males (Ellis et al., 1983). Chronic administration of an LHRH receptor antagonist suppressed the LH-testes axis in male sheep (Lincoln and Fraser, 1987). Based upon the more numerous studies in which LHRH has been immunoneutralized or antagonized in females, it appears that an induced deficiency of LHRH decreases the release of LH much more than FSH. The intracellular actions of LHRH that culminate in the exocytosis of LH and FSH secretory vesicles have been studied in detail (Jennes and Conn, 1988), but they will not be considered in the present discussion.

Secretion of FSH in the male is under inhibitory feedback of blood-borne inhibin secreted by FSH-stimulated Sertoli cells in the testes (Figure 11-2). The molecular structure of testes-derived inhibin appears identical to ovary-derived inhibin; Figure 11-3 summarizes information about the subunits that comprise inhibin in both males and females (Ying et al., 1988; deKretser and Robertson, 1989). The intact bioactive inhibin molecule consists of a dimer containing two

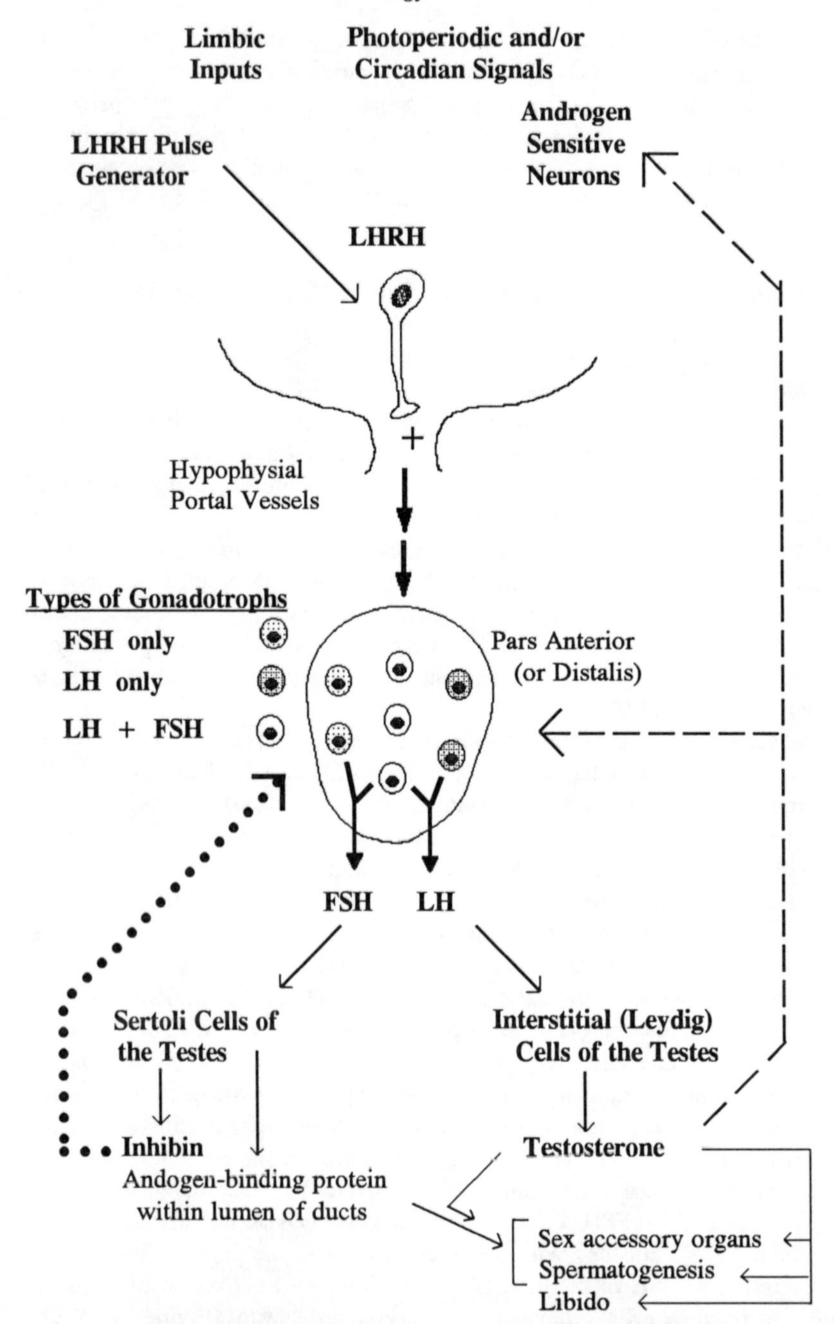

Figure 11-2. Neuroendocrine integration of LH and FSH release in males. Simplified diagram illustrating regulation of gonadotropin secretion in the male. Monohormonal and bihormonal gonadotroph cells in the adenohypophysis are depicted with different shading.

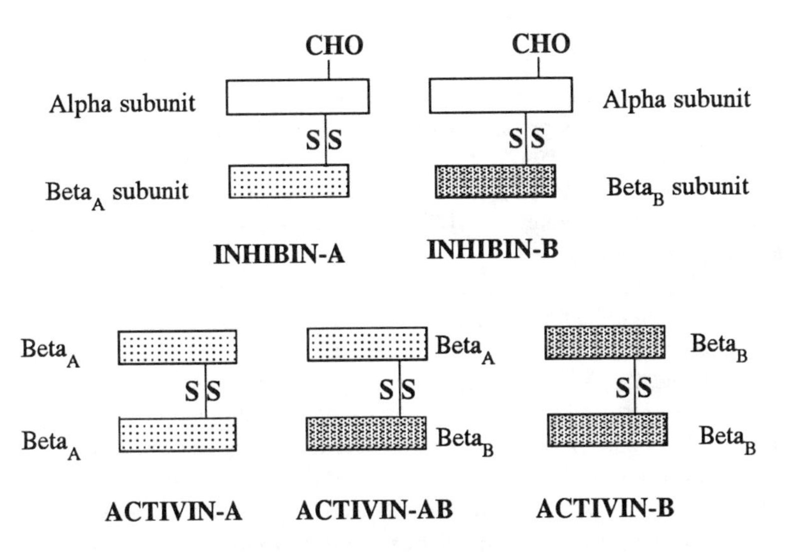

Figure 11-3. Subunits forming the dimeric molecules of inhibin and activin. Each different type of subunit is shaded differently. The multiple disulfide linkages between subunits are denoted SS. Glycosylation of the alpha subunit with carbohydrates is depicted as CHO.

dissimilar subunits (i.e., a heterodimer). The alpha subunit is always larger (about 20,000 Da) and is glycosylated at two sites. Because variation exists between species in the composition of the alpha subunit, evolutionary conservation of the molecule was apparently not necessary. In contrast, the non-glycosylated beta subunit of inhibin is smaller (about 13,000 Da) and is highly conserved across species. Species such as pig, rat, and human all possess both an A and B form of the beta subunit. These A and B forms of beta-inhibin share moderate sequence homology and are both bioactive when combined with the alpha subunit in a heterodimer. As denoted in Figure 11-3, combination of the alpha subunit with beta$_A$ subunit is called inhibin-A while a similar combination of alpha subunit with beta$_B$ subunit is called inhibin-B. The dimeric molecule is held together by numerous disulfide bonds between the many cysteine residues in both subunits.

During the isolation and identification of inhibin heterodimers, another fraction was found to possess bioactivity exactly opposite to that of inhibin. In other words, this opposing bioactivity stimulated the release of FSH rather than inhibiting FSH release as did inhibin. It was subsequently discovered that the molecule displaying FSH-releasing bioactivity consisted of two *similar* beta subunits of inhibin bound together in a homodimer that was named ***activin*** (Ying et al., 1988). The molecule denoted activin-AB contains one beta$_A$ and one beta$_B$ subunit while a homodimer consisting of two beta$_A$ subunits is called activin-A. The dimeric combination of two beta$_B$ subunits forms activin-B (Figure 11-3).

It should be noted that the expression of the activin subunits is not restricted to gonadal tissues, and the molecules may not be restricted to regulation of hypophysial FSH (DePaolo et al., 1991). Of particular interest, the gonadotroph cells of the adenohypophysis may produce the subunits of inhibin and activin (Roberts et al., 1989, 1992). This production raises the possibility of autocrine regulation in which locally produced inhibin suppresses or locally produced activin stimulates the release of FSH by these gonadotrophs (see Chapter 12 for more information).

Blood-borne inhibin secreted in response to FSH action on Sertoli cells exerts inhibitory feedback on FSH release in the male (see next paragraph). Testosterone secreted by the interstitial cells may also participate in FSH secretion, but this action is largely mediated by suppression of LHRH release that affects both LH and FSH. Another putative inhibitor of FSH secretion in males and females is a linear glycoprotein known as follistatin. Although follistatin can inhibit FSH as does inhibin, it also binds to inhibin and decreases its FSH-inhibiting potency. The mRNA encoding follistatin is present in various tissues including the adenohypophysis where it occurs in gonadotrophs as well as a specific cell type known as the folliculostellate cell (Kaiser et al., 1992). Because inhibin, activin, follistatin, and FSH have all been shown to coexist in gonadotrophs, complex paracrine/autocrine control of FSH release becomes a realistic possibility.

Blood-Borne Inhibin in the Male. Circulating concentrations of inhibin declined as immature rats and bulls approached puberty (Rivier et al., 1988; MacDonald et al., 1991). In contrast, serum inhibin increased as male monkeys attained puberty (Abeyawardene et al., 1989). Immunoneutralization of blood-borne inhibin in prepubertal rats and bulls increased the release of FSH, thus demonstrating a feedback role for blood-borne inhibin (Culler and Negro-Vilar, 1988; Martin et al., 1991). Similar effects of inhibin immunoneutralization were also observed in adult monkeys (Medhamurthy et al., 1991) but not in adult rats (Culler, 1990). It appears that endogenous inhibin may exert an inhibitory feedback on FSH secretion, particularly in those situations when circulating inhibin is elevated (e.g., prepubertal rat, prepubertal bull, and postpubertal monkey). In other situations, it is likely that CNS and adenohypophysial actions of testicular steroids exert a major inhibitory feedback on FSH as well as LH.

Inhibitory Feedback of Testosterone on LH Secretion. Blood-borne testosterone secreted by the interstitial cells of the testes in response to LH acts to inhibit further release of LH through actions on the gonadotroph cell and on androgen-sensitive neurons in the brain (Figure 11-2). When the testes are removed, episodic discharges of LH gradually become more frequent and of larger magnitude than in control males. At least part of this post-castration effect appears to result from reduced inhibition of the LHRH pulse generator that exists in the brain. However, there is some evidence that testosterone itself, or its *in vivo* metabolite, estradiol, can also directly act on gonadotroph cells to inhibit secretion of LH. Because of these dual feedback actions of testosterone, it is difficult to determine the relative *in vivo* contribution of each action.

One valuable investigative approach to answer this difficult question is known as the ***hypophysiotrophic clamp*** model. This approach involves experimental stabilization, but not elimination, of hypothalamic input to gonadotrophs followed by administration of testosterone. Stabilization of hypothalamic input to gonadotrophs is usually achieved by substitution of a constant exogenous source of LHRH for endogenous LHRH that has been blocked or prevented from reaching the adenohypophysis. It may be concluded that feedback acts in part on the hypophysis whenever administration of testosterone, or its removal by gonadectomy, alters the secretion of LH in such *clamped* males. When male rats prepared in this manner were gonadectomized, there occurred a rapid increase in LH secretion, indicating that testicular hormones (probably testosterone) were inhibitory to LHRH-induced release of LH (Strobl et al., 1989). In male sheep studied during the breeding season, chronic administration of testosterone, dihydrotestosterone, or estradiol to rams clamped as described above failed to inhibit LH when these same treatments did inhibit LH in intact males (Tilbrook et al., 1991). Therefore, inhibitory feedback of testicular steroids on LH secretion may occur primarily at the level of the hypothalamus in male sheep. Although the primary testicular steroid that inhibits LH in untreated males is testosterone, testes-derived estradiol may also exert some inhibition. Immunoneutralization of endogenous estradiol in intact rams also resulted in an increase of circulating LH (Schanbacher, 1984). This observation provides much stronger evidence for a physiological role of estradiol in the male than do the many other reports in which exogenous estradiol inhibited LH secretion in males.

Neural Input To LHRH Neurons. The neural elements listed at the top of Figure 11-2 influence the secretion of LHRH either through direct actions on LHRH neurons or indirectly by affecting the LHRH pulse generator. The peptidergic and monoaminergic compounds that mediate these effects have been studied less extensively in males than in females. However, there is good evidence for LHRH-inhibiting actions of opioidergic neural elements in the male. This evidence derives primarily from transient discharges of LH release following antagonism of opioid receptors by an injection of naloxone. In summary, circulating levels of LH in intact males, that were postpubertal or injected with androgen, increased abruptly after naloxone administration (Cicero et al., 1986; Rawlings et al., 1991). This naloxone-induced disinhibition of LH release in males was probably due to an induced discharge of LHRH into the hypophysial portal blood, but this neurohormone was measured in only a few studies (Caraty et al., 1987).

Exposure of intact males to stressful stimuli suppresses testicular function, and part of this effect is due to reduced secretion of LH (Krulich et al., 1974). Opioidergic mechanisms may mediate stress-induced suppression of LHRH in some but not all situations. Moreover, there may be differences between species and/or differences due to the intensity of stress in whether antagonism of opioid receptors prevents stress-induced decreases in LH release (Pontiroli et al., 1982; Gonzalez-Quijano et al., 1991).

Neurohormonal Regulation of Puberty

Mechanisms that determine the onset of puberty have been studied extensively in both sexes (Grumbach et al., 1990). Research in the female is often easier because occurrence of the pubertal ovulation is more readily determined than is the onset of male puberty (i.e., ejaculation of fertile spermatozoa). Nevertheless, much information is available on neurohormonal regulation of puberty in the male.

There is experimental support for at least three different neuroendocrine changes, each listed in Table 11-1, as having roles in initiating puberty. In humans and monkeys, there appears to be a period during prepubertal development when spontaneous secretion of LH (and presumably LHRH) is minimal, but this hyposecretion is *not* due to feedback of any signal from the testes (Plant, 1988). As these juvenile primates approach puberty, the LHRH pulse generator becomes more active and the LH-testes axis is stimulated. In human males, the spontaneous hypersecretion of LH is first observed during the period of sleep (Boyar et al., 1974) and then gradually extends to all periods.

In other species such as rats and sheep, the prepubertal testes appear to maintain effective inhibitory feedback on LH secretion until just before puberty occurs. During prepubertal maturation of the LHRH regulatory mechanisms, sensitivity to blood-borne testosterone *decreases* allowing puberty-related increases in the release of LHRH, LH, and testosterone to occur simultaneously. This hypothetical neuroendocrine mechanism for regulation of puberty is sometimes called the ***gonadostat hypothesis***, and changes in feedback sensitivity to gonadal hormones have been observed in pubertal rats and sheep (McCann et al., 1974; Olster and Foster, 1986).

The third puberty-related change listed in Table 11-1 involves maturation of the interstitial cells of the testes to secrete testosterone. In some ungulate species, spontaneous secretion of LH can be very prominent in prepubertal males, but there is no corresponding release of testosterone following each LH discharge (Foster et al., 1978; McCarthy et al., 1979). As these males approach puberty,

Table 11-1. Possible neuroendocrine changes that initiate puberty in juvenile males.

1. Increased CNS activation of LHRH pulse generator (*demonstrated only in humans and monkeys*).
2. Reduced sensitivity to testosterone inhibition of the secretion of LHRH by the hypothalamus and/or LH by the hypophysis.
3. Increased ability of interstitial cells of testes to secrete testosterone in reponse to blood-borne LH.

pulsatile discharges of LH still occur, and now each LH discharge is usually followed by an increase in circulating testosterone. This apparent maturation of testosterone secretory capacity prior to and sometimes immediately before puberty may be due to prolonged prepubertal exposure of the testes to LH and FSH, or alternatively the maturation may reflect other unknown age-related changes. Irrespective of the mechanism, it should be noted that this puberty-related change may precede and even coexist with other neuroendocrine changes that initiate puberty. In summary, the endogenous induction of puberty in the juvenile male may involve a combination of several neuroendocrine changes in the LHRH-LH-testosterone axis.

Although it is clear that the onset of puberty may be determined by maturational changes in several mechanisms, some mechanisms appear not be involved in the regulation of puberty. First, an increased sensitivity of the adenohypophysial gonadotrophs to modest dosages of LHRH does not initiate puberty. LH responses to very large dosages of LHRH may sometimes show puberty-related changes, but such changes mainly reflect increases in the quantity of releasable LH. One action of PRL in the male may be to synergize with testosterone in stimulating the sex accessory glands (Chapter 9). However, PRL does not appear to be involved in regulation of puberty because chronic suppression of PRL release did not affect onset of puberty in male sheep (Ravault et al., 1977).

Environmental Influences on Male Reproduction

Photoperiod is the environmental factor that has the greatest influence on male reproduction, and this influence differs among species. In long-day breeding species such as the hamster, the testes atrophy when the animal is exposed to short photoperiods. As discussed in Chapter 10, this effect is mediated by pineal secretion of melatonin and is closely related to the endogenous free-running circadian rhythm. As little as 1 sec of light occurring 8 h after the end of a once daily 6-h period of light could entrain the circadian rhythm of locomotor activity equivalent to a 14-h photoperiod and prevent the testes atrophy usually observed in 6-h photoperiods (Earnest and Turek, 1983).

Environmental photoperiod influences testes development in many species of birds, and this effect is also related to the endogenous circadian rhythm. When light occurs during the photoinducible period of the circadian rhythm, the testes are stimulated to develop as though the light occurred continuously since dawn (Farner, 1975). The major difference between birds and mammals is that blinded male birds can respond to photoperiod while blinded mammals cannot (Underwood and Menaker, 1970). As stated in Chapter 10, cells of the chicken pineal gland are directly sensitive to light, whereas those of the mammalian pineal gland are not. The avian hypothalamus may also be directly sensitive to light because local implantation of radioluminous material into the median eminence activated the LH-testes axis in quail (Oliver et al., 1979).

Male sheep also respond to light depending on when it occurs during the circadian rhythm. Because sheep are short-day breeders, the relationship between long photoperiods and testis development is not always clear. Exposure of prepubertal male sheep to a lighting regimen in which extra light occurred about 16 h after dawn consistently increased PRL secretion (Schanbacher et al., 1985) and sometimes also hastened pubertal development (Chemineau et al., 1988). In domestic pigs that breed all year, exposure of males to supplemental lighting hastened the onset of pubertal libido, but did not consistently affect secretion of LH or testosterone (Hoagland and Diekman, 1982).

REFERENCES

Abeyawardene, S.A., W.W. Vale, G.R. Marshall, and T.M. Plant. Circulating inhibin-alpha concentrations in infant, prepubertal, and adult male rhesus monkeys (*Macaca mulatta*) and in juvenile males during premature initiation of puberty with pulsatile gonadotropin-releasing hormone treatment. *Endocrinology* (1989) 125:250-256.

Allen, L.S., M. Hines, J.E. Shryne, and R.A. Gorski. Two sexually dimorphic cell groups in the human brain. *J. Neurosci.* (1989) 9:497-506.

Bastings, E., A. Beckers, M. Reznik, and J.F. Beckers. Immunological evidence for production of luteinizing hormone and follicle-stimulating hormone in separate cells in the bovine. *Biol. Reprod.* (1991) 45:788-796.

Batten, T.F.C., and C.R. Hopkins. Discrimination of LH, FSH, TSH and ACTH in dissociated porcine anterior pituitary cells by light and electron microscope immunocytochemistry. *Cell Tiss. Res.* (1978) 192:107-120.

Boyar, R.M., R.S. Rosenfeld, S. Kapen, J.W. Finkelstein, H.P. Roffwarg, E.D. Weitzman, and L. Hellman. Human puberty. Simultaneous augmented secretion of luteinizing hormone and testosterone during sleep. *J. Clin. Invest.* (1974) 54:609-618.

Breedlove, S.M. Sexual differentiation of the brain and behavior. In: Becker, J.B., S.M. Breedlove, and D. Crews (eds). *Behavioral Endocrinology* (The MIT Press, Cambridge, MA, 1992) pg 39-68.

Caraty, A., A. Locatelli, and B. Schanbacher. Augmentation par la naloxone de la frequence et de l'amplitude des pulses de LH-RH dans le sang porte hypothlamo-hypophysaire chez le belier castre. *C. R. Acad. Sci. III* (1987) 305:369-374.

Chemineau, P., J. Pelletier, Y. Guerin, G. Colas, J.P. Ravault, G. Toure, G. Almeida, J. Thimonier, and R. Ortavant. Photoperiodic and melatonin treatments for the control of seasonal reproduction in sheep and goats. *Reprod. Nutr. Devel.* (1988) 28:409-422.

Chen, W.P., J.W. Witkin, and A.J. Silverman. Sexual dimorphism in the synaptic input to gonadotropin releasing hormone neurons. *Endocrinology* (1990) 126:695-702.

Cicero, T.J., P.F. Schmoeker, E.R. Meyer, B.T. Miller, R.D. Bell, S.M. Cytron, and C.C. Brown. Ontogeny of the opioid-mediated control of reproductive endocrinology in the male and female rat. *J. Pharm. Exp. Ther.* (1986) 236:627-633.

Convey, E.M., E. Bretschneider, H.D. Hafs, and W.D. Oxender. Serum levels of LH, prolactin and growth hormone after ejaculation in bulls. *Biol. Reprod.* (1971) 5:20-24.

Courte, C., M. Hurault, C. Clary, P. de la Llosa, and M. Jutisz. Purification and physiochemical properties of bovine luteinizing hormone (LH) and comparison between bull and cow LH preparations. *Gen. Comp. Endocrinol.* (1972) 18:284-291.

Culler, M.D. Role of Leydig cells and endogenous inhibin in regulating pulsatile gonadotropin secretion in the adult male rat. *Endocrinology* (1990) 127:2540-2550.

Culler, M.D., and A. Negro-Vilar. Passive immunoneutralization of endogenous inhibin: sex-related differences in the role of inhibin during development. *Mol. Cell. Endocrinol.* (1988) 58:263-273.

Dada, M.O., G.T. Campbell, and C.A. Blake. A quantitative immunocytochemical study of the luteinizing hormone and follicle-stimulating hormone cells in the adenohypophysis of adult male rats and adult female rats throughout the estrous cycle. *Endocrinology* (1983) 113:970-984.

de Kretser, D.M., and D.M. Robertson. The isolation and physiology of inhibin and related proteins. *Biol. Reprod.* (1989) 40:33-47.

DePaolo,L.V., T.A. Bicsak, G.F. Erickson, S. Shimasaki, and N. Ling. Follistatin and activin: a potential intrinsic regulatory system within diverse tissues. *Proc. Soc. Exp. Biol. Med.* (1991) 198:500-512.

Dial, G.D., B.S. Wiseman, R.S. Ott, A.L. Smith, and J.E. Hixon. Absence of sexual dimorphism in the goat: induction of luteinizing hormone discharge in the castrated male and female and in the intersex with estradiol benzoate. *Theriogenology* (1985) 23:351-360.

Dohler, K.D. The special case of hormonal imprinting, the neonatal influence of sex. *Experientia* (1986) 42:759-769.

Earnest, D.J., and F.W. Turek. Effect of one-second light pulses on testicular function and locomotor activity in the golden hamster. *Biol. Reprod.* (1983) 28:557-565.

Ellendorff, F., N. Parvizi, D.K. Pomerantz, A. Hartjen, A. Konig, D. Smidt, and F. Elsaesser. Plasma luteinizing hormone and testosterone in the adult male pig: 24 hour fluctuations and the effect of copulation. *J. Endocrinol.* (1975) 67:403-410.

Ellis, G.B., C. Desjardins, and H.M. Fraser. Control of pulsatile LH release in male rats. *Neuroendocrino*logy (1983) 37:177-183.

Farner, D.S. Photoperiodic controls in the secretion of gonadotropins in birds. In: Barrington, E.J.W. (ed). *Trends in Comparative Endocrinology.* (Suppl. 1 to *Amer. Zool.*, Vol. 15, 1975), pg 117-135.

Foster, D.L., I.H. Mickelson, K.D. Ryan, G.A. Coon, R.A. Drongowski, and J.A. Holt. Ontogeny of pulsatile luteinizing hormone and testosterone secretion in male lambs. *Endocrinology* (1978) 102:1137-1146.

Gombe, S., W.C. Hall, K. McEnkee, W. Hansel, and B.W. Pickett. Regulation of blood levels of LH in bulls: influence of age, breed, sexual stimulation and temporal fluctuations. *J. Reprod. Fertil.* (1973) 35:493-503.

Gonzalez-Quijano, M.I., C. Ariznavarreta, A.I. Martin, J.A.F. Treguerres, and A. Lopez-Calderon. Naltrexone does not reverse the inhibitory effect of chronic restraint on gonadotropin secretion in the intact male rat. *Neuroendocrinology* (1991) 54:447-453.

Graham, J.M., and C. Desjardins. Classical conditioning: induction of luteinizing hormone and testosterone secretion in anticipation of sexual activity. *Science* (1980) 210:1039-1041.

Greenstein, B.D. Effects of rat α-fetoprotein administration on estradiol free fraction, the onset of puberty, and neural and uterine estrogen receptors. *Endocrinology* (1992) 130:3184-3190.

Grumbach, M.M., P.C. Sizonenko, and M.L. Aubert. *Control of the Onset of Puberty.* (Williams & Wilkins, Baltimore, 1990), pp 710.

Hammer, R.P. The sex hormone-dependent development of opiate receptors in the rat brain medial preoptic area. *Brain Res.* (1985) 360:65-74.

Harlan, R.E., J.H. Gordon, and R.A. Gorski. Sexual differentiation of the brain: implications for neuroscience. In: Schneider, D.M. (ed). *Reviews of Neuroscience, Vol. 4* (Raven Press, New York), 1979), pg 31-71.

Hoagland, T.A., and M.A. Diekman. Influence of supplemental lighting during increasing daylength on libido and reproductive hormones in prepubertal boars. *J. Anim. Sci.* (1982) 55:1483-1489.

Ishii, S., A. Miwa, T. Furusawa, K. Kawada, H. Migano, M. Umezu, and J. Masaki. Reproductive hormone levels and testicular inhibin content in adult rats after the unilateral and bilateral cryptorchism. *Jpn. J. Anim. Reprod.* (1990) 36:140-144.

Jacobson, C.D., V.J. Csernus, J.E. Shryne, and R.A. Gorski. The influence of gonadectomy, androgen exposure, or a gonadal graft in the neonatal rat on the volume of the sexually dimorphic nucleus of the preoptic area. *J. Neurosci.* (1981) 1:1142-1147.

Jennes, L., and P.M. Conn. Mechanisms of gonadotropin releasing hormone action. In: Cook, B.A., B.J.B. King, and H.J. van der Molen (eds) *Hormones and Their Actions. Part II* (Elsevier, Amsterdam, 1988) pg 135-154.

Kaiser, U.B., B.L. Lee, R.S. Carroll, G. Unabia, W.W. Chin, and G.V. Childs. Follistatin gene expression in the pituitary: localization in gonadotrophs and folliculostellate cells in diestrous rats. *Endocrinology* (1992) 130:3048-3056.

Karsch, F.J., D.J. Dierschke, and E. Knobil. Sexual differentiation of pituitary function: apparent difference between primates and rodents. *Science* (1973) 179:484-486.

Katongole, C.B., F. Naftolin, and R.V. Short. Relationship between blood levels of luteinizing hormone and testosterone in bulls and the effects of sexual stimulation. *J. Endocrinol.* (1971) 50:457-466.

Krulich, L., E. Hefco, P. Illner, and C.B. Read. The effects of acute stress on the secretion of LH, FSH, prolactin and GH in the normal male rat, comments on their statistical evaluation. *Neuroendocrinology* (1974) 16:293-311.

Lauber, A.H., G.J. Romano, and D.W. Pfaff. Sex difference in estradiol regulation of progestin receptor mRNA in rat mediobasal hypothalamus as demonstrated by in situ hybridization. *Neuroendocrinology* (1991) 53:608-613.

Lincoln, G.A., and H.M. Fraser. Compensatory response of the luteinizing hormone (LH)-releasing hormone (LHRH)/LH pulse generator after administration of a potent LHRH antagonist in the ram. *Endocrinology* (1987) 120:2245-2250.

Lloyd, J.M., and G.V. Childs. Differential storage and release of luteinizing hormone and follicle-stimulating hormone from individual gonadotropes separated by centrifugal elutriation. *Endocrinology* (1988) 122:1282-1290.

Lunstra, D.D., G.W. Boyd, and L.R. Corah. Effects of natural mating stimuli on serum luteinizing hormone, testosterone and estradiol-17β in yearling beef bulls. *J. Anim. Sci.* (1989) 67:3277-3288.

MacDonald, R.D., D.R. Deaver, and B.D. Schanbacher. Prepubertal changes in plasma FSH and inhibin in Holstein bull calves: responses to castration and (or) estradiol. *J. Anim. Sci.* (1991) 69:276-282.

Martin, T.L., G.L. Williams, D.D. Lunstra, and J.J. Ireland. Immunoneutralization of inhibin modifies hormone secretion and sperm production in bulls. *Biol. Reprod.* (1991) 45:73-77.

McCann, S.M., S. Ojeda, and A. Negro-Vilar. Sex steroid, pituitary and hypothalamic hormones during puberty in experimental animals. In: Grumbach, M.M., G.D. Grave, and F.E. Mayer (eds). *Control of the Onset of Puberty* (John Wiley & Sons, New York, 1974) pg 1-19.

McCarthy, M.S., H.D. Hafs, and E.M. Convey. Serum hormone patterns associated with growth and sexual development in bulls. *J. Anim. Sci.* (1979) 49:1012-1020.

Medhamurthy, R., M.D. Culler, V.L. Gay, A. Negro-Vilar, and T.M. Plant. Evidence that inhibin plays a major role in the regulation of follicle-stimulating hormone secretion in the fully adult male rhesus monkey (Macaca mulatta). *Endocrinology* (1991) 129:389-395.

Merchenthaler, I., F.J. Lopez, D.E. Lennard, and A. Negro-Vilar. Sexual differences in the distribution of neurons coexpressing galanin and luteinizing hormone-releasing hormone in the rat brain. *Endocrinology* (1991) 129:1977-1986.

Naftolin, F., K.J. Ryan, I.J. Davies, V.V. Ready, F. Flores, Z. Petro, M. Kuhn, R.J. White, Y. Takaoka,and L. Wolin. The formation of estrogens by central neuroendocrine tissues. *Rec. Prog. Horm. Res.* (1975) 31:295-319.

Oliver, J., M. Jallageas, and J.D. Bayle. Plasma testosterone and LH levels in male quail bearing hypothalamic lesions or radioluminous implants. *Neuroendocrinology* (1979) 28:114-122.

Olster, D.H., and D.L. Foster. Control of gonadotropin secretion in the male during puberty: A decrease in response to steroid inhibitory feedback in the absence of an increase in steroid-independent drive in the sheep. *Endocrinology* (1986) 118:2225-2234.

Phifer, R.F., A.R. Midgley, and S.S. Spicer. Immunohistologic and histologic evidence that follicle-stimulating hormone and luteinizing hormone are present within the same cell type in the human pars distalis. *J. Clin. Endocrinol. Metab.* (1973) 36:125-141.

Pierce, J.G. Gonadotropins: chemistry and biosynthesis. In: Knobil, E., and J. Neill (eds). *The Physiology of Reproduction* (Raven Press, New York, 1988), pg 1335-1348.

Plant, T.M. Puberty in primates. In: Knobil, E., and J. Neill (eds). *The Physiology of Reproduction* (Raven Press, New York, 1988), pg 1763-1788.

Pontiroli, A.E., G. Baio, L. Stella, A. Crescenti, and A.M. Girardi. Effects of naloxone on prolactin, luteinizing hormone, and cortisol responses to surgical stress in humans. *J. Clin. Endocrinol. Metab.* (1982) 55:378-380.

Raisman, G., and P.M. Field. Sexual dimorphism in the neuropil of the preoptic area of the rat and its dependence on neonatal androgen. *Brain Res.* (1973) 54:1-29.

Ravault, J.P., M. Courot, D. Garnier, J. Pelletier, and M. Terqui. Effect of 2-bromo-alpha-ergocryptine (CB-154) on plasma prolactin, LH and testosterone levels, accessory reproductive glands and spermatogenesis in lambs during puberty. *Biol. Reprod.* (1977) 17:192-197.

Rawlings, N.C., I.J. Churchill, W.D. Currie, and I.B.J.K. Joseph. Maturational changes in opioidergic control of luteinizing hormone and follicle-stimulating hormone in ram lambs. *J. Reprod. Fertil.* (1991) 93:1-7.

Rivier, C., S. Cajander, J. Vaughan, A.J.W. Hsueh, and W. Vale. Age-dependent changes in physiological action, content, and immunostaining of inhibin in male rats. *Endocrinology* (1988) 123:120-126.

Roberts, V., H. Meunier, J. Vaughan, J. Rivier, C. Rivier, W. Vale, and P. Sawchenko. Production and regulation of inhibin subunits in pituitary gonadotropes. *Endocrinology* (1989) 124:552-554.

Roberts, V.J., C.A. Peto, W. Vale, and P. Sawchenko. Inhibin/activin subunits are costored with FSH and LH in secretory granules of the rat anterior pituitary gland. *Neuroendocrinology* (1992) 56:214-224.

Schanbacher, B.D. Testosterone secretion in cryptorchid and intact bulls injected with gonadotropin-releasing hormone and luteinizing hormone. *Endocrinology* (1979) 104:360-364.

Schanbacher, B.D. Regulation of luteinizing hormone secretion in male sheep by endogenous estrogen. *Endocrinology* (1984) 115:944-950.

Schanbacher, B.D., and J.J. Ford. Gonadotropin secretion in cryptorchid and castrate rams and the acute effects of exogenous steroid treatment. *Endocrinology* (1977) 100:387-393.

Schanbacher, B.D., W. Wu, J.A. Nienaber, and G.L. Hahn. Twenty-four-hour profiles of prolactin and testosterone in ram lambs exposed to skeleton photoperiods consisting of various light pulses. *J. Reprod. Fertil.* (1985) 73:37-43.

Segarra, A.C., and B.S. McEwen. Estrogen increases spine density in ventromedial hypothalamic neurons of peripubertal rats. *Neuroendocrinology* (1991) 54:365-372.

Simerly, R.B. Hormonal control of neuropeptide gene expression in sexually dimorphic olfactory pathways. *Trends Neurosci.* (1990) 13:104-110.

Strobl, F.J., C.A. Gilmore, and J.E. Levine. Castration induces luteinizing hormone (LH) secretion in hypophysectomized pituitary-grafted rats receiving pulsatile LH-releasing hormone. *Endocrinology* (1989) 124:1140-1144.

Tilbrook, A.J., D.M. de Kretser, J.T. Cummins, and I.J. Clarke. The negative feedback effects of testicular steroids are predominately at the hypothalamus in the ram. *Endocrinology* (1991) 129:3080-3092.

Underwood, H., and M. Menaker. Photoperiodically significant photoreception in sparrows: Is the retina involved? *Science* (1970) 167:298-301.

Van der Werff ten Bosch, J.J., W.E. Tuinebreijer, and J.T.M. Vreeburg. The incomplete or delayed early-androgen syndrome. In: Hamburgh, M. and E.J.W. Barrington (eds). *Hormones in Development* (Appleton-Century-Crofts, New York, 1971), pg 669-675.

Van Eerdenburg, F.J.C.M., and D.F. Swaab. Increasing neuron numbers in the vasopressin and oxytocin containing nucleus of the adult female pig hypothalamus. *Neurosci. Lett.* (1991) 132:85-88.

Watson, R.E., G.E. Hoffmann, and S.J. Wiegand. Sexually dimorphic opioid distribution in the preoptic area: manipulation by gonadal steroids. *Brain Res.* (1986) 398:157-163.

Wilson, J.D., J.E. Griffin, F.W. George, and M. Leshin. The endocrine control of male phenotypic development. *Aust. J. Biol. Sci.* (1983) 36:101-128.

Wilson, P.R., and M.F. Tarttelin. Studies of sexual differentiation of sheep. I. Foetal and maternal modifications and post-natal plasma LH and testosterone content following androgenisation early in gestation. *Acta Endocrinol.* (1978) 89:182-189.

Witkin, J.W., M. Ferin, S.J. Popilskis, and A.J. Silverman. Effects of gonadal steroids on the ultrastructure of GnRH neurons in the rhesus monkey: synaptic input and glial apposition. *Endocrinology* (1991) 129:1083-1092.

Ying, S.Y., N. Ling, and R. Guillemin. Inhibins and activins. structures and radioimmunoassays. In: Jones, H.W., and C. Schrader (eds). *In Vitro Fertilization and Other Assisted Reproduction* (*Ann. N. Y. Acad. Sci.* Vol. 541, 1988). pg 143-152.

Chapter 12

GONADOTROPINS IN THE FEMALE

The present chapter will be devoted to those aspects of gonadotropin secretion, action, and neuroendocrine integration that are unique to female mammals. The preceding chapter, dealing with gonadotropins in the male, discussed the bihormonal nature of adenohypophysial gonadotrophs (LH and FSH) as well as the subjects of sexual differentiation and CNS correlates of that differentiation. Although these topics are also pertinent to the female, they will not be repeated in this chapter.

Biological Actions of Gonadotropins in the Female

The actions of LH and FSH on the female gonads are very complex, due mainly to the variation in ovarian cell types and differing hormone secretion by these cell types as the female progresses through the various phases of her reproductive cycle. Adenohypophysial stimulation of the growth and estrogen secretion by cells of the ovarian follicle involves both LH and FSH acting differentially on the ***granulosa*** and ***theca interna*** cells of that follicle. Rupture of the mature follicle and release of the ovum is known as ***ovulation*** and depends primarily on the action of circulating LH. The postovulatory formation of a ***corpus luteum*** in the cavity of the ruptured follicle represents one mammalian adaptation to viviparity, and it appears to require no further gonadotrophic stimulation beyond that which causes ovulation. The luteal tissue of the transient corpus luteum secretes progesterone, which is required to (1) suppress development of preovulatory ovarian follicles, (2) prevent sexual receptivity in non-primates, and (3) maintain a uterine environment suitable for embryonic development. The life span of the corpus luteum in most species depends on whether the ovum is fertilized after ovulation and whether pregnancy is initiated. Maintenance of progesterone secretion by the luteal tissue in pregnant females depends on the interactions among the following: (1) presence of adenohypophysial hormones, such as LH and PRL, which stimulate luteal cells (i.e., are luteotrophic), (2) absence of a hormonal signal from the uterus, such as prostaglandin-F_2 alpha, which regresses the corpus luteum (i.e., is luteolytic), and (3) various uterine and placental hormones that also influence the secretion of progesterone by the corpus luteum. In some mammals, the placenta secretes enough progesterone so that luteal tissue is not needed to sustain the latter part of pregnancy.

Figure 12-1 illustrates the complex actions of LH and FSH on theca interna and granulosa cells of ovarian follicles that result eventually in the secretion of estradiol into blood. As is true for other hormonal peptides, the action of LH or FSH depends on the target cells possessing cell-surface receptors specific for one of these blood-borne gonadotrophic hormones. The ovarian secretion of

SMALL FOLLICLES

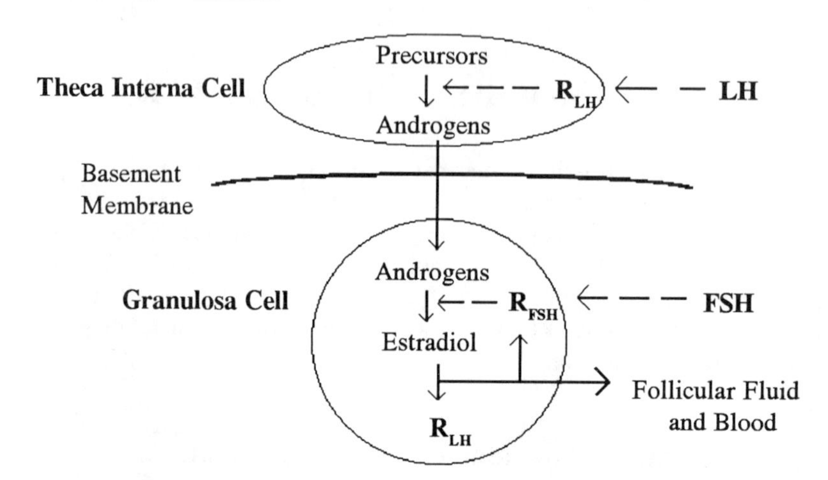

LARGER FOLLICLES

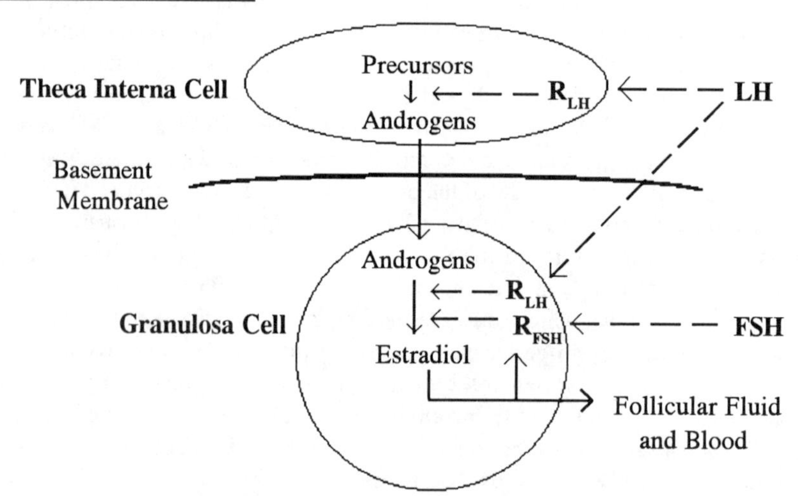

Figure 12-1. LH and FSH regulation of steroidogenesis in cell types of the ovarian follicle.

Biosynthetic steps in steroidogenesis and the presence of cell-surface receptors for gonadotropins (R_{LH} and R_{FSH}) are depicted in both cell types of the ovarian follicle. The upper diagram illustrates one theca interna cell and one granulosa cell in a small follicle whereas the lower diagram illustrates these same two cells after gonadotropin-induced growth and differentiation. ***Solid*** lines associated with steroid molecules denote either intracellular conversion and receptor activation or intercellular transport and secretion. ***Dashed*** lines associated with gonadotropins and their receptors denote binding and/or facilitation of steroid biosynthesis.

estradiol is very complex because the receptors differ among cell types and also vary with the degree of gonadotrophic stimulation of follicular growth. The wall of the follicle consists of an inner non-vascularized layer of granulosa cells that line the follicle wall and surround the oocyte. This inner layer of granulosa cells, also called membrana granulosa, is separated from the vascularized layer of theca interna cells by the basement membrane of the follicle. The cells of the membrana granulosa are not a homogeneous population probably due to location-dependent specializations. Granulosa cells located in the cumulus region surrounding the oocyte appear to have specialized contacts with the oocyte. The layers of granulosa cells located just inside the basement membrane are classified as pseudostratified because processes from the cells in several layers have contact with the basement membrane (Lipner and Cross, 1968). However, some granulosa cells located in the interior of the follicle lack cellular contact with the basement membrane.

Development of knowledge about steroidogenesis in ovarian follicles was derived primarily from *in vitro* studies of granulosa and theca interna cells. When cultured together and exposed to LH and FSH, these two cell types synthesize androgenic steroids (e.g., testosterone and androstenedione) as well as estradiol. By physical separation of the cell types prior to culture and *in vitro* exposure to only one of the gonadotropins, the specific hormonal responses of each cell type have been determined as illustrated in Figure 12-1 (Moor and Seamark, 1986; Gore-Langton and Armstrong, 1988). The recent development of techniques to localize and quantify the mRNA for FSH receptor (FSH_R) and LH receptor (LH_R) in follicles of different size and maturity (Camp et al., 1991) has confirmed much of the earlier *in vitro* data. In summary, theca interna cells of small follicles respond to LH stimulation with increased synthesis of androgens as well as of LH_R. The secreted androgens diffuse across the basement membrane and are taken up by granulosa cells. Under the influence of FSH, these granulosa cells in small follicles convert the androgens into estradiol that is secreted into follicular fluid and blood. This estradiol also stimulates the synthesis of additional FSH_R in these cells, and it initiates the synthesis of LH_R in granulosa cells as the follicle enlarges. These induced changes in gonadotropin receptors, as the follicle enlarges, create the situation depicted in the lower part of Figure 12-1. Theca interna cells in large follicles possess an abundance of LH_R that facilitates LH-induced stimulation of androgen production by these cells for export to the granulosa cells. The abundance of FSH_R in granulosa cells and their acquisition of LH_R combine to facilitate joint LH- and FSH-induced stimulation of estradiol production from the theca-derived androgenic precursors. The overall concept illustrated in Figure 12-1 appears to hold for most species, and it is sometimes called the two-cell, two-gonadotropin model. As with most unifying concepts, there are some aspects that are not fully understood. Although the transport of steroid compounds is mainly unidirectional (i.e., theca interna to granulosa), there may also be transport of granulosa-derived progesterone to the theca interna to provide substrate for androgen synthesis in bovine follicles (Fortune, 1986). Also, rat granulosa cells transiently produce progesterone receptor mRNA during preovulatory events (Park and Mayo, 1991). Such an intraovarian action of progesterone on granulosa cells is not accounted for in the established concepts of ovarian function.

Pure populations of granulosa cells are readily harvested from ovarian follicles, and these cells have been widely studied *in vitro*. However, care must be taken in the interpretation of results obtained from long-term cultures of granulosa cells because these cells may change markedly during culture. These cells undergo a process called luteinization in which they become like luteal cells and secrete large amounts of progesterone upon appropriate gonadotrophic stimulation. However, luteal cells that form *in vivo* are derived from a mixture of theca interna and granulosa cells, and the *in vitro* luteinization of granulosa cells in the absence of theca interna may create an unphysiological group of luteal cells.

In addition to the production of steroidogenic compounds, FSH-stimulated granulosa cells synthesize and secrete inhibin (see structure in Figure 11-3). The production of inhibin by cultured granulosa cells in response to FSH was enhanced by the addition of either androstenedione or estradiol, both of which would be present *in vivo* (Ying et al., 1987). The inhibitory feedback of granulosa-derived inhibin on FSH secretion in the female will be discussed later in this chapter. A variety of other peptides can be produced by granulosa cells, but their physiological relevance is not fully understood. These ovarian peptides include the neuropeptides oxytocin and β-endorphin as well as insulin-like growth factor-I, which mediates biologic effects of somatotropin in most tissues of the body.

The developmental pattern of ovarian follicles differs among species depending primarily on the number of ova to be shed at any one time (Greenwald and Terranova, 1988). In litter-bearing species such as the rat and hamster, many follicles will grow, develop, and ovulate at the same time. In monotocous mammals (single births), mechanisms exist for a single follicle to become dominant and suppress any further development of competing follicles until the dominant follicle either wanes or ovulates. There are two or more separate waves of follicular development during the 21-day estrous cycle of non-pregnant cattle (Ginther et al., 1989; Fortune et al., 1991). The dominant follicle of every wave, except the final one of the estrous cycle, regresses without rupture (i.e., becomes atretic). Studies of such dominant follicles in sheep indicate that a very large proportion of total estradiol secretion is derived from the dominant follicle, and that a lesser proportion of inhibin is derived from the dominant follicle (Mann et al., 1992). Therefore, these results demonstrate a dissociation between the production of estradiol and inhibin from the granulosa cells in follicles of different sizes and physiological maturities.

Cellular Actions of Gonadal Steroids. The gonadal hormones estradiol, progesterone, and testosterone each bind to a receptor molecule that is a member of the receptor superfamily mentioned earlier in Chapters 6 and 7 (Evans, 1988; Parker, 1988). These gonadal steroids are transported from their cells of origin to their target tissues primarily in blood, although paracrine actions on adjacent cells are also possible. During blood-borne transit, they may be bound to carrier proteins, but the low affinity of such binding allows dissociation of the hormone when the complex arrives in target tissues. The molecular structure of these hormones readily allows their entry into cells, and they are retained and concentrated in those cells that contain the specific hormone receptor. The intracellular binding of the hormone to its receptor is illustrated in Figure 12-2. Early research on these steroid receptors suggested that unbound receptors are located exclusively in the cytoplasm, and that only after binding to hormone is the resulting complex translocated into the nucleus of the target cell. More recent information depicted as solid lines in Figure 12-2 indicates that steroid hormones bind to their receptors only in the nucleus. Binding of the hormone to its receptor activates the receptor, but the details of this activation remain controversial. Ligand-induced activation of the receptor may initiate receptor dimerization and/or enhanced binding to the DNA regulatory element on the hormone-

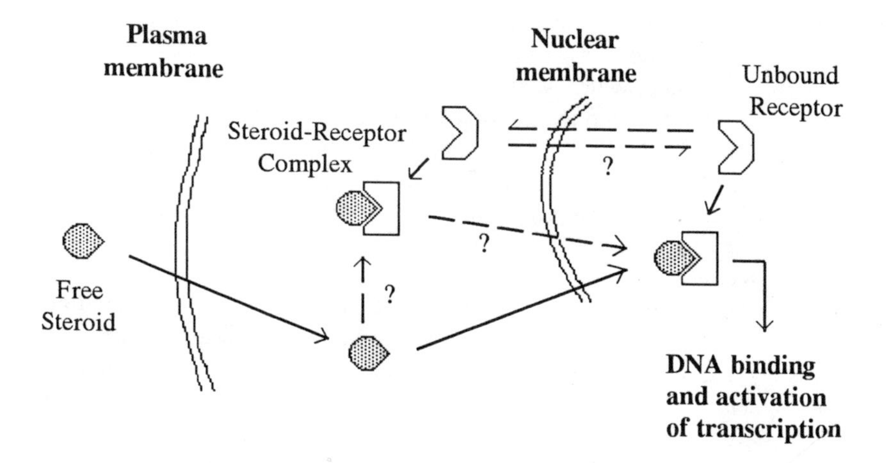

Figure 12-2. Intracellular movement and action of estradiol and other steroid hormones in target tissues.

Free molecules of the steroid hormone diffuse across the plasma membrane. These molecules were once thought to bind with their receptors present in the cytoplasm, but it is more likely that they move into the nucleus before binding to their receptor. Such binding activates the receptor, perhaps altering its conformation and DNA-binding affinity as denoted by the change in the shape of the present symbol. If the ligand-receptor complex does form initially in the cytoplasm, it is rapidly transported into the nucleus (shown as dashed lines). Irrespective of where it initially forms, the ligand-receptor complex in the nucleus activates transcription of specific hormone-responsive genes.

responsive gene. Regardless of the specific details, activation of the ligand-occupied receptor stimulates transcription of the gene that is sensitive to the hormone (Tsai and O'Malley, 1991). Figure 12-2 illustrates the intracellular movement and action of estradiol and other steroid hormones in target cells, because similar mechanisms seem to exist for all members of this receptor superfamily. However, there are different receptors for each of these hormones and different hormone-responsive genes in their target cells.

Luteotrophic Actions of LH. Progesterone secretion by luteal cells of the corpus luteum is stimulated by LH in all species. As discussed in Chapter 9, LH and PRL both possess luteotrophic actions, but the relative importance of each endogenous hormone differs among species (Niswender and Nett, 1988). In primates and most ungulates, LH acts as the primary luteotrophic hormone, with little or no PRL being required. In rats, mice, and hamsters, PRL is the primary luteotrophic hormone, but normal luteal function requires some gonadotropin secretion acting either directly on the luteal cells or indirectly to sustain estradiol secretion from the follicles. Rabbits are unique because progesterone secretion is indirectly sustained by the action of estrogen produced by ovarian follicles in response to the action of gonadotropins (Bill and Keyes, 1983). Despite the adequate secretion of whichever adenohypophysial hormones are luteotrophic in a particular species, uterine secretion of prostaglandin-F_2 alpha decreases progesterone and terminates the life span of the corpus luteum. Various factors (some of which have been identified) from the uterus and conceptus are able to prevent the secretion of prostaglandin-F_2 alpha and/or protect luteal cells from its luteolytic effects (Silvia et al., 1991).

The secretion of progesterone, whether from luteal or placental tissue, continues throughout pregnancy and parturition occurs when this secretion of progesterone wanes. Mechanisms controlling the species-specific timing of this pregnancy-ending event have been thoroughly investigated. Readers interested in additional details are referred to a recent review (Challis and Olson, 1988). In some species (e.g., rat and pig), there is a large discharge of the hormone ***relaxin*** from the corpora lutea of pregnancy as they undergo luteolysis and their secretion of progesterone decreases (Sherwood, 1988). In other species (e.g., guinea pig and horse), it is uterine and placental tissues that secrete relaxin during pregnancy, and this secretion is maximal at parturition. In pregnant women, the corpus luteum is the primary source of relaxin, but placental and decidual tissues may also contribute. It seems clear that the tissue source of relaxin varies among species, but this variation is not related to whether luteal or placental tissue is the source of the progesterone required to maintain pregnancy to term (Sherwood, 1988). The primary actions of relaxin are to promote relaxation of the uterine cervix and to maintain the uterine myometrium in a quiescent state. Concurrent with the decline in progesterone and the rise in relaxin at the end of pregnancy, there are also increases in estrogen, oxytocin, and prostaglandin secretion that probably facilitate normal evacuation of the pregnant uterus.

Patterns of LH and FSH Secretion in the Female

During the prepubertal period in female mammals, the secretion of LH and FSH undergoes a biphasic change. Circulating concentrations of both gonadotropins are usually somewhat elevated during the first few weeks, months, or years of life. The duration of this hypergonadotrophic period depends on whether the prepubertal period in a particular species is measured in weeks, months, or years. During the subsequent period of prepubertal development in non-primate species, secretion of LH and FSH declines in association with development of inhibitory feedback signals from the ovary (e.g., estrogen and inhibin). A similar decline occurs in juvenile female monkeys, but inhibitory feedback from the gonad is not responsible for the reduction in gonadotropin secretion until the period several months before puberty (Plant, 1988). The best way to determine experimentally the ages at which ovarian secretions suppress gonadotropin secretion is to compare blood concentrations of LH or FSH between immature females that have been ovariectomized and age-matched females with intact ovaries. As stated above, the early hypergonadotrophic period of life is characterized by a lack of ovarian feedback, but such feedback inhibition of LH, and in some cases of FSH, invariably develops during the period prior to puberty in all mammalian females. Despite this inhibitory feedback from ovarian secretions, the secretion of gonadotropins increases sufficiently at puberty to stimulate ovarian follicular development as well as the production of estradiol sufficient to trigger the pubertal preovulatory surge of LH.

It is important to note that secretion of LH during the prepubertal period consists of basal secretion plus pulsatile discharges occurring at irregular intervals. As noted in the male, FSH secretion is less pulsatile than LH secretion. Although studied in much detail, increased secretion of immunoreactive LH has not usually been observed until puberty occurs, or in some cases, just before it occurs (Diekman et al., 1983; Dodson et al., 1988b; Ryan et al., 1991). Because of the ovarian stimulation that precedes puberty, it has been hypothesized that the biopotency of blood-borne LH may be increased before puberty. However, no such increased LH biopotency could be demonstrated in cattle (Dodson et al., 1988a).

The occurrence of puberty is greatly influenced by nutrition, body weight, and body condition in female mammals. Mechanisms mediating such influences may involve suppression of hypothalamic release of LHRH in undernourished females (Foster et al., 1989). Female sheep became adapted to chronic undernutrition and were able to initiate puberty at body weights lower than normal (Suttie et al., 1991). When severely undernourished prepubertal female sheep were provided additional dietary nutrients, secretion of both gonadotropins increased rapidly (Padmanabhan et al., 1992).

Cyclic Females. The secretory profiles of LH and FSH in postpubertal females that are not pregnant are invariably linked to the various phases of their reproductive cycles (e.g., estrous cycle, menstrual cycle, or copulation-induced cycle). Estrous and menstrual cycles are both characterized by spontaneous ovulations followed sequentially by luteal development, luteal regression, and a proestrous period. The major difference between estrous cycles and menstrual cycles is that menstruation occurs between the end of luteal regression and the start of proestrous development (also called follicular phase in female primates) in menstrual cycles. Estrous cycles in non-primate species vary greatly in length, being as short as 4 days in some rodents. Those species with very short estrous cycles have evolved a mechanism through which the life span of the corpus luteum can be prolonged by copulation with a male. Such a prolongation is essential to allow sufficient development of the fetoplacental unit so that it can provide for maintenance of the corpus luteum during pregnancy. This copulation-induced prolongation of luteal life span is called ***pseudopregnancy***, and the profiles of prolactin secretion associated with it were presented in Figure 9-2.

In other mammalian species, copulation induces ovulation and the subsequent reproductive cycle because ovulation either does not occur spontaneously or its spontaneous occurrence is not optimum for fertilization of the ovum. Those species in which ovulation *only* occurs after copulation are sometimes called ***reflex ovulators***, and they include such diverse species as rabbit, ferret, cat, and llama. The secretion of LH in mated females of these species is characterized by an initial large copulation-induced surge as illustrated in Figure 12-3 for rabbits. This discharge of LH in female rabbits lasts for several hours after a single brief copulation and is followed about 10 h later by ovulation. In female cats which may copulate as many as 30 times in 36 h, the induced surge of LH peaked at about 3 h after the first copulation, with further mating activity having no apparent stimulatory effect on LH release (Concannon et al., 1989). The refractoriness of LH release to subsequent copulations apparently involved the neurohormone LHRH because there were still releasable amounts of LH in the adenohypophysis of these female cats. Figure 12-3 also depicts FSH profiles in mated rabbits, and a biphasic profile is evident. The initial period of increased FSH coincided with the post-copulatory surge of LH, but a secondary increase of FSH secretion occurred between 10 and 30 h after copulation, a period when LH secretion was very low.

Females that ovulate spontaneously during estrous or menstrual cycles release a large preovulatory surge of LH shortly before ovulation. Figure 12-4 presents hormone profiles for a representative species (e.g., cow, ewe, pig, mare, guinea pig) during the estrous cycle. The preovulatory LH surge usually coincides with the onset of sexual receptivity by the female (estrous behavior). Hormonal determinants of sexual behavior will be discussed in Chapter 14. The period preceding the LH surge is often called the proestrous period, and it is characterized by (1) an initial abrupt decline in progesterone secretion from the corpus luteum formed at the previous ovulation, (2) more pulsatile discharges of LH secretion than occurred when progesterone was elevated, and (3) increasing secretion of estradiol from rapidly growing ovarian follicles (all illustrated in Figure 12-4). The pattern of FSH secretion during the estrous cycle consists of two peaks, one coincident with the LH surge and a secondary peak about 1 day

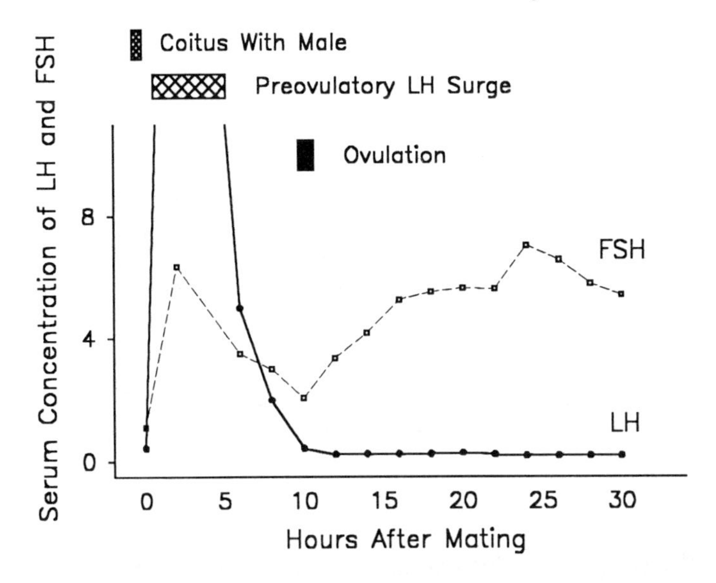

Figure 12-3. Serum concentrations of LH and FSH in female rabbits after copulation.

These concentrations of LH and FSH (ng/ml) following copulation were estimated and redrawn from Mills et al. (1981). The concentration of LH at 2 h after mating was >30 ng/ml (not plotted). Differentially shaded boxes denote the timing of various reproductive events.

later. During the period after ovulation and after the secondary peak of FSH, the corpus luteum begins to increase its output of progesterone, which is maintained until luteolysis occurs several days before the next estrus. Profiles of LH, FSH, and estradiol during the entire period of corpus luteum function remain at basal levels although LH secretion is also characterized by pulsatile discharges at irregular intervals. In those species that have several waves of follicular development during each estrous cycle, estradiol secretion may be increased during the period in which one follicle of a midcycle wave becomes dominant and estrogen-active (Ireland, 1987). Primate females undergoing menstrual cycles have hormone profiles similar to those in Figure 12-4 except that the proestrous period (follicular phase) is lengthened from 2-3 days to 10-14 days.

In those rodents such as rats, mice, and hamsters that ovulate spontaneously every 4-5 days, the progesterone profile is different from that depicted in Figure 12-4. Unless mating occurs to activate the corpora lutea, these structures secrete very modest quantities of progesterone, with much of it being converted into a 20-hydroxylated metabolite with reduced bioactivity. However, these rodent females possess a unique LHRH regulatory system that only allows secretion of

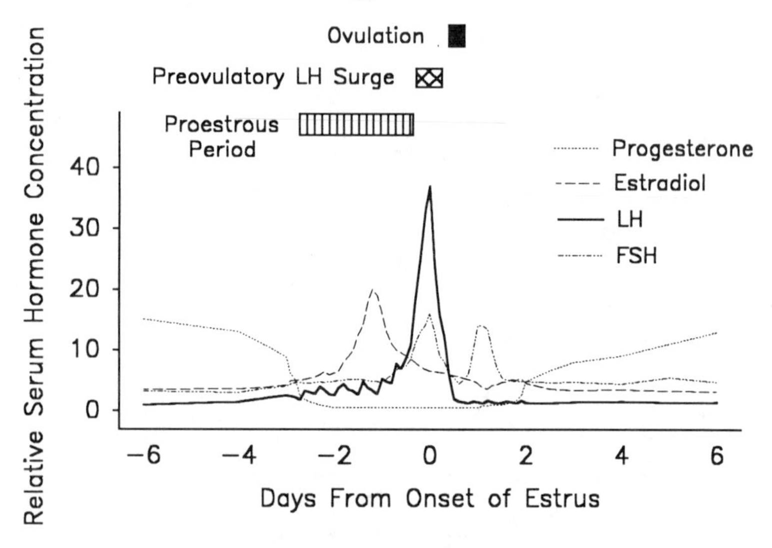

Figure 12-4. Representative hormone profiles in cyclic females around the time of estrus.

Hormone concentrations are plotted in arbitrary units due to large differences among the various hormones depicted. Shaded boxes denote various periods around estrus.

a preovulatory surge of LH at one time of the day. The evolutionary association between short-lived corpora lutea (i.e., short estrous cycles) and the circadian-dependent secretion of preovulatory LH surges is suggestive of an adaptive advantage. When female rodents copulate or receive vaginocervical stimulation, the corpora lutea are activated to secrete more progesterone and to have a life span approaching that of species with longer estrous cycles. In this case, the hormone profiles depicted in Figure 12-4 are approximately correct for such mated but non-pregnant rodents at the end of one pseudopregnancy and the start of another pseudopregnancy. Furthermore, the selective secondary increase of FSH about one day after the preovulatory LH/FSH surge occurs during every 4-day estrous cycle. The function of this FSH increase may be to initiate development of the next wave of ovarian follicles because immunoneutralization of this blood-borne FSH surge disrupted follicular development necessary for subsequent ovulations (Schwartz et al., 1973).

Circulating concentrations of inhibin are not included in Figure 12-4 because multispecies generalizations are presently not possible. In many species, inhibin appears to derive exclusively from granulosa cells of developing follicles, both dominant and non-dominant (Mann et al., 1992), whereas the corpus luteum is a major source of blood-borne inhibin in primates (Basseti et al., 1990). Therefore, luteolysis in primate species would be associated with a decrease of circulating inhibin that does not occur in other species.

Establishment of pregnancy in cyclic females prolongs the life span of the corpus luteum and continues its secretion of progesterone by preventing the secretion of luteolytic compounds by the uterus (e.g., prostaglandin-F_2 alpha). The secretion of LH and FSH during pregnancy remains at basal levels with irregular pulsatile discharges, similar to those that occurred during the luteal period of recurring estrous cycles. When the pregnancy ends with parturition, the secretion of gonadotropins in many species must undergo a period of recovery before reinitiation of postpartum follicular growth, ovulation, and fertility. However, no such recovery is needed in female rats, which secrete enough LH and FSH to grow follicles and ovulate them within 1-2 days after parturition (Hoffman and Schwartz, 1965). Reflex ovulation can also be induced in female rabbits by copulation occurring a few days after parturition (Harned and Casida, 1969). In contrast, the secretion of gonadotropins (mainly LH) is deficient during the first weeks or months after parturition in many ungulate species. Moreover, females that are suckled by their offspring and/or are thin or poorly nourished require an even longer period of recovery before normal LH secretion resumes.

Secretion of LH and FSH during the period of anestrus in seasonal breeding species is characterized by low basal secretion with pulsatile discharges at irregular intervals. The ovaries of such females are devoid of corpora lutea and progesterone secretion is absent from the circulation. Blood-borne estradiol is present in low concentrations, but these low levels are apparently able to exert a potent suppression of gonadotropin secretion. This conclusion is based upon the consistent observation that surgical removal of the quiescent ovaries from anestrus females promptly disinhibits the secretion of LH. This post-ovariectomy secretion of LH occurs as highly regular pulsatile discharges that are nearly equivalent to those observed after ovariectomy during the annual breeding season.

Neurohormonal Control of LH and FSH in the Female

Important aspects of the neurohormonal control of LH and FSH have already been discussed in Chapters 4 and 11. These subjects include (1) the central role of hypothalamic LHRH controlling adenohypophysial release of LH and FSH and (2) a mixed population of monohormonal and bihormonal gonadotrophs in the adenohypophysis of many species. Figure 12-5 summarizes these previously discussed subjects and others in the neuroendocrine integration of gonadotropin secretion in the female. Blood-borne FSH and LH act upon various cells in the ovary, as described earlier in the present chapter, resulting in the variable production of three hormones (inhibin, estradiol, and progesterone) that can regulate the further secretion of FSH and LH through inhibitory feedback.

Immunoneutralization of blood-borne LHRH results in consistent suppression of LH secretion in ovariectomized females and in intact females about to express a preovulatory surge (McCormack et al., 1977; Blake and Kelch, 1981).

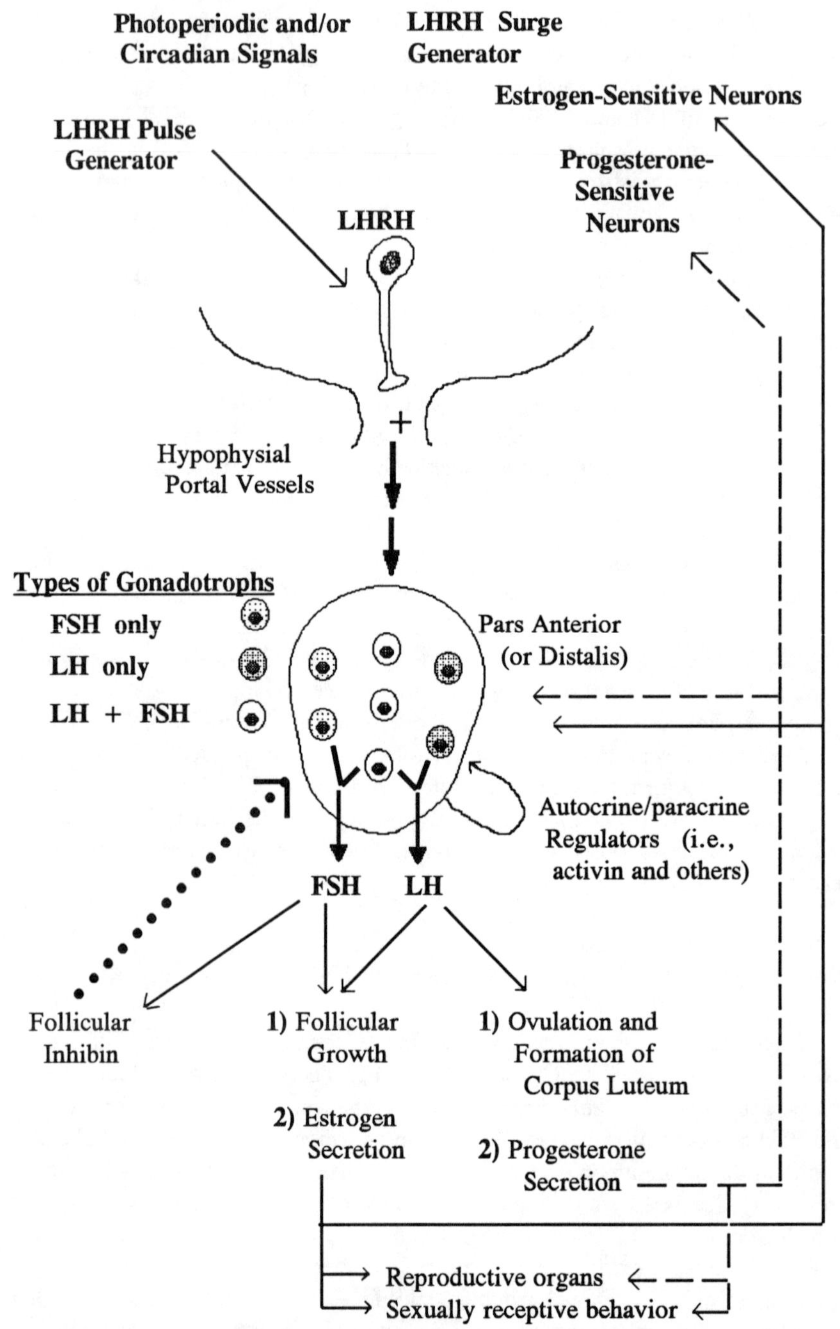

Figure 12-5. Neuroendocrine integration of LH and FSH release in females. This simplified diagram illustrates regulation of gonadotropin secretion in females. Monohormonal and bihormonal cells in the adenohypophysis are depicted with different shading.

Suppression of FSH following immunoneutralization of LHRH is less consistently observed. For example, the surge of FSH occurring 1 day after the preovulatory LH surge (see Figure 12-4) was not blocked by immunoneutralization of LHRH, whereas the FSH surge coincident with the LH surge was inhibited (Blake and Kelch, 1981).

FSH-stimulated granulosa cells in ovarian follicles produce and release the same heterodimer inhibin molecules as found in the male (see Figure 11-3). Blood-borne inhibin acts directly on gonadotrophs to suppress FSH release in adult cyclic females because immunoneutralization of inhibin in rats and sheep increased the secretion of FSH as well as some aspects of LH release (Culler and Negro-Vilar, 1989; Rivier et al., 1990; Mann et al., 1990). Inhibin bioactivity was defined over 50 years ago as suppression of FSH secretion without any effect on LH secretion. Following the discovery of heterodimer molecules of inhibin as well as the bihormonal nature of gonadotroph cells, various reports of granulosa-derived inhibin suppressing some aspects of LH secretion required a revision of the original definition. As depicted in Figure 12-5, follicle-derived inhibin acts mainly on the adenohypophysis in a closed-loop inhibitory feedback of further FSH release. In addition to inhibin-mediated suppression, FSH secretion is also regulated by inhibitory feedback of estrogen acting on the adenohypophysis (Baird et al., 1991).

Feedback of Estradiol and Progesterone. Neuroendocrine regulation of LH and FSH in the female is complicated greatly by the ability of the ovarian steroids, estradiol and progesterone, to exert inhibitory as well as stimulatory feedback. An additional complication is that feedback of these steroids can occur at both the adenohypophysis and the hypothalamus (Figure 12-5).

Tonic levels of blood-borne estradiol appear to maintain a general suppression of LH secretion through inhibitory actions on the gonadotrophs as well as on the LHRH systems in the hypothalamus. As stated earlier, the secretion of LH in ovariectomized females consists of large pulsatile discharges occurring at regular intervals. Immunoneutralization of endogenous estrogens in sheep resulted in a pattern of pulsatile LH secretion similar to that in ovariectomized sheep (Pant et al., 1978). Rising levels of estrogen secretion during the proestrous period exert a stimulatory feedback on LH secretion resulting in the preovulatory surge (see Figure 12-4). Immunoneutralization of endogenous estradiol in cyclic female rats abolished the circadian-dependent surge of LH and prolactin that normally occurs on the afternoon of proestrus (Neill et al., 1971). Replacement with an estrogenic molecule that could not be immunoneutralized restored the preovulatory LH surge. Furthermore, exogenous estrogen consistently induces a surge of LH secretion in appropriately treated females. As discussed in Chapter 11, the ability of estrogen to induce such a surge of LH secretion in rodents and many ungulates is lost during masculine sexual differentiation.

The sites of estradiol feedback on LH secretion have been thoroughly investigated in many species and using many different experimental models, and the conclusions drawn from these investigations sometimes differ. The occurrence of inhibitory and stimulatory feedback also complicates interpretation. Briefly, there is evidence in one or more species for LH-inhibitory actions of estradiol in the adenohypophysis and in the hypothalamus. Similarly, there is evidence for LH-stimulatory actions of estradiol in the adenohypophysis and in the hypothalamus. As discussed in Chapter 11 regarding testosterone feedback, experimental stabilization of hypothalamic input to gonadotrophs represents a valuable technique for investigating direct feedback on gonadotrophs. This experimental model is sometimes referred to as a ***hypophysiotrophic clamp***. In all applications of this model, the female receives repetitive intermittent injections of LHRH, or a synthetic analogue, while endogenous secretion of LHRH is experimentally blocked by any one of several techniques (e.g., transection of the hypophysial stalk; destruction of the median eminence; immunoneutralization of LHRH; pharmacologic suppression of LHRH release). Release of LH is monitored before, during, and after administration of estradiol (or sometimes progesterone) to these *clamped* females. Invariably, there is an initial estrogen-induced decline in LH secretion that must be due to direct inhibitory feedback on the gonadotrophs that are being exposed to unvarying pulses of exogenous LHRH. After the initial inhibition of LH release, there is an increase in release of LH to levels greater than those existing before estradiol treatment. This secondary increase in LH secretion is due to the increased synthesis of LH in gonadotrophs and greater sensitivity to blood-borne LHRH, both of which reflect stimulatory feedback of estradiol on gonadotrophs. The species differences, and in some cases the controversy, from the use of the clamp model are related to the magnitude of the secondary increase in LH release. In monkeys, this estrogen-induced surge of LH in clamped females usually approximates the LH surge occurring in non-clamped females (Knobil, 1980). Therefore, an estrogen-induced surge of endogenous LHRH may not be necessary. By contrast, similar studies in clamped females of several ungulate species (e.g., sheep and pig) indicated that the estrogen-induced LH surge was subnormal when the gonadotrophs were only exposed to unvarying pulses of exogenous LHRH (Clarke et al., 1989b; Kesner et al., 1989; Britt et al., 1991). In summary, stimulatory feedback of estradiol occurs at the adenohypophysis of all species, and estradiol-induced increases of LHRH release appear to be required in many non-primate species for normal preovulatory LH surges.

As stated above, the initial phase of estradiol-induced suppression of LH occurs on the gonadotrophs. Although the LH-inhibitory effect in normal females persists for a long time, the direct suppression of gonadotrophs wanes after only a few hours in clamped females. In addition, exogenous estradiol tonically inhibits the release of LHRH for delivery to the adenohypophysis (Karsch et al., 1987), and it also inhibits the volleys of neuronal activity that correlate with putative discharge of the LHRH pulse generator (Kesner et al., 1987). Therefore, estradiol also exerts a persistent inhibitory feedback on the hypothalamus to decrease the release of LHRH.

Based on studies in clamped females, progesterone antagonizes the estradiol-induced surge of LH release through mechanisms located central to the adenohypophysis (Wildt et al., 1981; Clarke and Cummins, 1984). However, progesterone also inhibits the basal (non-surge) secretion of LH, often in synergy with a similar action of estradiol (Goodman, 1978). There is evidence from a single species (sheep) that this tonic inhibition of LH by progesterone occurs both at the hypothalamus to decrease LHRH (Karsch et al., 1987) and at the gonadotroph to decrease LH responses to exogenous LHRH in clamped females (Clarke et al., 1989a). In summary, progesterone-induced suppression of the LH surge occurs in the hypothalamus while suppression of basal LH release may occur at both hypothalamus and adenohypophysis.

Progesterone exerts a stimulatory feedback on the timing of the LH surge in rodents. Demonstration of this effect of progesterone requires an estrogen-primed gonadotrophic axis in a female that will shortly initiate a LH surge. However, administration of progesterone facilitates or hastens the occurrence of the LH surge presumably though actions on the hypothalamus to stimulate early release of LHRH (Kalra and McCann, 1975; Levine and Ramirez, 1980). Metabolites of progesterone released immediately after copulation enhanced the preovulatory surge of LH in rabbits (Hilliard et al., 1967), perhaps through direct stimulation of LHRH release (Lin and Ramirez, 1990).

Steroid-Sensitive Neurons. As described in the preceding section, there is evidence that both estradiol and progesterone act on the hypothalamus to modify the secretion of LHRH. However, LHRH neurons lack intracellular receptors for either estradiol or progesterone. Therefore, estrogen-sensitive and progesterone-sensitive neurons exist in the brain to mediate the effects of these ovarian hormones on the secretion of LHRH, as illustrated in Figure 12-5. Learning about those steroid-sensitive neurons that influence LHRH is complicated by the fact that the brain contains a large number of steroid-sensitive neurons, many of which probably do not influence LHRH release. Although steroid-sensitive neurons can be identified by their uptake of radiolabeled steroid or by immunocytochemical staining for a particular steroid receptor, it is not currently possible to prove that an identified steroid-sensitive neuron also regulates the release of LHRH. Nevertheless, the following neuropeptides have been demonstrated in estrogen-sensitive neurons of the hypothalamus: β-endorphin, dynorphin, and neuropeptide Y (Morrell et al., 1985; Sar et al., 1990). The number of progesterone-sensitive neurons in the hypothalamus was increased greatly by estrogen administration and receptors for both steroids sometimes occurred in the same neuron (Blaustein and Turcotte, 1989). Progesterone receptors have been stained in neurons that also contain β-endorphin or the enzymes necessary for synthesis of catecholamines or gamma-aminobutyric acid (Olster and Blaustein, 1990; Fox et al., 1990; Leranth et al., 1992). Besides being present in the steroid-sensitive neurons that participate in neuroendocrine integration (Figure 12-5), receptors for both estradiol and progesterone are found in gonadotrophic cells of the adenohypophysis.

Most of the neuroendocrine actions of estradiol and progesterone on the hypothalamus to regulate LHRH involve actions on the genome, as illustrated in Figure 12-2. However, some effects of progesterone and its metabolites occur too rapidly to be explained by effects on the genome. Moreover, covalently linking progesterone to a large molecule that prevented its entry inside the neuron did not abolish all neuroendocrine effects of progesterone (Ramirez et al., 1990). Therefore, progesterone may influence release of LHRH through both genomic and non-genomic mechanisms.

Autocrine/Paracrine Regulation of Gonadotrophs. Adenohypophysial gonadotrophs may also be regulated by intraglandular mechanisms, as illustrated in Figure 12-5. As mentioned in Chapter 5, gonadotrophic cells also contain dynorphin and calcitonin gene-related peptide, although there is little evidence that these compounds influence the release of LH and/or FSH. However, the presence of the molecule known as ***activin*** within gonadotrophic cells is another situation (Roberts et al., 1989, 1992). As discussed in Chapter 11, this molecule is a homodimer consisting of two beta-subunits of inhibin (see Figure 11-3). Because activin can stimulate the selective release of FSH, it might explain the selective increase of FSH that is usually observed about 1 day after the preovulatory surge of LH/FSH (see Figures 12-3 and 12-4). It has long been known that this secondary increase of FSH does not require blood-borne LHRH (Blake and Kelch, 1981). It has recently been reported that immunoneutralization of activin-B in cyclic female rats attenuated this secondary increase of FSH. Immunoneutralization of activin-B also antagonized ovariectomy-induced increases of FSH secretion in hypophysectomized rats bearing ectopic adenohypophysial grafts on the kidney (DePaolo et al., 1992). Immunoneutralization of activin-B in dispersed adenohypophysial cells also increased spontaneous secretion of FSH (Corrigan et al., 1991). These results suggest that production of activin-B by gonadotroph cells or perhaps other tissues may stimulate the secretion of FSH independent of LHRH and LH. In normal female rats, ovariectomy increased the transcription of the gene for $beta_B$-activin in the adenohypophysis, whereas exogenous estrogen inhibited transcription (Roberts et al., 1989). It seems possible that intrahypophysial mechanisms for selective secretion of FSH may be regulated by signals originating in the ovary (e.g., inhibin or estrogen).

Regulation of LHRH receptors in gonadotrophs constitutes another intraglandular mechanism that affects secretion of LH and FSH. Receptors for LHRH appear to be controlled primarily by the actions of LHRH. The ability of gonadotrophs to bind radiolabeled LHRH can be decreased (i.e., down-regulated) by continuous exposure to exogenous LHRH. In contrast, pulsatile exposure to exogenous and presumably endogenous LHRH increases LHRH binding capacity, and estradiol also directly increases binding of LHRH by gonadotrophs (Clayton, 1989).

LHRH Pulse Generator. After inhibitory feedback from ovaries or testes has been eliminated, LH secretion consists of large pulsatile discharges occurring at very regular intervals ranging from 20 min to 90 min, depending on species. Each discharge of LH follows immediately a discharge of LHRH into the portal vasculature (Clarke and Cummins, 1982). During the intervals between successive LH discharges, release of LH from the adenohypophysis was not detectable by arteriovenous methods (Rasmussen and Malven, 1982). Because of such observations, the concept of a LHRH pulse generator has developed. The hypothetical pulse generator regulates the rhythmic discharges of LHRH observed after castration. However, it exists and is suppressed to varying degrees during the female reproductive cycle by estradiol, progesterone, and environmental factors. It should be noted that the LHRH pulse generator (Lincoln et al., 1985) is a well-established hypothetical concept and not necessarily a well-defined physical entity. There may be endogenously pulsatile neurons dedicated to controlling the synchronous discharge of LHRH neurons. Alternatively, complex synaptic inputs (stimulatory and inhibitory) to LHRH neurons may coordinate the firing of LHRH neurons and release of their neurohormone. Also, LHRH neurons may possess an endogenous rhythmicity that can be temporarily suppressed by various synaptic inputs (Wetsel et al., 1992).

Efforts to locate and understand the LHRH pulse generator have utilized primarily electrophysiology or neurosurgical deafferentation as research approaches. Synchronous firing of neurons coincident with the regular discharges of LH (and presumably LHRH) in ovariectomized females has been reported in several species (monkey, rat, sheep, goat). These electrical discharges appear to be functionally very important because pharmacologic suppression of them with aminergic or opioidergic drugs produces concurrent changes in the pulsatile discharges of LH (Kaufman et al., 1985; Williams et al., 1990; Nishihara et al., 1991). The sites in which these electrical correlates of the LHRH pulse generator could be recorded have been dispersed generally within the hypothalamus, and it not known whether LHRH neurons were the only cells firing in synchronous bursts.

Other efforts to localize the pulse generator have involved neurosurgical deafferentation of hypothalamic areas from the rest of the brain leaving intact the portal vein linkage to the adenohypophysis. In summary, neural input from rostral structures into the mediobasal hypothalamus seems essential for pulsatile release of LH to occur in ovariectomized females. The exact position of the frontal deafferentation appears to determine whether pulsatile discharges continue. If the deafferentation was located sufficiently rostral, the pulse generator remained functional. The neuroanatomical area that must remain in synaptic contact with the median eminence was the rostral-most part of the arcuate nucleus in rats and the retrochiasmatic area in sheep (Jackson et al., 1978; Soper and Weick, 1980).

Preovulatory LH Surge. The large increase in LH release, sustained for several hours and accompanied by a modest increase in FSH, is induced by copulation in some species (Figure 12-3) and occurs spontaneously after luteolysis in other species (Figure 12-4). In both situations, a concurrent increase in LHRH release has been observed, and this LHRH increase appears to be necessary for the preovulatory surge in all non-primate species (see earlier section on estradiol feedback). Except for reflex ovulators, prior exposure to increased estradiol is necessary to produce a normal preovulatory surge. This estradiol acts on the gonadotrophs to increase the quantity of LH available for surge-like release, and it also acts on the brain to produce a sustained increase of LHRH release. The pattern of this sustained release of LHRH consists of very frequent pulsatile discharges that contribute to a steadily increasing baseline and, together with that baseline, constitute a surge of LHRH secretion. Although the ascending phase of LHRH secretion coincides very closely with the ascending phase of LH release in sheep, the descending phase of LH secretion occurs while LHRH release still remains elevated (Moenter et al., 1991). Therefore, the termination of the preovulatory LH surge may result from a depletion of releasable LH or other inhibitory mechanisms.

One of the early and major advances in the study of ovulation was the discovery that neuroactive drugs could block ovulation in the rat only when administered before a critical period on the afternoon of proestrus (Everett et al., 1949). During the subsequent decades, the CNS location of a hypothetical LHRH surge generator (Figure 12-5) has been investigated in great detail. As discussed in Chapter 4, neuronal perikarya that contain LHRH are distributed from rostral areas, such as the preoptic area and diagonal band of Broca, back to the mediobasal hypothalamus depending on species (Silverman, 1988). Lesions of neural structures such as the median eminence or hypothalamic areas through which LHRH axons project on their way to the median eminence may compromise the LH surge and ovulation, but they do not necessarily involve the surge generator. The only lesion experiments with the potential to determine the location of surge-generating mechanisms are those in which basal pulsatile LH secretion persists after the lesion, but no LH surges occur even when challenged with a bolus of estradiol. Neurosurgical deafferentation of rostral inputs to the rat hypothalamus was initially applied to such investigations, and the occurrence of ovulation was used as an indirect measure of the LH surge (Halasz and Gorski, 1967). Subsequent quantification of LH in blood confirmed that deafferentation of rostral inputs to the hypothalamus prevented the LH surge while allowing enough basal gonadotropin secretion to maintain polyfollicular ovaries (Blake et al., 1972). The neuroanatomical location of the effective ***hypothalamic deafferentation*** is illustrated in Figure 12-6. This same figure depicts the location of a ***preoptic deafferentation*** that was located more rostrally and that allowed ovulation to recur at irregular intervals (Koves and Halasz, 1970; Kaasjager et al., 1971). These results are consistent with LHRH surge-generating mechanisms being located in the neural tissue between the two deafferentation sites (i.e., the preoptic area and adjacent structures). When studied in other species, a similar deafferentation through the suprachiasmatic nucleus in sheep allowed pulsatile LH but partially compromised the LH surge (Jackson et al., 1978; Pau et al., 1982). In contrast, deafferentation through the anterior hypothalamus of the female monkey did not antagonize the estrogen-induced surge of LH (Cogen et al., 1980), but as noted in Chapter 4, LHRH neurons are more concentrated in the mediobasal hypothalamus of primate species.

The functional importance of the preoptic area in generating the preovulatory surge of LH in rats is also emphasized by the differences in preoptic morphology that have been associated with sexual differentiation of the ability to release a surge of LH (see Chapter 11). In addition, electrical stimulation of

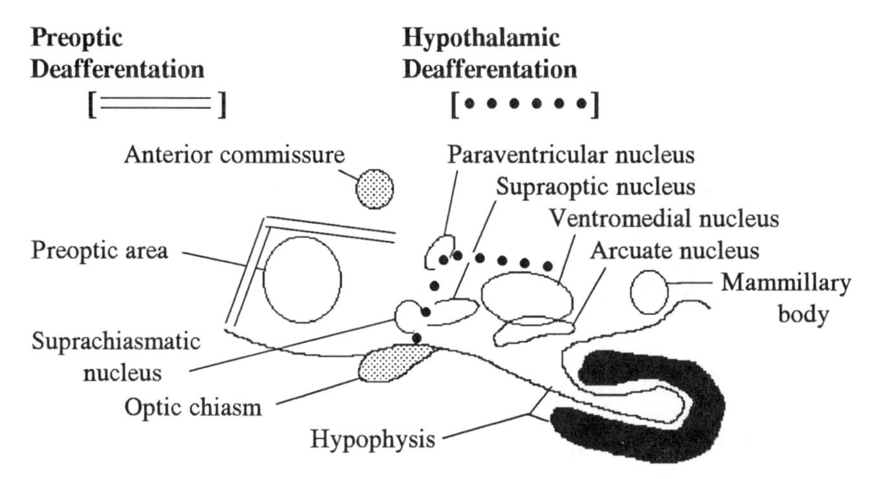

Figure 12-6. Location of neurosurgical deafferentations of the hypothalamus in female rats.

Diagrammatic representation of two types of neurosurgical deafferentations in the female rat. **Preoptic deafferentation** (double lines) of the preoptic area from its rostral and dorsal inputs allowed irregular ovulations to occur (Koves and Halasz, 1970; Kaajager et al., 1971) whereas **hypothalamic deafferentation** (dotted line) of the anterior hypothalamus from its rostral and dorsal inputs completely blocked ovulation (Halasz and Gorski, 1967) and preovulatory release of LH (Blake et al., 1972).

the preoptic area in rats induced ovulation through an increase in secretion of LHRH and LH (Everett et al., 1973; Fink and Jamieson, 1976). Electrical stimulation of the preoptic area in species other than rats have sometimes increased LH release, but the magnitude of these increases have not been surge-like as occurs in the rat. Perhaps the LHRH surge-generating neural center, or its projections to the median eminence, are more widely dispersed in these other species and consequently more difficult to simultaneously activate by localized electrical stimulation.

As stated earlier, there is a circadian component involved in the generation of a preovulatory LH surge in those rodent species with very short estrous cycles; females can generate a surge of LH *only* during the afternoon hours. However, electrical stimulation of the preoptic area is effective at all times of the day. Exposure of the female to estradiol sufficiently early in the day can induce a preovulatory surge that afternoon. If estrogen exposure occurs too late in the day, the occurrence of the LH surge is delayed until the next afternoon. In non-rodent species, administration of estrogen can induce a surge of LH at any time of day, usually a specific interval after the initial exposure to estrogen.

Aminergic and Peptidergic Input to LHRH Neurons. There exists a large body of evidence that the secretion of LH (basal and surge-like) is regulated by aminergic and peptidergic neurotransmitters that probably act on LHRH neurons. The following aminergic compounds have been implicated in some type of LHRH regulation: norepinephrine, dopamine, serotonin, and gamma-aminobutryic acid. Generalizations regarding the type of control exerted by each compound are very difficult because some neurotransmitters appear to stimulate LHRH neurons when basal release is being suppressed by estrogens but to inhibit LHRH neurons when basal release is elevated following elimination of ovarian feedback. More comprehensive discussions about the roles of these various compounds during specific reproductive states can be found in two excellent review articles (Weiner et al., 1988; Thiery and Martin, 1991).

Experimental information about the role of peptidergic neurotransmitters in regulating LHRH and LH is even more limited than for aminergic compounds. Moreover, a major obstacle to understanding the function of endogenous neuropeptides is that receptor antagonists are rarely available, whereas they are readily available for aminergic receptors. Nevertheless, administration of the following neuropeptides has been reported to influence the secretion of LH and/or LHRH: galanin, vasoactive intestinal peptide, neuropeptide Y, neurotensin, and various opioid peptides such as enkephalin, β-endorphin and dynorphin (Weiner et al., 1988). As occurred with aminergic neurotransmitters, administration of some neuropeptides (e.g., neuropeptide Y) produces opposite effects on LH depending on the degree of ovarian feedback. These differential effects greatly complicate generalizations about the physiological effects of any particular neuropeptide.

Because it is possible to antagonize the action of endogenous opioid peptides (EOP) with naloxone, there is greater understanding of their role in regulating LHRH (see Chapter 5). In summary, stress-induced and suckling-induced suppression of LH secretion appears to be mediated by EOP because naloxone administration temporarily relieves the suppression (Haynes et al., 1989). Females in which LH secretion is low as a result of either (1) progesterone feedback or (2) postpartum hyposecretion of LH also respond to naloxone injection with transient increases in LH release (Trout and Malven, 1987; Malven and Hudgens, 1987). Identification of which specific EOP may be involved in each experimental model is a difficult task. However, localized immunoneutralizations of endogenous enkephalin in the mediobasal hypothalamus and endogenous β-endorphin in the preoptic area and other rostral structures both transiently relieve the LH suppression of luteal-phase sheep (Weesner and Malven, 1990).

Regulation of Puberty in the Female. The occurrence of the pubertal ovulation and the subsequent initial secretion of progesterone are readily monitored events, which makes the study of puberty easier in females than males. The neuroendocrine changes listed in the previous chapter (Table 11-1) as possible initiators of puberty in males are also possibilities for females (Grumbach et al., 1990). As occurs in male primates, females of these species experience a period during their juvenile life when the secretion of LH and FSH is low despite the absence of inhibitory feedback from the ovary. As the females approach puberty, ovarian secretions do appear to suppress gonadotropins, but there is a selective hypersecretion of gonadotropins during the period of sleep in human females as they initiate puberty just as occurs in peripubertal human males.

Puberty-related decreases in the LH-inhibitory potency of ovarian estradiol have also been observed in many species (Ojeda et al., 1983), which supports the gonadostat hypothesis. In seasonal breeders such as sheep, this reduction in the LH-inhibitory feedback is linked to seasonal changes in sensitivity to LH-inhibitory feedback (Foster, 1988). In addition, ovarian steroidogenic capacity increases as females approach puberty. Although this ovarian maturation can be stimulated by administration of exogenous gonadotropins, the prepubertal ovaries of untreated females gradually acquire the ability to generate a surge of estradiol in response to physiological concentrations of endogenous LH and FSH. Such puberty-related maturation of gonadal steroidogenesis may be significantly more important in females than in males because the preovulatory surge of LH at puberty requires an antecedent surge of estradiol secretion.

Environmental Effects on Female Reproduction

As occurs in the male, photoperiod exerts a major influence on female reproduction in some species. Because these influences were discussed thoroughly in Chapter 10 on the pineal gland, they will only be summarized here. Exposure of female hamsters to long photoperiods and female sheep to short photoperiods leads to follicular development and ovulation due to the decreased potency of blood-borne estradiol to suppress gonadotropin secretion. As presented in Chapter 10, the duration of the dark-related hypersecretion of pineal melatonin conveys information to the LHRH-LH axis about the ambient photoperiod. The LHRH-LH axis may become photorefractory to specific melatonin patterns, but the pineal gland always correctly reflects ambient photoperiods.

In female mammals, secretion of sufficient gonadotropin for initiation of puberty or the onset of seasonal ovulations can sometimes be enhanced by exposure to mature males (Martin et al., 1986; Hughes et al., 1990). It appears that odor from the male contributes greatly to the enhanced secretion of gonadotropin, and in some cases, odor exposure alone is sufficient to stimulate LH secretion (Over et al., 1990). In addition to these stimulatory effects of male odor, there is an inhibitory effect generally known as the ***Bruce*** effect (Bruce, 1961). During the first few days after mating, exposure of the mated female mouse to the urinary odor of a strange male prevents implantation by blocking the required secretion of prolactin and progesterone. One of the interesting unresolved questions about the Bruce effect concerns the olfactory recognition that permits the female to distinguish between the odor of a male that is familiar and one that is not (Brennan et al., 1990).

Stressful Environments. Reproductive activities of both females and males are suppressed during chronic exposure to stressors (Ramaley, 1981). Preventing ovulation and the possible onset of pregnancy when females reside in stressful environments has beneficial evolutionary consequences. Stress-activated mechanisms to inhibit female reproduction include peripheral effects on the ovary and uterus as well as CNS effects to prevent secretion of LHRH and LH (Rivier and Rivest, 1991).

Undernutrition constitutes a specific type of stressor that can antagonize the secretion of LH in many species, including prepubertal heifers and ewe lambs (Kurz et al., 1990; Ebling et al., 1990). In adult cows that are thin and/or underfed, postpartum resumption of the LH secretion necessary for initiation of ovarian cycles is suppressed (Wright et al., 1987). Moreover, suckling stimuli appear to exert a greater inhibition of LH secretion in sows and cows that are thin and/or underfed (Booth, 1990; Williams, 1990). Postpartum reinitiation of ovulation in ewes does not appear to be suppressed by suckling (Malven and Hudgens, 1987), perhaps because seasonal regulation of ovulation evolved in this species. In postpartum dairy cows, which are milked rather than suckled, ovulation occurs within a few weeks after parturition, but in this species inadequate nutrition and/or thin body condition are associated with a failure to express appropriate estrous behaviors during these early ovulations (Butler and Smith, 1989). In summary, thin and/or undernourished females of many ungulate species are delayed in their attainment of puberty and in their resumption of fertile ovulations after parturition.

Biotechnology of Female Reproduction

Suppression, enhancement, and more precise control of reproduction in the female have all, at one time or another, been objectives of biotechnology. The development of orally active estrogen-like and progesterone-like compounds led to their successful use as ovulation-blocking contraceptive agents in women and animals. Much neuroendocrine research was directed toward understanding the site and mechanism of action of these contraceptive steroids, and as stated earlier in this chapter, there is evidence for LH-inhibitory and LHRH-inhibitory actions at the levels of the gonadotroph and the hypothalamus, respectively.

Orally active progesterone-like compounds have been widely used to control ovulation times in farm animals. Administration of these compounds to a group of females for several weeks prevents ovulation until the compound is withdrawn. After removal of the exogenous progesterone-like compound, all the females initiate together a proestrous period followed by a relatively synchronous ovulation. Fertility at the synchronized ovulation is sometimes equivalent to that observed at non-synchronized ovulations. Another successful method to produce synchronous ovulations is to inject a group of cyclic females with prostaglandin-F_2 alpha. This compound produces luteolysis of all corpora lutea irrespective of their age and synchronous removal of endogenous progesterone from the circulation. The resulting proestrous development of ovarian follicles and ovulation occurs synchronously just as it occurs after removal of exogenous progesterone-like compounds.

Enhancement of female reproductive capacity is an important objective whether it applies to individual women or groups of agricultural animals. Exogenous gonadotropins can be used in females to stimulate follicular development, and mature oocytes can be collected either before or after fertilization. Embryos resulting from either *in vivo* or *in vitro* fertilization of the stimulated oocyte can be transferred into the uterus of individual recipients (e.g., surrogates or oocyte donors with damaged oviducts) for subsequent development to term.

Increases in the number of offspring per female may also be achieved by techniques that do not involve embryo transfer. For example, immunoneutralization of blood-borne androstenedione or inhibin in ewes increased the number of follicles ovulated at each estrus as well as the number of offspring born (Martensz and Scaramuzzi, 1979; Meyer et al., 1991). There are also substantial differences in ovulation rates between breeds of sheep and pigs, and many researchers have sought to transfer this biologic advantage to other breeds using traditional and non-traditional methods of breeding.

The suckled beef cow constitutes a particularly difficult problem because resumption of fertile ovulations after parturition is often delayed, especially when the cows are undernourished. The wide variety of treatments, which have been tested and/or are being commercially used, probably reflects the fact that no single solution has been discovered. Obvious solutions such as improved nutrition and weaning of the calf have economic disadvantages. Pulsatile administration of LHRH or one of its analogues has only sometimes hastened the onset of fertile ovulations, and practical delivery systems for frequent pulsatile administration remain to be perfected. Exogenous gonadotropins require less frequent injections than LHRH, but the cost/benefit relationship of such treatments needs to be improved. Another approach has been to suppress the LHRH-LH axis of anestrous cows through temporary subcutaneous implantation of compounds with the biological activity of progesterone. Removal of this progestational implant followed by one injection of estrogen often results in activation of gonadotropin secretion and resumption of ovarian cycles. In summary, termination of postpartum anestrus in suckled beef cows can be hastened by several treatments, but additional research is required for a full and practical solution to this problem.

REFERENCES

Baird, D.T., B.K. Campbell, G.E. Mann, and A.S. McNeilly. Inhibin and oestradiol in the control of FSH secretion in the sheep. *J. Reprod. Fertil.* (1991) Suppl. 43:125-138.

Basseti, S.G., S.J. Winters, H.S. Keeping, and A.J. Zeleznik. Serum immunoreactive inhibin levels before and after lutectomy in the cynomologus monkey (*Macaca fascicularis*). *J. Clin. Endocrinol. Metabol.* (1990) 70:590-594.

Bill, C.H., and P.L. Keyes. 17β-estradiol maintains normal function of corpora lutea throughout pseudopregnancy in hypophysectomized rabbits. *Biol. Reprod.* (1983) 28:608-617.

Blake, C.A., and R. Kelch. Administration of antiluteinizing hormone-releasing hormone serum to rats: Effects on periovulatory secretion of luteinizing hormone and follicle-stimulating hormone. *Endocrinology* (1981) 109:2175-2179.

Blake, C.A., R.I. Weiner, R.A. Gorski, and C.H. Sawyer. Secretion of pituitary luteinizing hormone and follicle stimulating hormone in female rats made persistently estrous or diestrous by hypothalamic deafferentation. *Endocrinology* (1972) 90:855-860.

Blaustein, J.D., and J.C. Turcotte. Estradiol-induced progestin receptor immunoreactivity is found only in estrogen receptor-immunoreactive cells in guinea pig brain. *Neuroendocrinology* (1989) 49:454-461.

Booth, P.J. Metabolic influences on hypothalamic-pituitary-ovarian function in the pig. In: Cole, D.J.A., G.R. Foxcroft, and B.J. Weir (eds). *Control of Pig Reproduction III.* (*J. Reprod. Fertil.* Suppl. 40, 1990) pg 89-100.

Brennan, P., H. Kaba, and E.B. Keverne. Olfactory recognition: A simple memory system. *Science* (1990) 250:1223-1226.

Britt, J.H., K.L. Esbenshade, and A.J. Ziecik. Roles of estradiol and gonadotropin-releasing hormone in controlling negative and positive feedback associated with the luteinizing hormone surge in ovariectomized pigs. *Biol. Reprod.* (1991) 45:478-485.

Bruce, H.M. Time relations in the pregnancy-block induced in mice by strange males. *J. Reprod. Fertil.* (1961) 2:138-142.

Butler, W.R., and R.D. Smith. Interrelationships between energy balance and postpartum reproductive function in dairy cattle. *J. Dairy Sci.* (1989) 72:767-783.

Camp, T.A., J.O. Rahal, and K.E. Mayo. Cellular localization and hormonal regulation of follicle-stimulating hormone and luteinizing hormone receptor messenger RNAs in the rat ovary. *Mol. Endocrinol.* (1991) 5:1405-1417.

Challis, J.R.G., and D.M. Olson. Parturition. In: Knobil, E., and J.D. Neill (eds). *The Physiology of Reproduction, Vol. 2* (Raven Press, New York, 1988), pg 2177-2216.

Clarke, I.J., and J.T. Cummins. The temporal relationship between gonadotropin releasing hormone (GnRH) and luteinizing hormone (LH) secretion in ovariectomized ewes. *Endocrinology* (1982) 111:1737-1739.

Clarke, I.J., and J.T. Cummins. Direct pituitary effects of estrogen and progesterone on gonadotropin secretion in the ovariectomized ewe. *Neuroendocrinology* (1984) 39:267-274.

Clarke, I.J., J.T. Cummins, M.E. Crowder, and T.M. Nett. Long-term negative feedback effects of oestrogen and progesterone on the pituitary gland of the long-term ovariectomized ewe. *J. Endocrinol.* (1989a) 120:207-214.

Clarke, I.J., J.T. Cummins, M. Jenkin, and D.J. Phillips. The oestrogen-induced surge of LH requires a signal pattern of gonadotropin-releasing hormone input to the pituitary gland in the ewe. *J. Endocrinol.* (1989b) 122:127-134.

Clayton, R.N. Gonadotropin-releasing hormone: its actions and receptors. *J. Endocrinol.* (1989) 120:11-19.

Cogen, P.H., J.L. Antunes, K.M. Louis, I. Dyrenfurth, and M. Ferin. The effects of anterior hypothalamic disconnection on gonadotropin secretion in the female rhesus monkey. *Endocrinology* (1980) 107:677-683.

Concannon, P.W., D.H. Lein, and B.G. Hodgson. Self-limiting reflex luteinizing hormone release and sexual behavior during extended period of unrestricted copulatory activity in estrous domestic cats. *Biol. Reprod.* (1989) 40:1179-1187.

Corrigan, A.Z., L.M. Bilezikjian, R.S. Carroll, L.N. Bald, C.H. Schmelzer, B.M. Fendly, A.J. Mason, W.W. Chin, R.H. Schwall, and W. Vale. Evidence for an autocrine role of activin B within rat anterior pituitary cultures. *Endocrinology* (1991) 128:1682-1684.

Culler, M.D., and A. Negro-Vilar. Endogenous inhibin suppresses only basal follicle-stimulating hormone secretion but suppresses all parameters of pulsatile luteinizing hormone secretion in the diestrous female rat. *Endocrinology* (1989) 124:2944-2953.

DePaolo, L.V., L.N. Bald, and B.M. Fendly. Passive immunoneutralization with a monoclonal antibody reveals a role for endogenous activin-B in mediating FSH hypersecretion during estrus and following ovariectomy of hypophysectomized, pituitary-grafted rats. *Endocrinology* (1992) 130:1741-1743.

Diekman, M.A., W.E. Trout, and L.L. Anderson. Serum profiles of LH, FSH and prolactin from 10 weeks of age until puberty in gilts. *J. Anim. Sci.* (1983) 56:139-145.

Dodson, S.E., M.P. Abbott, and W. Haresign. Comparison of bioassay and radioimmunoassay data for study of changes in the pattern of LH secretion from birth to puberty in the heifer. *J. Reprod. Fertil.* (1988a) 82:539-543.

Dodson, S.E., B.J. McLeod, W. Haresign, A.R. Peters, and G.E. Lamming. Endocrine changes from birth to puberty in the heifer. *J. Reprod. Fertil.* (1988b) 82:527-538.

Ebling, F.J.P., R.I. Wood, F.J. Karsch, L.A. Vannerson, J.M. Suttie, D.C. Bucholtz, R.E. Schall, and D.L. Foster. Metabolic interfaces between growth and reproduction. III. Central mechanisms controlling pulsatile luteinizing hormone secretion in the nutritionally growth-limited female lamb. *Endocrinology* (1990) 126:2719-2727.

Evans, R.M. The steroid and thyroid hormone receptor superfamily. *Science* (1988) 240:889-895.

Everett, J.W., L.C. Krey, and L. Tyrey. The quantitative relationship between electrochemical preoptic stimulation and LH release in proestrous versus late-diestrous rats. *Endocrinology* (1973) 93:947-953.

Everett, J.W., C.H. Sawyer, and J.E. Markee. A neurogenic timing factor in control of the ovulating discharge of luteinizing hormone in the cyclic rat. *Endocrinology* (1949) 44:234-250.

Fink, G., and M.G. Jamieson. Immunoreactive luteinizing hormone releasing factor in rat stalk blood: effects of electrical stimulation of the medial preoptic area. *J. Endocrinol.* (1976) 68:71-87.

Fortune, J.E. Bovine theca and granulosa cells interact to promote androgen production. *Biol. Reprod.* (1986) 35:292-299.

Fortune, J.E., J. Sirois, A.M. Turzillo, and M. Lavoir. Follicle selection in domestic ruminants. *J. Reprod. Fertil.* (1991) Suppl. 43:187-198.

Foster, D.L. Puberty in the female sheep. In: Knobil, E., and J.D. Neill (eds). *The Physiology of Reproduction, Vol.* 2. (Raven Press, New York, 1988), pg 1739-1762.

Foster, D.L., F.J.P. Ebling, L.A. Vannerson, J.M. Suttie, T.D. Landefeld, V. Padmanabhan, A.M. Micka, D.C. Bucholtz, R.I. Wood, and D.E. Fenner. Toward an understanding of interfaces between nutrition and reproduction: the growth-restricted lamb as a model. In: Pirke, L., W. Wuttke, and U. Schweiger (eds). *Behavioral Influences on the Menstrual Cycle in Health and Disease.* (Springer-Verlag, Heidelberg, 1989), pg 50-65.

Fox, S.R., R.E. Harlan, B.D. Shivers, and D.W. Pfaff. Chemical characterization of neuroendocrine targets for progesterone in the female rat brain and pituitary. *Neuroendocrinology* (1990) 51:276-283.

Ginther, O.J., L. Knopf, and J.P. Kastelic. Temporal associations among ovarian events in cattle during oestrous cycles with two and three follicular waves. *J. Reprod. Fertil.* (1989) 87:223-230.

Goodman, R.L. A quantitative analysis of the physiological role of estradiol and progesterone in the control of tonic and surge secretion of luteinizing hormone in the rat. *Endocrinology* (1978) 102:142-150.

Gore-Langton, R.E., and D.T. Armstrong. Follicular steroidogenis and its control. In: Knobil, E., and J.D. Neill (eds). *The Physiology of Reproduction, Vol. 1.* (Raven Press, New York, 1988), pg 331-385.

Greenwald, G.S., and P.F. Terranova. Follicular selection and its control. In: Knobil, E., and J.D. Neill (eds). *The Physiology of Reproduction, Vol. 1.* (Raven Press, New York, 1988), pg 387-445.

Grumbach, M.M., P.C. Sizonenko, and M.L. Aubert. *Control of the Onset of Puberty.* (Williams & Wilkins, Baltimore, 1990), pp 710.

Halasz, B., and R.A. Gorski. Gonadotropic hormone secretion in female rats after partial or total interruption of neural afferents to the medial basal hypothalamus. *Endocrinology* (1967) 80:608-622.

Harned, M.A., and L.E. Casida. Some postpartum reproductive phenomena in the domestic rabbit. *J. Anim. Sci.* (1969) 28:785-788.

Haynes, N.B., G.E. Lamming, K.P. Yang, A.N. Brooks, and A.D. Finnie. Endogenous opioid peptides and farm animal reproduction. *Oxford Rev. Reprod. Biol.* (1989) 11:111-145.

Hilliard, J., R. Penardi, and C.H. Sawyer. A functional role for 20α-hydroxy-preg-4-en-3-one in the rabbit. *Endocrinology* (1967) 80:901-909.

Hoffmann, J.C., and N.B. Schwartz. Timing of post-partum ovulation in the rat. *Endocrinology* (1965) 76:620-625.

Hughes, P.E., G.P. Pearce, and A.M. Paterson. Mechanisms mediating the stimulatory effects of the boar on gilt reproduction. In: Cole, D.J.A., G.R. Foxcroft, and B.J. Weir (eds). *Control of Pig Reproduction III.* (*J. Reprod. Fertil.* Suppl. 40, 1990) pg 323-341.

Ireland, J.J. Control of follicular growth and development. In: Niswender, G.D., D.T. Baird, and J.K. Findlay (eds). *Reproduction in Domestic Ruminants.* (*J. Reprod. Fertil.* Suppl. 34, 1987) pg 39-54.

Jackson, G.L., D. Kuehl, K. McDowell, and A. Zaleski. Effect of hypothalamic deafferentation on secretion of luteinizing hormone in the ewe. *Biol. Reprod.* (1978) 18:808-819.

Kaasjager, W.A., D.M. Woodbury, J.A.M.J. Van Dieten, and G.P. Van Rees. The role played by the preoptic region and the hypothalamus in spontaneous ovulation and ovulation induced by progesterone. *Neuroendocrinology* (1971) 7:54-64.

Kalra, P.S. and S.M. McCann. The stimulatory effect on gonadotropin release of implants of estradiol and progesterone in certain sites in the central nervous system. *Neuroendocrinology* (1975) 19:289-302.

Karsch, F.J., J.T. Cummins, G.B. Thomas, and I.J. Clarke. Steroid feedback inhibition of pulsatile secretion of gonadotropin-releasing hormone in the ewe. *Biol. Reprod.* (1987) 36:1207-1218.

Kaufman, J.M., J.S. Kesner, R.C. Wilson, and E. Knobil. Electrophysiological manifestation of luteinizing hormone-releasing hormone pulse generator activity in the rhesus monkey: influence of alpha-adrenergic and dopaminergic blocking agents. *Endocrinology* (1985) 116:1327-1333.

Kesner, J.S., M.J. Estienne, R.R. Kraeling, and G.B. Rampacek. Luteinizing hormone and prolactin secretion in hypophysial-stalk-transected pigs given estradiol and pulsatile gonadotropin-releasing hormone. *Neuroendocrinology* (1989) 49:502-508.

Kesner, J.S., R.C. Wilson, J.M. Kaufman, J. Hotchkiss, Y. Chen, H. Yamamoto, R.R. Pardo, and E. Knobil. Unexpected responses of the hypothalamic gonadotropin-releasing hormone "pulse-generator" to physiological estradiol inputs in the absence of the ovary. *Proc. Natl. Acad. Sci.* (1987) 84:8745-8749.

Knobil, E. The neuroendocrine control of the menstrual cycle. *Rec. Prog. Horm. Res.* (1980) 36:53-88.

Koves, K., and B. Halasz. Location of the neural structures triggering ovulation in the rat. *Neuroendocrinology* (1970) 6:180-193.

Kurz, S.G., R.M. Dyer, Y. Hy, M.D. Wright, and M.L. Day. Regulation of luteinizing hormone secretion in prepubertal heifers fed an energy-deficient diet. *Biol. Reprod.* (1990) 43:450-456.

Leranth, C., N.J. MacLusky, T.J. Brown, E.C. Chen, D.E. Redmond, and F. Naftolin. Transmitter content and afferent connections of estrogen-sensitive progestin receptor-containing neurons in the primate hypothalamus. *Neuroendocrinology* (1992) 55:667-682.

Levine, J.E., and V.D. Ramirez. In vivo release of luteinizing hormone-releasing hormone estimated with push-pull cannulae from the mediobasal hypothalami of ovariectomized, steroid-primed rats. *Endocrinology* (1980) 107:1782-1790.

Lin, W.W., and V.D. Ramirez. Infusion of progestins into the hypothalamus of female New Zealand White rabbits: Effect on in vivo luteinizing hormone-releasing hormone release as determined with push-pull perfusion. *Endocrinology* (1990) 126:261-272.

Lincoln, D.W., H.M. Fraser, G.A. Lincoln, G.B. Martin, and A.S. McNeilly. Hypothalamic pulse generators. *Rec. Prog. Horm. Res.* (1985) 41:369-411.

Lipner, H., and N.L. Cross. Morphology of the membrana granulosa of the ovarian follicle. *Endocrinology* (1968) 82:638-641.

Malven, P.V., and R.E. Hudgens. Naloxone-reversible inhibition of luteinizing hormone in postpartum ewes: effects of suckling and season. *J. Anim. Sci.* (1987) 65:196-202.

Mann, G.E., B.K. Campbell, A.S. McNeilly, and D.T. Baird. Effects of passively immunizing ewes against inhibin and oestradiol during the follicular phase of the oestrous cycle. *J. Endocrinol.* (1990) 125:417-424.

Mann, G.E., A.S. McNeilly, and D.T. Baird. Hormone production in vivo and in vitro from follicles at different stages of the oestrous cycle in the sheep. *J. Endocrinol.* (1992) 132:225-234.

Martensz, N.D., and R.J. Scaramuzzi. Plasma concentrations of luteinizing hormone, follicle-stimulating hormone and progesterone during the breeding season in ewes immunized against androstenedione or testosterone. *J. Endocrinol.* (1979) 81:249-259.

Martin, G.B., C.M. Oldham, Y. Cognie, and D.T. Pearce. The physiological responses of anovulatory ewes to the introduction of rams - A review. *Livestock Prod. Sci.* (1986) 15:219-247.

McCormack, J.T., T.M. Plant, D.L. Hess, and E. Knobil. The effect of luteinizing hormone releasing hormone (LHRH) antiserum administration on gonadotropin secretion in the rhesus monkey. *Endocrinology* (1977) 100:663-667.

Meyer, R.L., K.M. Carlson, J. Rivier, and J.E. Wheaton. Antiserum to an inhibin alpha-chain peptide neutralizes inhibin bioactivity and increases ovulation rate in sheep. *J. Anim. Sci.* (1991) 69:747-754.

Mills, T., A. Copland, and K. Osteen. Factors affecting the postovulatory surge of FSH in the rabbit. *Biol. Reprod.* (1981) 25:530-535.

Moenter, S.M., A. Caraty, A. Locatelli, and F.J. Karsch. Pattern of gonadotropin-releasing hormone (GnRH) secretion leading up to ovulation in the ewe: existence of preovulatory GnRH surge. *Endocrinology* (1991) 129:1175-1182.

Moor, R.M., and R.F. Seamark. Cell signaling, permeability, and microvasculatory changes during antral follicle development in mammals. *J. Dairy Sci.* (1986) 69:927-943.

Morrell, J.I., J.F. McGinty, and D.W. Pfaff. A subset of β-endorphin- or dynorphin-containing neurons in the medial basal hypothalamus accumulates estradiol. *Neuroendocrinology* (1985) 41:417-426.

Neill, J.D., M.E. Freeman, and S.A. Tillson. Control of the proestrus surge of prolactin and luteinizing hormone secretion by estrogens in the rat. *Endocrinology* (1971) 89:1448-1453.

Nishihara, M., H. Hiruma, and F. Kimura. Interactions between the noradrenergic and opioid peptidergic systems in controlling the electrical activity of luteinizing hormone-releasing hormone pulse generator in ovariectomized rats. *Neuroendocrinology* (1991) 54:321-326.

Niswender, G.D., and T.M. Nett. The corpus luteum and its control. In: Knobil, I., and J. Neill (eds). *The Physiology of Reproduction* (Raven Press, New York, 1988), pg 489-525.

Ojeda, S.R., L.I. Aguado, and S. Smith. Neuroendocrine mechanisms controlling the onset of female puberty: the rat as a model. *Neuroendocrinology* (1983) 37:306-313.

Olster, D.H., and J.D. Blaustein. Immunocytochemical colocalization of progestin receptors and β-endorphin or enkephalin in the hypothalamus of female guinea pigs. *J. Neurobiol.* (1990) 21:768-780.

Over, R., J. Cohen-Tannoudji, M. Dehnhard, R. Clase, and J.P. Signoret. Effect of pheromones from male goats on LH-secretion in anoestrous ewes. *Physiol. Behav.* (1990) 48:665-668.

Padmanabhan, V., K.M. Reno, M. Borondy, T.D. Landfeld, F.J.P. Ebling, D.L. Foster, and I.Z. Beitins. Effect of nutritional repletion on pituitary and serum follicle-stimulating hormone isoform distribution in growth-retarded lambs. *Biol. Reprod.* (1992) 46:964-971.

Pant, H.C., H. Dobson, and W.R. Ward. Effect of active immunization against oestrogens on plasma gonadotropins in the ewe and the response to synthetic oestrogen or LH. *J. Reprod. Fertil.* (1978) 53:241-248.

Park, O.K., and K.E. Mayo. Transient expression of progesterone receptor messenger RNA in ovarian granulosa cells after the preovulatory luteinizing hormone surge. *Mol. Endocrinol.* (1991) 5:967-978.

Parker, M.G. Gene regulation by steroid hormones. In: Cook, B.A., R.J.B. King, and H.J. van der Molen (eds). *Hormones and Their Actions. Part I* (Elsevier, Amsterdam, 1988), pg 39-59.

Pau, K.Y.F., D.E. Kuehl, and G.L. Jackson. Effects of frontal hypothalamic deafferentation on luteinizing hormone secretion and seasonal breeding in the ewe. *Biol. Reprod.* (1982) 27:999-1009.

Plant, T.M. Puberty in primates. In: Knobil, I., and J. Neill (eds). *The Physiology of Reproduction, Vol. 1.* (Raven Press, New York, 1988), pg 1763-1788.

Ramaley, J.A. Stress and fertility. In: Gilmore, D., and B. Cook (eds). *Environmental Factors in Mammal Reproduction* (University Park Press, Baltimore, 1981), pg 127-141.

Ramirez, V.D., D.E. Dluzen, and F.C. Ke. Effects of progesterone and its metabolites on neuronal membranes. In: Chadwick, D., and K. Widdows (eds). *Steroids and Neuronal Activity.* (Ciba Foundation Symp. 153; John Wiley & Sons, Chichester, UK, 1990), pg 125-141.

Rasmussen, D.D., and P.V. Malven. Characterization of cephalic arteriovenous LH differences by continuous sampling in ovariectomized sheep. *Neuroendocrinology* (1982) 34:415-420.

Rivier, C., H. Meunier, V. Roberts, and W. Vale. Inhibin: role and secretion in the rat. *Rec. Prog. Horm. Res.* (1990) 46:231-257.

Rivier, C., and S. Rivest. Effect of stress on the activity of the hypothalamic-pituitary-gonadal axis: peripheral and central mechanisms. *Biol. Reprod.* (1991) 45:523-532.

Roberts, V. Meunier, J. Vaughan, J. Rivier, C. Rivier, W. Vale, and P. Sawchenko. Production and regulation of inhibin subunits in pituitary gonadotrophs. *Endocrinology* (1989) 124:552-554.

Roberts, V.J., C.A. Peto, W. Vale, and P. Sawchenko. Inhibin/activin subunits are costored with FSH and LH in secretory granules of the rat anterior pituitary gland. *Neuroendocrinology* (1992) 56:214-224.

Ryan, K.D., R.L. Goodman, F.J. Karsch, S.J. Legan, and D.L. Foster. Patterns of circulating gonadotropins and ovarian steroids during the first periovulatory period in the developing sheep. *Biol. Reprod.* (1991) 45:471-477.

Sar, M., A. Sahu, W.R. Crowley, and S.P. Kalra. Localization of neuropeptide-Y immunoreactivity in estradiol-concentrating cells in the hypothalamus. *Endocrinology* (1990) 127:2752-2756.

Schwartz, N.B., K. Krone, W.L. Talley, and C.A. Ely. Administration of antiserum to ovine FSH in the female rat: failure to influence immediate events of the cycle. *Endocrinology* (1973) 92:1165-1174.

Sherwood, O.D. Relaxin. In: Knobil, E., and J.D. Neill (eds). *The Physiology of Reproduction, Vol. 1* (Raven Press, New York, 1988), pg 585-673.

Silverman, J.A. The gonadotropin-releasing hormone (GnRH) neuronal systems: immunocytochemistry. In: Knobil, E., and J. Neill (eds). *The Physiology of Reproduction, Vol. 1.* (Raven Press, New York, 1988), pg 1283-1304.

Silvia, W.J., G.S. Lewis, J.A. McCracken, W.W. Thatcher, and L. Wilson. Hormonal regulation of uterine secretion of prostaglandin F_2alpha during luteolysis in ruminants. *Biol. Reprod.* (1991) 45:655-663.

Soper, B.D., and R.F. Weick. Hypothalamic and extrahypothalamic mediation of pulsatile discharges of luteinizing hormone in the ovariectomized rat. *Endocrinology* (1980) 106:348-355.

Suttie, J.M., D.L. Foster, B.A. Veenvliet, T.R. Manley, and I.D. Corson. Influence of food intake but independence of body weight on puberty in female sheep. *J. Reprod. Fertil.* (1991) 92:33-39.

Thiery, J.C., and G.B. Martin. Neurophysiological control of the secretion of gonadotropin-releasing hormone and luteinizing hormone in the sheep - A review. *Reprod. Fertil. Devel.* (1991) 3:137-173.

Trout, W.E., and P.V. Malven. Effects of exogenous estradiol-17β and progesterone on naloxone-reversible inhibition of the release of luteinizing hormone in ewes. *J. Anim. Sci.* (1987) 65:1602-1609.

Tsai, M.J., and B.W. O'Malley. Mechanisms of gene transcription by steroid receptors. In: Cohen, P., and J.G. Foulkes (eds). *The Hormonal Control of Gene Transcription.* (Elsevier, Amsterdam, 1991), pg 101-116.

Weesner, G.D., and P.V. Malven. Intracerebral immunoneutralization of beta-endorphin and met-enkephalin disinhibits release of pituitary luteinizing hormone in sheep. *Neuroendocrinology* (1990) 52:382-388.

Weiner, R.I., P.R. Findell, and C. Kordon. Role of classic and peptide neuromediators in the neuroendocrine regulation of LH and prolactin. In: Knobil, E., and J.D. Neill (eds). *The Physiology of Reproduction, Vol. 1.* (Raven Press, New York, 1988), pg 1235-1281.

Wetsel, W.C., M.M. Valencia, I. Merchenthaler, Z. Liposits, F.J. Lopez, R.I. Weiner, P.L. Mellon, and A. Negro-Vilar. Intrinsic pulsatile secretory activity of immortalized luteinizing hormone-releasing hormone-secreting neurons. *Proc. Natl. Acad. Sci.* (1992) 89:4149-4153.

Wildt,L., J.S. Hutchison, G. Marshall, C.R. Pohl, and E. Knobil. On the site of action of progesterone in the blockade of the estradiol-induced gonadotropin discharge in the rhesus monkey. *Endocrinology* (1981) 109:1293-1294.

Williams, G.L. Suckling as a regulator of postpartum rebreeding in cattle: a review. *J. Anim. Sci.* (1990) 68:831-852.

Williams, C.L., M. Nishihara, J.C. Thalabard, K.T. O'Byrne, P.M. Grosser, J. Hotchkiss, and E. Knobil. Duration and frequency of multiunit electrical activity associated with the hypothalamic gonadotropin releasing hormone pulse generator in the rhesus monkey: differential effects of morphine. *Neuroendocrinology* (1990) 52:225-228.

Wright, I.A., S.M. Rhind, A.J.F. Russel, T.K. Whyte, A.J. McBean, and S.R. McMillen. Effects of body condition, food intake and temporary calf separation on the duration of the post-partum anoestrous period and associated LH, FSH and prolactin concentrations in beef cows. *Anim. Prod.* (1987) 45:395-402.

Ying, S.Y., J. Czvik, A. Becker, N. Ling, N. Ueno, and R. Guillemin. Secretion of follicle-stimulating hormone and production of inhibin are reciprocally related. *Proc. Natl. Acad. Sci.* (1987) 84:4631-4635.

Chapter 13

ADRENAL MEDULLA

The ***chromaffin*** cells of the adrenal medulla are innervated neuroendocrine cells that secrete catecholaminergic neurohormones into the blood. Because of the anatomical isolation of the adrenal glands and the easily observed physiological responses to its medullary hormones, the neuroendocrine status of the adrenal medulla has been appreciated since the beginning of this century. Neurally derived chromaffin cells are innervated by sympathetic preganglionic axons, and they constitute a neuroendocrine transducer converting the neural information of those axons into hormonal output by chromaffin cells. These neuroendocrine cells of the adrenal medulla are specifically represented as example A of secretomotor innervation in Figure 1-1 (Chapter 1). Knowledge about the adrenal medulla developed during the early part of this century in parallel with the concept of ***homeostasis***, which can be defined as maintenance of a steady internal state in the face of various internal and external challenges. The sympathetic nervous system, including the adrenal medulla, was also found to initiate and coordinate the ***fight-or-flight*** responses to an external threat. Although these responses evolved in order to improve the chances of individual animals surviving the threat, their role in today's animals and humans is not uniformly beneficial. It has also been difficult to determine exactly which biological responses depend entirely on blood-borne hormones from the adrenal medulla or what is the relative role of these same molecules produced by widely distributed sympathetic axons for local action or secretion into blood. The present chapter will address some of these questions as well as neuroendocrine integration involving the adrenal medulla.

Secretions of the Adrenal Medulla

The primary secretory products of adrenal medullary cells are the catecholamines, ***norepinephrine*** and ***epinephrine*** (referred to in the British literature as noradrenaline and adrenaline). The biosynthetic steps for synthesis of these catecholamines are presented in Figure 13-1 (Kirshner, 1975). The chromaffin cells take up tyrosine from the blood and, using the enzyme ***tyrosine hydroxylase,*** convert tyrosine into dihydroxyphenylalanine (abbreviated DOPA). This conversion is probably the rate-limiting step in the entire biosynthesis, and it is highly regulated by acetylcholine released from innervating sympathetic axons. This step is also subject to feedback inhibition by end-products of catecholamine biosynthesis. In addition, the activity of tyrosine hydroxylase in chromaffin cells may be stimulated by ACTH or ACTH-stimulated compounds from the adrenal cortex. The DOPA formed by tyrosine hydroxylase is readily converted into dopamine by the nonspecific enzyme, aromatic L-amino acid decarboxylase (formerly called DOPA decarboxylase). Dopamine is converted into norepinephrine by the enzyme dopamine β-hydroxylase, a mixed function oxidase enzyme that adds one hydroxyl group to dopamine. This enzyme may also be stimulated by ACTH or alternatively by ACTH-stimulated compounds originating in adrenocortical cells. Interestingly, the conversion of dopamine into

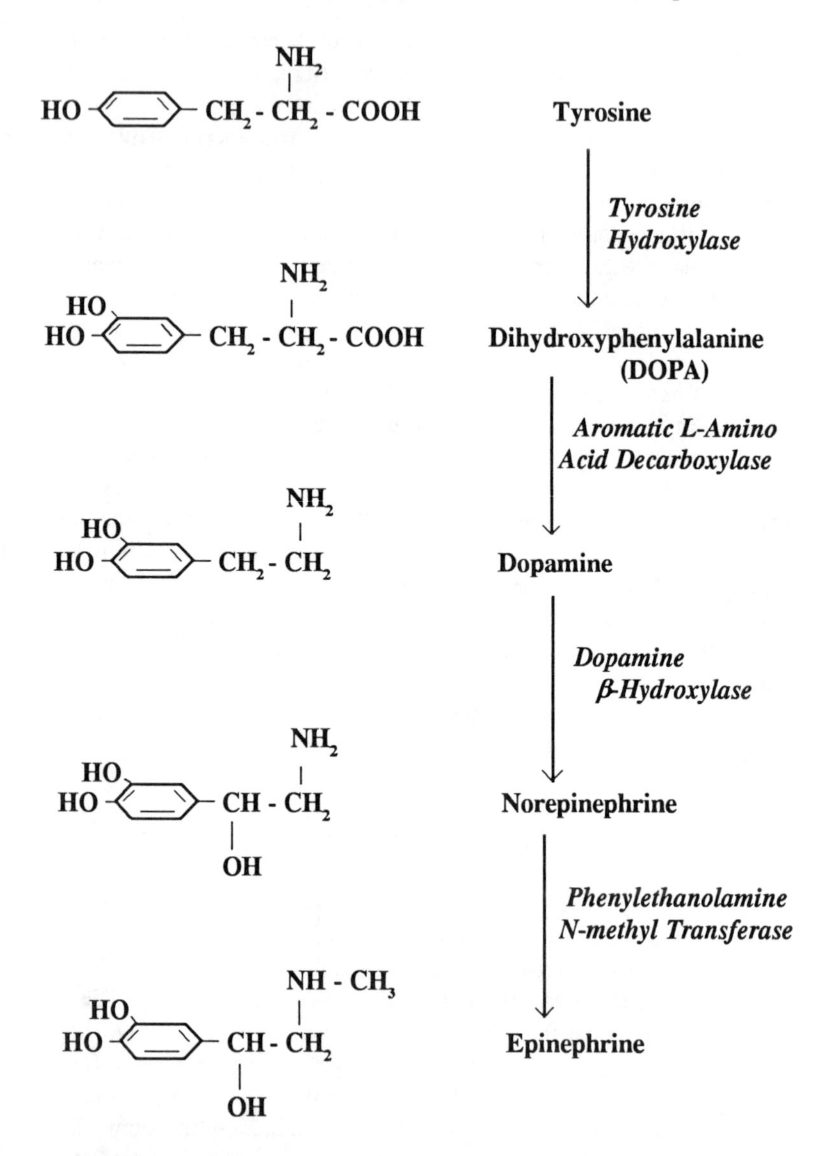

Figure 13-1. Biosynthesis of norepinephrine and epinephrine.
The chemical structures (left column) and the names of the compounds and the enzymes (right column) involved in biosynthesis of norepinephrine and epinephrine.

norepinephrine occurs within intracellular secretory vesicles (also known as chromaffin granules) resulting in the cosecretion of dopamine β-hydroxylase with the other contents of the secretory vesicles. The biosynthetic steps leading to norepinephrine occur in *all* the chromaffin cells of the adrenal medulla. However, a subset of these chromaffin cells *also* possess the enzyme phenylethanolamine N-methyl transferase (PNMT) that converts norepinephrine into epinephrine by adding a methyl group. This methylation uses methionine as a donor molecule and occurs in the cytoplasm, rather than within the secretory vesicle where norepinephrine is formed. In mammals, the PNMT enzyme is almost exclusively localized in the adrenal medulla with only small quantities in the brain and in the sympathetic nerves of the heart.

The subset of chromaffin cells that contain PNMT and synthesize epinephrine are regionally localized in a species-specific manner. In ungulate species, these cells are located in the periphery of the medulla, often in finger-like interdigitations with adrenocortical cells. In other species, the distribution of epinephrine-containing chromaffin cells is more diffuse and variable throughout the medulla (Weiner, 1975; McMillen et al., 1988). However, there does appear to be an association between development of epinephrine-containing medullary cells and adrenocortical function across species. First, the relative size of the adrenal cortex is directly related to the proportion of medullary cells that synthesize epinephrine. Second, very few medullary cells contain epinephrine in fetal rats and rabbit, but these cells begin to appear postnatally as adrenocortical function increases (Weiner, 1975). In fetal sheep, the degree of interdigitation between peripherally located epinephrine-containing cells of the medulla and the inner cells of the adrenal cortex could be enhanced in hypophysectomized fetuses by the administration of ACTH (Coulter et al., 1991).

In addition to catecholamines, the secretory vesicles of chromaffin cells contain a variety of peptides, some of which may be secreted into the blood. Opioid neuropeptides derived from pre-proenkephalin (see Figure 5-1) are abundant in some chromaffin cells. In newborn and adult sheep, these enkephalin-related peptides are found only in epinephrine-containing cells. However, a different situation exists earlier in fetal development (days 80-110) when centrally located epinephrine-lacking cells also contain enkephalin. During development to term, these chromaffin cells lose the ability to synthesize enkephalin-related peptides (McMillen et al., 1988). Interestingly, hypophysectomy retarded the disappearance of enkephalin from these cells, but exogenous ACTH failed to have any effect on enkephalin distribution in the medullary cells of hypophysectomized fetuses (Coulter et al., 1991). In other species (rat, cat, and dog), enkephalin is found in the PNMT-lacking chromaffin cells of adults. In summary, the functional role of adrenomedullary enkephalin is unclear, but in adults of some species (sheep, cow, and hamster), it is restricted to those chromaffin cells that synthesize epinephrine (McMillen et al., 1988).

Secretomotor Control of Chromaffin Cells. The sympathetic preganglionic axons that innervate the chromaffin cells originate from perikarya located in the thoracic spinal cord. These terminals release acetylcholine that acts through cholinergic receptors on the chromaffin cells to stimulate catecholamine release. The cholinergic receptors that are mainly responsible for this release are of the nicotinic subtype, although those of the muscarinic subtype also play a minor role (Ungar and Phillips, 1983). Those chromaffin cells that lack PNMT release mostly norepinephrine and perhaps some dopamine. Those cells that possess PNMT to synthesize epinephrine probably release all three catecholamines, although the exact amount of dopamine and norepinephrine release from these cells is not known.

Sympathetic preganglionic axons reach the adrenal medulla through the splanchnic nerve, which also contains sympathetic preganglionic axons projecting to other tissues. Electrophysiological investigation of antidromically identified perikarya that send axons through the splanchnic nerves failed to demonstrate any major differences between perikarya that project their axons to the adrenal medulla and those that project to other tissues (Backman et al., 1990). Among the adrenal projections, it is not known if there is anything unique about those specific axons which innervate the subset of chromaffin cells synthesizing epinephrine. However, the ratio between blood-borne norepinephrine and epinephrine is highly variable and can be altered by different provocative stimuli (Ungar and Phillips, 1983).

The mechanisms by which cholinergic activation of nicotinic receptors of chromaffin cells result in the discharge of catecholamines have been thoroughly studied (Douglas, 1975). The action of acetylcholine does not trigger action potentials in the chromaffin cells as it does in neurons but does allow movement of extracellular calcium into the cell. As depicted in Figure 3-5 for neurohypophysial axon terminals, this intracellular calcium promotes exocytosis of the secretory vesicles and release of their contents (catecholamines, neuropeptides, and dopamine β-hydroxylase). It should be noted that those aspects of Figure 3-5 pertaining to action potentials and restoration of the depolarized potential of the plasma membrane do not apply to secretion from chromaffin cells.

Neuroendocrine Integration of Adrenal Medulla

Biological Actions. Norepinephrine and epinephrine are well recognized as activators of physiological responses to emergency situations (i.e., fight-or-flight). Although epinephrine derives mainly from the adrenal medulla, norepinephrine is also produced by sympathetic postganglionic neurons and released at their widely distributed terminals either for local action or entry into blood. The biological actions of norepinephrine and epinephrine appear to be antagonistic in the arterioles of skeletal muscle and skin where blood-borne epinephrine causes vasodilation while norepinephrine (probably locally released) causes vasoconstriction. The sympathetic nervous system activation of the heart and its pumping capacity appears to be accomplished mainly through postganglionic innervation of the necessary elements, with possible reinforcement by circulating hormones from the adrenal medulla.

One particular response to emergency situations depends almost exclusively on blood-borne, adrenal-derived epinephrine. This emergency response is the induced breakdown of glycogen stored in the liver and in skeletal muscle to generate glucose needed as a metabolic substrate. This process is called ***glycogenolysis***, and it has been observed in liver cells that lack sympathetic innervation. Moreover, norepinephrine does not promote breakdown of glycogen in skeletal muscle, even though it does exert this effect in the liver (Lewis, 1975). The induced hyperglycemia resulting from glycogenolysis in the liver provides immediate benefits to the organism responding to the emergency. Mobilization of fatty acids due to breakdown of stored triglycerides (i.e., lipolysis) in adipose tissue represents another emergency response mediated by the sympathetic nervous system. However, lipolysis appears to be due primarily to postganglionic innervation of adipose tissue.

Biological actions of neuropeptides secreted by the adrenal medulla are poorly understood. Enkephalin-related peptides are secreted into adrenal effluent blood in response to stimuli that also release medullary catecholamines. Such stimuli reported to date include (1) electrical stimulation of the splanchnic nerve and (2) hypovolemia (Farrell et al., 1983; Edwards and Jones, 1989). However, cosecretion of enkephalin and catecholamine by chromaffin cells is not obligatory because they can be dissociated under certain circumstances (Edwards and Jones, 1989). A hypothesized local action of medullary enkephalin on the adjacent adrenocortical cells is inconsistent with the observation of stimulation-induced increases in release of enkephalin and catecholamines, whereas there is no change in basal or ACTH-induced secretion of cortisol in calves (Edwards et al., 1986). Some types of stress-induced analgesia in rodents require an intact adrenal medulla. Because antagonism of opioid receptors decreases the same types of stress-induced analgesia, it has been suggested that enkephalins from the adrenal gland contribute to stress-induced analgesia (Lewis et al., 1982). However, definitive evidence for hormonal actions of adrenal-derived enkephalin-related peptides is lacking.

Other neuropeptides present in substantial quantities in the secretory vesicles of chromaffin cells include chromogranin A, chromogranin B, and neuropeptide Y. Chromogranin A was first discovered in chromaffin cells, but it was later found in many different types of secretory cells. Chromogranin A may be a precursor for biologically important peptides in some types of secretory cells, but the protein remains intact in chromaffin cells (Watkinson et al., 1991). Therefore, its function in chromaffin cells is not known. Other peptides occurring in lesser concentrations in chromaffin cells include neurotensin, dynorphin, and substance P (Winkler et al., 1986). The neuropeptide, galanin, was recently identified in chromaffin cells from hamsters and guinea pigs, but not rats (Zentel et al., 1990).

Stimuli that Provoke Adrenomedullary Secretion. A wide variety of stressors (listed across the top of Figure 13-2) can provoke the secretion of epinephrine and norepinephrine by the adrenal medulla. Environments that trigger anxiety in animals and humans consistently increase adrenal medullary

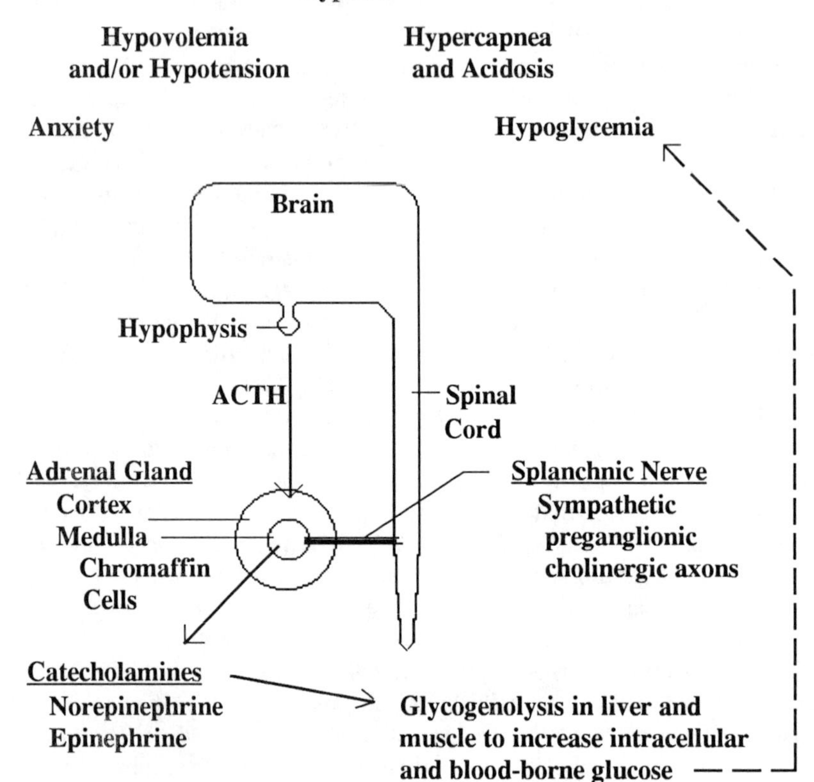

Figure 13-2. Neuroendocrine integration of catecholamine secretion from the adrenal medulla.

Schematic diagram illustrating the CNS control of the adrenal medulla via the splanchnic nerve. Factors that activate the adrenal medulla are listed across the top of the diagram.

secretion, probably through hypothalamic activation of the sympathetic nervous system. Electrical stimulation of hypothalamic areas can stimulate release of both norepinephrine and epinephrine although the ratio between the two compounds depends on the site and strength of stimulation.

Alterations in blood gases due to respiratory distress can also activate the release of epinephrine. A relative lack of oxygen in the blood (hypoxia) acts via the brain to increase sympathetic discharges to the adrenal medulla. An excess of carbon dioxide in the blood (hypercapnea) and/or low blood pH (acidosis) can also act centrally to increase sympathetic outflow to the adrenal medulla, but the stimulus of hypercapnea/acidosis may also act directly on the chromaffin cells, especially in fetal mammals (Lewis, 1975). Extreme physical exertion also stimulates adrenomedullary secretion, but this stimulation may be mediated by the induced changes in blood gases described above. Although adrenal-derived norepinephrine and epinephrine do not play important roles in the maintenance of blood pressure, a decline in blood pressure (hypotension) or blood volume (hypovolemia) will stimulate the release of both adrenomedullary hormones.

In the discussion on biological actions of catecholamines, the induced breakdown of glycogen to yield glucose was one important action that could only be accomplished by adrenomedullary secretions. Therefore, low blood glucose (hypoglycemia) represents one of the most important provocative stimuli of such secretion. Chromaffin cells are directly stimulated by hypoglycemia that is perceived in the CNS and increases sympathetic outflow to the adrenal medulla. Another possible mechanism by which hypoglycemia may stimulate chromaffin cells is through increases in the secretion of the hormone glucagon from the pancreas. Hypoglycemia is known to stimulate glucagon release, and this hormone can directly stimulate norepinephrine and epinephrine release from chromaffin cells (Lewis, 1975).

Applied Aspects of Adrenal Catecholamines

A small proportion (<1%) of humans with hypertension have a tumor (pheochromocytoma) composed of adrenomedullary chromaffin cells (Lightman, 1979). These tumorous cells may be located in the adrenal medulla or elsewhere, but they produce excessive amounts of catecholamines that create either persistent or variable hypertension. All hypertension should be controlled by a variety of available pharmacological agents, but when definitive diagnosis of pheochromocytoma is made, surgical removal may be attempted. All sources of catecholamine hypersecretion should be located because therapeutic benefits depend largely on all sources being removed.

In contrast to hypersecretion of tumor-derived catecholamines, another category of human patients suffer from a deficiency of the catecholamine dopamine in certain parts of the brain. The largest category of these dopamine-related disorders is Parkinson's disease which causes severe movement dysfunction. Although these patients benefit from drugs that enhance intracerebral dopamine, the progressive nature of Parkinson's disease and its debilitating symptoms have stimulated efforts to find an alternative or supplemental therapy. Experimentally induced movement disorders in dopamine-deficient laboratory animals can sometimes be corrected by autotransplantation of one adrenal medulla of the afflicted animal from its normal site directly into the brain (Freed et al., 1990). This procedure has also been utilized experimentally in humans with severe Parkinson's disease. Temporary reduction in the movement disorders is sometimes observed, but long-term benefits are smaller and observed less frequently (Dunnett and Richards, 1990). The technique remains controversial, and the reduction of symptoms immediately after autotransplantation may be due to dopamine secreted from the chromaffin cells. However, adrenal chromaffin cells do not survive in the brain for extended periods after autotransplantation.

Therefore, any long-term clinical benefits, which sometimes occur, cannot be due to secretions from these chromaffin cells (Quinn, 1990). Potentially longer survival of the grafted cells in the recipient brain may be possible using fetal tissue derived from the human CNS.

One application of catecholamine biotechnology in food-producing animals is based on the discovery of a unique catecholamine agonist that is orally active and possesses extreme selectivity for inducing lipolysis. A number of different molecules sharing these two important properties have been discovered, and they act upon the beta subtype of the adrenergic receptor in adipose tissue (i.e., called beta-agonists). Food-producing animals administered these compounds during the period of active growth come to have a much reduced proportion of body fat relative to muscle (Williams, 1987). Energetic efficiency (i.e., defined as conversion of dietary input into animal products) is also improved in growing animals fed beta-agonists because less energy is required per gram to synthesize muscle than fat. Human consumers of animal products also prefer those products with reduced quantities of fat as long as other qualities are maintained. Therefore, economic advantages may be realized provided that the issues of food safety can be adequately resolved.

REFERENCES

Backman, S.B., H. Sequeira-Martinho, and J.L. Henry. Adrenal versus non-adrenal sympathetic preganglionic neurones in the lower thoracic intermediolateral nucleus of the cat: Physiological properties. *Can. J. Physiol. Pharmacol.* (1990) 68:1447-1456.

Coulter, C.L., I.R. Young, C.A. Browne, and I.C. McMillen. Different roles for the pituitary and adrenal cortex in the control of enkephalin peptide localization and cortio-medullary interaction in the sheep adrenal during development. *Neuroendocrinology* (1991) 53:281-286.

Douglas, W.W. Secretomotor control of adrenal medullary secretion: synaptic, membrane, and ionic events in stimulus-secretion coupling. In: *Handbook of Physiology. Endocrinology, Section 7. Volume VI. Adrenal Gland* (Amer. Physiol. Soc., Washington, DC, 1975) pg 367-388.

Dunnett, S.B., and S.J. Richards (eds). *Neural Transplantation: From Molecular Basis to Clinical Applications.* (Prog. Brain Res. Vol. 82; Elsevier, Amsterdam, 1990), pp 743.

Edwards, A.V., D. Hansell, and C.T. Jones. Effects of synthetic adrenocorticotrophin on adrenal medullary responses to splanchnic nerve stimulation in conscious calves. *J. Physiol.* (1986) 379:1-16.

Edwards, A.V., and C.T. Jones. Adrenal responses to splanchnic nerve stimulation in conscious calves given naloxone. *J. Physiol.* (1989) 418:339-351.

Farrell, L.D., T.S. Harrison, and L.M. Demers. Immunoreactive met-enkephalin in the canine adrenal: response to acute hypovolemic stress. *Proc. Soc. Exp. Biol. Med.* (1983) 173:515-518.

Freed, W.J., M. Poltorak, and J.B. Becker. Intracerebral adrenal medulla grafts: a review. *Exp. Neurol.* (1990) 110:139-166.

Kirshner, N. Biosynthesis of the catecholamines. In: *Handbook of Physiology. Endocrinology Section 7. Volume VI. Adrenal Gland* (Amer. Physiol. Soc., Washington, DC, 1975) pg 341-355.

Lewis, G.P. Physiological mechanisms controlling secretory activity of adrenal medulla. In: *Handbook of Physiology. Endocrinology Section 7. Volume VI. Adrenal Gland.* (Amer. Physiol. Soc., Washington, DC, 1975) pg 309-319.

Lewis, J.W., M.G. Tordoff, J.E. Sherman, and J.C. Liebeskind. Adrenal medullary enkephalin-like peptides may mediate opioid stress analgesia. *Science* (1982) 217:557-559.

Lightman, S. Adrenal medulla. In: James, V.H.T. (ed). *The Adrenal Gland* (Raven Press, New York, 1979) pg 283-307.

McMillen, I. C., H. M. Mulvogue, C. L. Coulter, C. A. Browne, and P.R.C. Howe. Ontogeny of catecholamine-synthesizing enzymes and enkepha-lins in the sheep adrenal medulla: An immunocytochemical study. *J. Endocrinol.* (1988) 118:221-226.

Quinn, N.P. The clinical application of cell grafting techniques in patients with Parkinson's disease. In: Dunnet, S.B. and S.J. Richards (eds). *Neural Transplantation: from Molecular Basis to Clinical Applications* (Elsevier, Amsterdam, 1990), pg 619-625.

Ungar, A., and J.H. Phillips. Regulation of the adrenal medulla. *Physiol. Rev.* (1983) 63:787-843.

Watkinson, A., A.C. Jonsson, M. Davison, J. Young, C.M. Lee, S. Moore, and G.J. Dockray. Heterogeneity of chromogranin A-derived peptides in bovine gut, pancreas and adrenal medulla. *Biochem. J.* (1991) 276:471-479.

Weiner, N. Control of the biosynthesis of adrenal catecholamines by the adrenal medulla. In: *Handbook of Physiology. Endocrinology Section 7. Volume VI. Adrenal Gland* (Amer. Physiol. Soc., Washington, DC, 1975) pg 357-366.

Williams, P.E.V. The use of β-agonists as a means of altering body composition in livestock species. *Nutrition Abst. Rev. Ser. B* (1987) 57:453-464.

Winkler, H., D.K. Apps, and R. Fischer-Colbrie. The molecular function of adrenal chromaffin granules: Established facts and unresoved topics. *Neuroscience* (1986) 18:261-290.

Zentel, H.J., D. Nohr, S. Muller, N. Yanaihara, and E. Weihe. Differential occurrence and distribution of galanin in adrenal nerve fibers and medullary cells in rodent and avian species. *Neurosci. Lett.* (1990) 120:167-170.

Chapter 14

HORMONES AND BEHAVIOR

This chapter will deal with hormonally influenced behaviors. Most of these behaviors are related to reproduction or sexuality, but a few others will be briefly considered. Actions of blood-borne hormones on the adult CNS include (1) ***feedback*** regulation of those neuroendocrine elements that influence, through one or more subsequent steps, the secretion of the particular hormone acting on the CNS (covered in previous chapters) and (2) ***activation*** of behaviors that, when coordinated with the other actions of the hormone, either maintain homeostasis or enhance reproduction. Such hormonal action on the CNS to promote specific behaviors constitutes an important element of ***neuroendocrine integration*** as introduced in Chapter 1 (Figure 1-2).

Sexually Dimorphic Behaviors

Specific behaviors that differ between the sexes include copulatory behavior, aggressive behavior, and maternal behavior. Sexually dimorphic behaviors are affected by the following three factors: (1) ***organizational*** effects of gonadal hormones produced perinatally, (2) ***activational*** effects of gonadal hormones produced during adulthood, and (3) ***sexually dimorphic anatomy***. Morphological differences between the sexes obviously determine the ability to perform certain specific components of copulatory behavior (e.g., intromission) as well as maternal behavior (e.g., delivery of milk to neonate). Aggressive behaviors are also affected greatly by morphological differences between the sexes. Males of certain mammalian species possess antlers, horns, and other unique structures that are often utilized in aggression. Body size is also sexually dimorphic in ways that probably contribute to aggression. At least for rats and mice, it has been demonstrated that aggressive behavior of castrated males is activated by acute administration of testosterone. However, the presence of testicular secretions during the neonatal period just after birth also organizes the neural substrate through which androgens secreted in adulthood activate aggression (Feder, 1981). In summary, sexually dimorphic aggressive behaviors are influenced by all three factors listed above as potential modulators of behavior.

Copulatory Behaviors. All three factors listed in the previous paragraph are also involved in sexually dimorphic copulatory behaviors. The adult secretion of gonadal hormones plays a major and well-recognized role in the activation of copulation in all non-primate species. However, past sexual experience of the male, especially in certain carnivores, can diminish the immediate requirement for testicular androgens. However, a decrease in male-like copulatory behavior eventually occurs in all castrated males.

A brief description of copulatory behaviors is needed to understand fully the experimental results to follow. Female sexual behavior involves ***proceptivity*** and ***receptivity***. Proceptive behavior in rodent females involves a series of specific solicitational behaviors, usually in response to stimuli received from the male and that appear to increase the interest of the male in copulation. In laboratory rats, behaviors known as hopping, darting, and ear wiggling are major elements of proceptive behaviors. In other species, proceptive behavior may involve the female initiating physical contact with the male, sometimes including mounting activity by the female. Sexual receptivity of the female consists of those behaviors which facilitate (or allow) copulation; immobility is an important element of sexual receptivity in most species. A reflexive dorsiflexion of the caudal vertebral column known as ***lordosis*** occurs in female rats, mice, and guinea pigs during copulation. Lordosis occurs in response to somatosensory stimuli received from the male and facilitates intromission by making the perineal region of the female more accessible to the male.

The organizational effects of gonadal steroids during the perinatal period were recognized first in the female offspring of testosterone-treated pregnant guinea pigs. The androgenization of these females *in utero* resulted in reduced lordosis behavior when these females were castrated and given estrogen-progesterone therapy in adulthood. These androgenized female guinea pigs also displayed increased male-like mounting behavior as adults when compared to control females. Subsequent research with guinea pigs demonstrated that the period of maximum susceptibility to the organizing effects of androgen occurs on Days 30-35 of embryonic development (term = 68 days). However, in rats the period of susceptibility to behavioral organization by androgen occurs around the time of birth and for a few days thereafter. In hamsters, this period occurs during the last few days of embryonic development, whereas in sheep it occurs during the first half of gestation (Feder, 1981). These apparent differences between various species are much less different when considered relative to the time at which testicular differentiation occurs in each species.

Activation of Female Copulatory Behavior. Estradiol, and in some cases progesterone, secreted by the adult ovary activate those neural substrates that were organized in the absence of perinatal androgen and thereby initiate female copulatory behavior. In those species that ovulate spontaneously and have recurring estrous cycles, the period during which the female is receptive to copulatory attempts by the male is known as ***estrus***. This name derives from the Greek name for the gadfly, an insect that could cause an excitement in cows similar to the observed sexual excitement. In species such as the cat and rabbit that only ovulate in response to copulation, sexual receptivity can occur whenever the season is appropriate and there is no corpus luteum present from a previous ovulation. In primate species, female copulatory behavior can occur at any time, but seasonal factors and the stage of the menstrual cycle both influence the incidence of such behaviors. It is generally inferred that female copulatory behavior in primates is less dependent on ovarian hormones, but it may still be influenced by these hormones.

The action of estradiol on neural structures constitutes the major activational effect leading to female copulatory behavior. A period of 24-48 h of estradiol action is usually required before expression of this behavior, and autoradiographic studies show that the molecule of estradiol has been metabolized or cleared some time before initiation of the estrous behavior. Estradiol-induced changes in copulatory behavior probably involve the intracellular mechanisms discussed in Chapter 12 (Figure 12-2), which involve activated transcription of specific genes in selected populations of neurons.

Differences exist among species regarding whether the action of estradiol alone is fully sufficient to induce female copulatory behavior or whether the behavior is facilitated by transient exposure to progesterone (Clemens and Weaver, 1985). There do not appear to be differences between the hormonal requirements for proceptive and receptive components of the estrous behavior. In other words, if progesterone facilitates estrus in a particular species, it facilitates both components of the behavior. A requirement for progesterone to initiate female copulatory behavior in an estrogenized female occurs in hamsters and guinea pigs. In the female rat, administration of estradiol alone can induce estrous behavior, but progesterone injections activate the behavior to occur earlier and also increase its intensity. In spontaneously ovulating ungulates and in reflex ovulators, the action of progesterone in estrogenized females does *not* appear to facilitate copulatory behavior. In all species, including those in which progesterone facilitates estrus, prolonged administration of progesterone inhibits the occurrence of copulatory behavior even when estradiol is administered again.

Due in part to its extreme potency, crystalline estradiol can be implanted locally into the CNS of ovariectomized females and some aspects of copulatory behavior will occur. In some experiments progesterone was administered systemically to enhance the implant-induced copulatory behavior in rodents. The neural site(s) at which the local action of estradiol readily induces copulatory behavior differs among species. The anterior hypothalamic area was an effective site in both rats and hamsters. However, estradiol implants in the preoptic area were effective in rats but not hamsters (Pfaff and Modianos, 1985). The ventromedial hypothalamus was also an effective site in ovariectomized rats. In ovariectomized rabbits, the only site at which estradiol implants enhanced copulation was the premammillary area (Palka and Sawyer, 1966). Implants of estradiol into the ventromedial hypothalamus of ovariectomized sheep induced copulatory behavior, whereas implants in the preoptic area did not (Blanche et al., 1991). In summary, localized action of estradiol at species-specific sites within the hypothalamus can induce female copulatory behavior, suggesting that this action is of major importance. However, concurrent actions of estradiol at other CNS sites or at peripheral sites may facilitate full expression of estrous behavior (Pfaff, 1983).

The sites at which progesterone acts to facilitate female copulatory behavior in rodent species have also been studied. Effective sites of implantation include the ventromedial hypothalamus as well as the midbrain reticular formation (Ross et al., 1971; Clemens and Weaver, 1985). The facilitatory action of progesterone on female sexual behavior may be mediated by oxytocin acting as a neuropeptide modulator because intracerebral administration of a oxytocin receptor antagonist decreased both proceptive and receptive behaviors in estrogenized rats injected with progesterone (Witt and Insel, 1991).

As stated earlier, progesterone inhibits female copulatory behavior when administered in a chronic manner. This effect can also be demonstrated when estrogen and progesterone are injected concurrently into ovariectomized females. The neural site of this progesterone action has not been definitively localized, perhaps because multiple sites of action are necessary. Even in those species in which progesterone does not facilitate estradiol-induced copulatory behavior (e.g., ungulates), a period of progesterone exposure appears to be required before a second estradiol treatment can induce estrus (Fabre-Nys and Martin, 1991).

Neuroendocrine integration of female copulatory behavior is illustrated on the left side of Figure 14-1. Under the influence of adenohypophysial hormones (see Figure 12-5), the adult ovary secretes estrogen and progesterone, which act

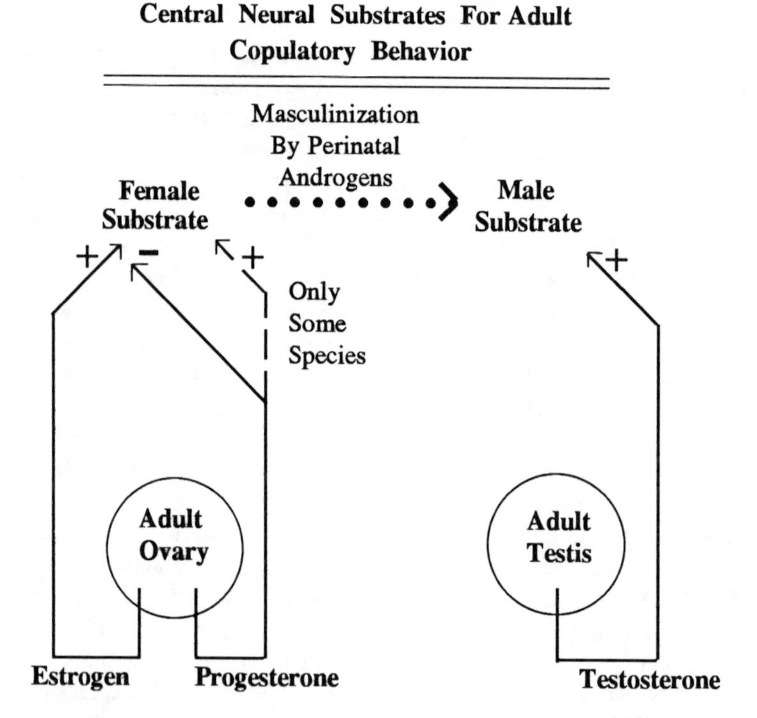

Figure 14-1. Neuroendocrine integration of female and male copulatory behaviors.

This diagram illustrates the organizational effects of perinatal androgens and the activational effects of hormones from the adult gonads on the CNS substrates for copulatory behavior.

upon a neural substrate that has not been masculinized by perinatal androgens. Progesterone initially facilitates female behavior in some species, but chronic exposure to progesterone is inhibitory to copulatory behavior in all species.

Activation of Male Copulatory Behavior. Sexual behavior of the male involves the interaction between social cues received from the female and a testosterone-activated neural substrate that developed under the influence of perinatal androgens (see Chapter 11 and Figure 14-1). Copulatory behavior consists of a male seeking and/or identifying a female that is potentially receptive followed by mounting, attempted intromission, and ultimately ejaculation. Blood-borne testosterone during the period prior to male copulatory behavior appears to act on many neural structures that contribute to the behavior. These neural structures include the hypothalamus, spinal cord, and sensory nerves from the genital organs (Hart and Leedy, 1985). Nevertheless, some aspects of male copulatory behavior can be induced in castrated male rats by localized implants of testosterone into the hypothalamus, including the preoptic area, anterior hypothalamic area, and posterior hypothalamus (Davidson, 1980).

Although circulating testosterone is essential for the activation of male copulatory behavior, there remains controversy about whether the molecule must be metabolically converted before it can activate sexual behavior (Whalen et al., 1985). Figure 14-2 summarizes different types of metabolic conversions of the androgenic compounds, androstendione and testosterone. Androstendione may be secreted by the testis or formed peripherally by oxidation of blood-borne testosterone. Except for certain species (pig and horse), testicular synthesis of estrogen is quite low. However, many tissues possess the aromatase enzymes necessary to convert testosterone and androstendione into estradiol and estrone, respectively. Testosterone may also be enzymatically converted into dihydrotestosterone by 5-alpha reductase which exists in androgen-sensitive tissues.

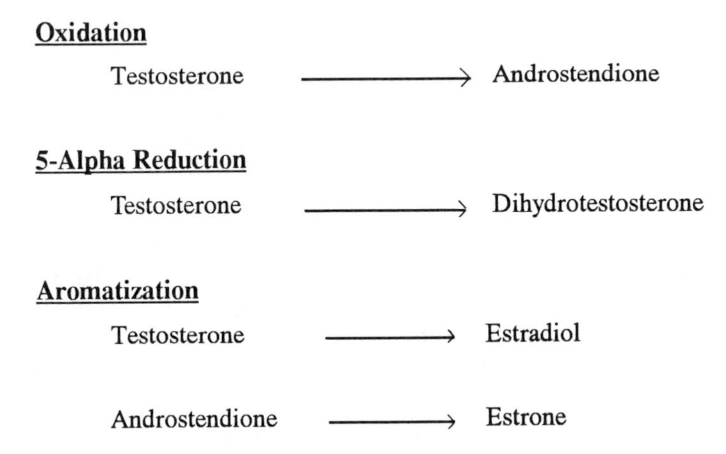

Figure 14-2. Metabolic conversions of androgenic compounds in blood and/or target tissues.

In peripheral tissues, this conversion into dihydrotestosterone is often required for androgen-dependent development. As discussed in Chapter 11, a deficiency of the required enzyme, 5-alpha reductase, in a genetic group of human males demasculinizes the prostate gland and the external genitalia (Wilson et al., 1983). The behavioral effects of androgen, at least in most experimental animals, do not depend on 5-alpha reduction of the molecule because administration of dihydrotestosterone at equivalent dosages cannot be substituted for testosterone without compromising the androgen-induced behavior (Whalen et al., 1985; Crichton et al., 1991). However, there are some data obtained in guinea pigs and rhesus monkeys that are consistent with endogenous testosterone being converted to dihydrotestosterone for behavioral action.

There is somewhat stronger evidence for aromatization of androgen to estrogen being required in order to induce male copulatory behavior. First, the aromatization reactions listed in Figure 14-2 are known to occur in hypothalamic and/or brain tissues from male rats and monkeys but not guinea pigs (Whalen et al., 1985). Therefore, estrogen could mediate the CNS effects of endogenous androgen, except in male guinea pigs. Testing this hypothesis has involved (1) dose-response comparisons of estrogen with various androgens, including those that cannot be aromatized (dihydrotestosterone) and (2) blocking androgen-induced behaviors with anti-estrogens or aromatase inhibitors. Although the results appear to depend greatly on specific conditions and the species being studied, administration of estrogenic compounds can induce some components of male copulatory behavior. The dosages required are often quite large and some behavioral deficits may still be observed. However, some of these deficits in hormone-treated castrates may relate to the failure of exogenous estrogens to maintain penile morphology and sensitivity required for normal copulatory behavior. In addition, the possible contribution of endogenous androgens secreted by the adrenal gland of castrated males must be considered.

Results obtained from administration of various synthetic androgens which may or may not be aromatized are somewhat complicated. In part, this may be due to use of placental tissues to determine whether a particular androgen can be aromatized. Metabolic conversions in neural tissue may be different, and neural enzymes catalyzing aromatization may even be induced when castrated males are administered high doses of a particular compound (Whalen et al., 1985).

Certain synthetic anti-estrogens are able to antagonize some aspects of androgen-induced copulatory behavior in male rats, but the results vary greatly among compounds and testing paradigms. Drugs that antagonize the aromatization of androgen into estrogen also decrease some components of testosterone-induced male copulatory behavior, but again results differ among compounds and specific conditions of testing (Whalen et al., 1985). Immunoneutralization of estrogen in castrated male sheep decreased many testosterone-induced behaviors (Crichton et al., 1991). In summary, neural action of estrogens resulting from aromatization of androgens probably contributes to some components of male copulatory behavior, but it seems unlikely that all effects of androgen on the brain are mediated by estrogens and certainly very few of the other effects (neural and genital) are dependent on aromatization to estrogen.

Precopulatory Behaviors in the Male. As discussed earlier in this chapter, females may exhibit certain proceptive behaviors that enhance the sexual interest of the male. Estrous females may also produce chemicals that attract the male and/or signify her potential receptivity. These sexual attractants are volatile and air-borne in many invertebrate species, but they are not as volatile in mammals and also appear to be present in urine. Male precopulatory behaviors may exist in part to optimize the detection of these sexual attractants in female urine. The ***vomernasal organ*** is a blind pouch located above the hard palate and communicates with the nasal and/or oral cavity via the nasopalatine duct. The mucosa of the vomernasal organ is similar to that in the main olfactory system, but its neural connections enter the accessory olfactory system. There is much evidence that the ability of rodent and ungulate males to detect and/or respond to sexual attractants in female urine depends on access of the urine to a normally innervated vomernasal organ. Delivery of urine to the vomernasal organ is also facilitated by the flehmen (i.e. lip-curl) behavioral response which usually occurs when ungulate males come in contact with urine (Hart and Leedy, 1985). Interestingly, the vomernasal organ is absent in primate species where copulatory behavior is less dependent on estrogen-mediated receptivity of the female.

Maternal Behaviors. For purposes of this discussion, the definition of maternal behavior includes all aspects of interaction of the female with offspring. Interaction of the male with offspring (i.e., parental behavior) will not be covered even though it may be important in some mammalian species. The antecedents of maternal behavior occur during pregnancy when nest building occurs in pigs and many rodent species. The timing of nest building varies among species, but in all cases the placental, ovarian, and/or hypophysial hormones of pregnancy (e.g., estrogen, progesterone, and prolactin) contribute to the behavior. The experience of previous maternal behavior can facilitate onset of the behavior and perhaps even decrease the requirement for gestational hormones (Rosenblatt et al., 1985). Periparturient females also display aggression against intruders which often develops even before delivery of her offspring. Under experimental conditions, pregnant rats of certain genetic strains show normal maternal behaviors even before parturition. Other periparturient maternal behaviors in rodents include self-licking of the mammary regions of the body that may facilitate lactogenic development of the mammary glands as well as eating of the placenta (i.e., placentophagia) immediately after its delivery.

Although specific aspects of maternal care and suckling of offspring differ widely among species, hormonal regulation of these behaviors may be more general. Through actions on neural systems primed by the hormones of pregnancy, blood-borne prolactin and intracerebral oxytocin both appear to facilitate maternal behavior toward offspring. As discussed in Chapter 9, secretion of prolactin from the adenohypophysis increases just prior to parturition (Figure 9-3) and at every suckling episode (Figure 9-1). Although these increases of prolactin act on the mammary gland to promote synthesis of milk, there is evidence that prolactin also acts on the rat brain to promote maternal behavior (Bridges et al., 1990). In periparturient sheep, sensory stimuli from the genital tract are required for full development of maternal behavior (Levy et al., 1992). In addition, intracerebral administration of oxytocin can stimulate such behavior in estrogen-primed ovariectomized ewes (Kendrick et al., 1987) and in parturient ewes deprived of sensory input from the genital tract (Levy et al., 1992).

As discussed in Chapter 5, oxytocin can be (1) released within the CNS to act as a neuropeptide modulator and (2) secreted from the pars nervosa into blood to act as a neurohormone. During parturition, large quantities of oxytocin are released into blood, but a functional requirement for this endogenous oxytocin to facilitate parturition has not been established (see Chapter 3). Release of oxytocin within the brain is also increased greatly at parturition in rats and sheep (Neumann et al., 1991; Kendrick et al., 1991). These results suggest that periparturient increases in the release of oxytocin may be more important in the induction of maternal behavior than in the facilitation of parturition. Moreover, the results obtained from sheep raise the possibility of differential regulation of oxytocin release within the CNS as compared to release into blood. The results plotted in Figure 14-3 illustrate this possibility. The concentration of oxytocin in CSF of pregnant sheep was only about one half that in blood plasma before the onset of labor. When labor began, oxytocin levels in both blood and CSF

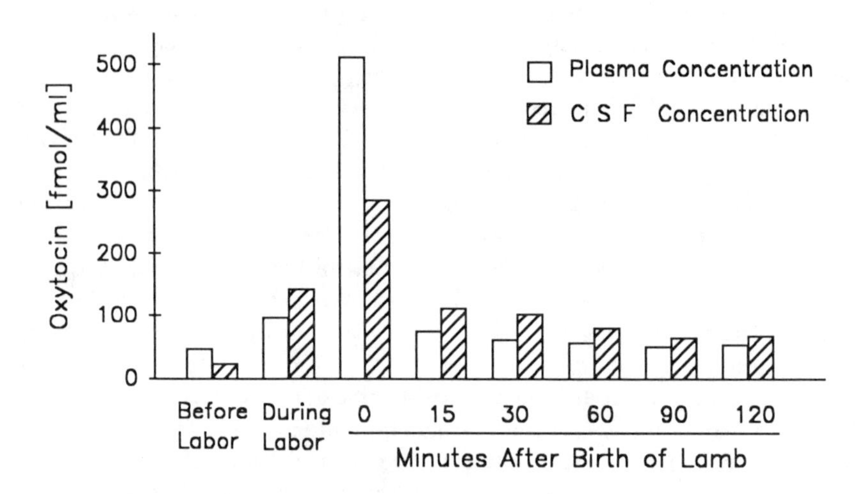

Figure 14-3. Periparturient concentrations of oxytocin in blood plasma and ventricular cerebrospinal fluid (CSF) of sheep.

These data are redrawn from Kendrick et al. (1991) and represent immunoreactive oxytocin levels in jugular venous plasma and CSF from the lateral ventricle as ewes delivered single lambs.

increased. The observed increase of oxytocin in CSF did not result from the transfer of blood-borne oxytocin because negligible quantities of this peptide are transported from blood into CSF. Rather, oxytocin in CSF probably reflected central release as also observed during tissue microdialysis of several brain regions (Kendrick et al., 1988a, 1988b). The data in Figure 14-3 illustrate the so-called Ferguson reflex in which dilation of cervix and vagina during parturition stimulates the release of oxytocin into blood, but these data also demonstrate that central release of oxytocin increases concurrently. In fact, the ratio of oxytocin concentrations in CSF to those in blood plasma was greater than 1.0 during labor and at 15 to 120 min after birth of the lamb. Only before labor and immediately at the time of birth were blood levels of oxytocin greater than CSF levels. Maternal acceptance of the lamb by its mother occurs during the first few hours after birth and requires sensory input from the genital tract (Levy et al., 1992). Moreover, maternal behavior can be induced in nonpregnant ewes by vaginocervical stimulation to release oxytocin centrally or by intracerebral administration of oxytocin. Parturition also increased local release of oxytocin in neural tissue of the olfactory bulb (Kendrick et al., 1988a) that may facilitate the formation of olfactory recognition memories that are involved in the subsequent identification of each lamb by her mother (Signoret, 1990).

Ingestive Behaviors

Primary control of feeding and drinking behaviors appears to reside in the brain, but these CNS mechanisms may be influenced by blood-borne compounds as well as by neural input from the periphery. Most of the blood-borne compounds that influence ingestive behaviors are either metabolic compounds that affect feeding or osmotic molecules that affect drinking. The present discussion will not attempt to cover the numerous CNS neuropeptides and amines regulating ingestive behavior, but rather it will briefly cover only blood-borne hormones, which have an effect on but appear not to be major regulators of ingestive behaviors. There is some evidence for two blood-borne hormones (insulin and cholecystokinin) as satiety signals to suppress feeding behavior (Gibbs and Smith, 1986; Baskin et al., 1987). Although systemic administration of both hormones can decrease feeding behavior (Anika et al., 1980; Parrott et al., 1991), such actions do not necessarily reflect physiological satiety produced by endogenous hormones. Both insulin and cholecystokinin are also neuropeptides (see Table 5-1) that may be released locally within the brain to enhance satiety and decrease feeding behavior. Investigations of blood-borne satiety signals are complicated by the fact that administration of any compound causing discomfort or anxiety will decrease feeding behavior nonspecifically. On the other hand, exogenous compounds that stimulate feeding behavior, such as neuropeptide Y, galanin, and norepinephrine, are less likely to produce nonspecific effects. However, these compounds are much more effective stimuli of feeding behavior when administered into the brain where they probably activate endogenous neural systems that normally respond only to locally released neurochemicals rather than to blood-borne compounds.

Other Behaviors

The interrelated processes of motivation, learning, and memory clearly involve CNS mechanisms independent of neuroendocrinology. However, some compounds with proven neuroendocrine functions have been reported to influence these processes. Therefore, a brief description of these influences seems warranted. Motivation, learning, and memory are assessed in experimental animals through examination of conditioned behavior. The conditioning can involve the experimental pairing of an unconditioned stimulus (either positive or negative to the animal) with a conditioned stimulus that formerly was neutral in its meaning to the animal. As a result of repeated pairing of the two stimuli, the animal learns to associate the unconditioned stimulus with the conditioned stimulus such that it will then respond to the conditioned stimulus by itself. The animal's response to the conditioned stimulus will be negative or positive depending on the nature of the original unconditioned stimulus. This process is often called classical conditioning or sometimes Pavlovian conditioning after the scientist who discovered it. When the unconditioned stimulus is unpleasant (i.e., aversive), the animal learns to avoid the unpleasant consequences of the unconditioned stimulus by responding to the conditioned stimulus prior to occurrence of the unconditioned stimulus. When this response involves active performance of a particular behavioral response, it is called ***active*** avoidance behavior. When the response necessary to avoid the unpleasant unconditioned stimulus is nonperformance of a usually pleasant behavior, it is called ***passive*** avoidance behavior.

Another type of conditioning used to assess motivation, learning, and memory is often called ***operant*** conditioning. In this process, animals learn to perform a specific instrumental task in order to obtain food or other types of reward. Because the instrumental task is not a natural behavior, the animal appears to learn by trial and error, but once the instrumental (i.e., operant) behavior is learned, the motivation to obtain the reward as well as memory of the instrumental task can both be thoroughly studied.

Administration of neuropeptides derived from pro-opiomelanocortin (Figures 5-1 and 5-3) and pro-vasopressin (Figure 3-1) have been shown to improve one or more aspects of motivation, learning, and memory (Le Moal et al., 1992). As stated in Chapter 6, administration of the seven amino acid sequence common to $ACTH_{4\text{-}10}$ (also found in the three forms of MSH) improves the ability and/or motivation to learn conditioned behaviors. Because hypophysectomy causes deficits in learning ability, it has been suggested that the learning improvements resulting from exogenous $ACTH_{4\text{-}10}$ represent a physiological mechanism. However, the behavioral depression and occasional poor health of hypophysectomized rats appears more likely to be responsible for their learning deficits. Therefore, physiological benefits in motivation or learning that may result from endogenous $ACTH_{4\text{-}10}$ or related compounds are probably due to intracerebral action of these neuropeptides that occur widely in the brain (see Chapter 5).

Peripheral or central administration of vasopressin as well as central administration of des-glycinamide$_9$vasopressin, which has little vasopressin bioactivity, facilitates learning of passive avoidance behavior in rodents (Koob et al., 1989). It has also been suggested that there exists in the CNS a unique subtype of vasopressin receptor that mediates these effects of exogenous vasopressinergic compounds on learning and memory (De Wied et al., 1991). However, it is not clear whether administration of vasopressin or its synthetic analogues simulates actions normally performed by endogenous vasopressin. One method to investigate the possible role of endogenous vasopressin is to use rats of the genetic strain that cannot synthesize vasopressin and therefore have diabetes insipidus (Brattleboro strain, discussed in Chapter 3). This vasopressin-deficient strain of rats learns passive avoidance behavior less efficiently than control strains (Van Wimersma Greidanus et al., 1983). However, other workers attribute the deficits in passive avoidance behavior to a reduced emotionality in Brattleboro rats and fail to observe consistent learning deficits (Lorenzini et al., 1991). To investigate the role of endogenous vasopressin in learning and/or retention of passive avoidance behavior in normal rats, endogenous vasopressin has been immunoneutralized by intracerebral administration of anti-vasopressin antibodies. Intraventricular or intrahippocampal administration in rats subjected to one-trial passive avoidance impaired retention of the learned behavior (Velduis et al., 1987).

Neonatal Experience, Emotionality, and Adrenocortical Function. In species whose offspring are immature and helpless at birth (so-called altricial species such as rodents and carnivores), it has been demonstrated that daily handling or mild stress of the neonate produces certain effects observable in adulthood (Hart, 1985). Such experimental treatments applied to neonates appear to decrease their adult emotionality when exposed to unfamiliar environments or to attempts to produce conditioned behaviors. Stress-induced adrenocortical secretion and circadian profiles of glucocorticoids also occurred at younger ages in rats that were neonatally handled or mildly stressed. When tested as adults, the adrenocortical responses to stressors seemed more appropriate because (1) prestress levels were lower, and (2) stress caused larger glucocorticoid increases which subsided more quickly in those rats that had been stressed as neonates. The physiological and developmental mechanisms responsible for the observed effects of neonatal experience are not understood. However, there may be a maternal component to the effect. Removal of the neonate from its mother undoubtedly activates her adrenocortical system as well as that of the neonate. There is evidence that maternal glucocorticoids present in milk may influence the development of glucocorticoid receptors in the adult hippocampus as well as the ability to learn conditioned behavior (Angelucci et al.,

1983). In later work from this same laboratory, the rearing of infant rats on milk that was deficient in glucocorticoids (i.e., adrenalectomized foster mothers) decreased their adult emotionality in test situations (Catalani et al., 1989). However, these apparent effects of maternal glucocorticoids in milk consumed by neonates were observed in adult female offspring but not male offspring. In summary, experimental stressors applied during infancy in altricial species produce developmental changes in emotionality and adrenocortical function that can be observed during adulthood, but the mechanisms that produce these changes are not understood.

REFERENCES

Angelucci, L., F.R. Patacchioli, C. Chierichetti, and S. Laureti. Perinatal mother-offspring pituitary-adrenal interrelationship in rats: corticosterone in milk may affect adult life. *Endocrinol. Exp.* (1983) 17:191-205.

Anika, S.M., T.R. Houpt, and K.A. Houpt. Insulin as a satiety hormone. *Physiol. Behav.* (1980) 25:21-23.

Baskin, D.G., D.P. Figlewicz, S.C. Woods, D. Porte, and D.M. Dorsa. Insulin in the brain. *Annu. Rev. Physiol.* (1987) 49:335-347.

Blanche, D., C.J. Fabre-Nys, and G. Venier. Ventromedial hypothalamus as a target for oestradiol action on proceptivity, receptivity and luteinizing hormone surge of the ewe. *Brain Res.* (1991) 546:241-249.

Bridges, R.S., M. Numan, P.M. Ronsheim, P.E. Mann, and C.E. Lupini. Central prolactin infusions stimulate maternal behavior in steroid-treated, nulliparous female rats. *Proc. Natl. Acad. Sci.* (1990) 87:8003-8007.

Catalani, A., E. Toth, A. Gambini, A. Giuliani, G. Lorentz, and L. Angelucci. Maternal adrenalectomy and adult offspring in a conflict situation in the rat. *Pharmacol. Biochem. Behav.* (1989) 32:323-329.

Clemens, L.G., and D.R. Weaver. The role of gonadal hormones in the activation of feminine sexual behavior. In: Adler, N., D. Pfaff, and R.W. Goy (eds). *Handbook of Behavioral Neurobiology. Vol. 7. Reproduction* (Plenum Press, New York, 1985) pg 183-227.

Crichton, J.S., A.W. Lishman, M. Hundley, and C. Amies. Role of dihydrotestosterone in the control of sexual behavior in castrated male sheep. *J. Reprod. Fertil.* (1991) 93:9-17.

Davidson, J.M. Hormones and sexual behavior in the male. In: Krieger. D.T., and J.C. Hughes. (eds). *Neuroendocrinology* (Sinauer Associates, Inc., Sunderland, MA, 1980), pg 232-238.

De Wied, D., J. Elands, and G. Kovacs. Interactive effects of neurohypophyseal neuropeptides with receptor antagonists on passive avoidance behavior: mediation by a cerebral neurohypophyseal receptor? *Proc. Natl. Acad. Sci.* (1991) 88:1494-1498.

Fabre-Nys, C., and G.B. Martin. Roles of progesterone and oestradiol in determining the temporal sequence and quantitative expression of sexual receptivity and the preovulatory LH surge in the ewe. *J. Endocrinol.* (1991) 130:367-379.

Feder, H.H. Perinatal hormones and their role in the development of sexually dimorphic behaviors. In: Adler, N.T. (ed). *Neuroendocrinology of Reproduction* (Plenum Press, New York, 1981), pg 127-157.

Gibbs, J., and G.P. Smith. Satiety: the roles of peptides from the stomach and the intestine. *Fed. Proc.* (1986) 45:1391-1395.

Hart, B.L. Early experience and behavior. Chapter 7. In: Hart, B.L. *The Behavior of Domestic Animals* (W.H. Freeman & Co., New York, 1985), pg 242-263.

Hart, B.L., and M.G. Leedy. Neurological bases of male sexual behavior: a comparative analysis. In: Adler, N., D. Pfaff, and R.W. Goy (eds). *Handbook of Behavioral Neurobiology. Vol. 7. Reproduction* (Plenum Press, New York, 1985) pg 373-422.

Kendrick, K.M., E.B. Keverne, and B.A. Baldwin. Intracerebroventricular oxytocin stimulates maternal behavior in the sheep. *Neuroendocrinology* (1987) 46:56-61.

Kendrick, K.M., E.B. Keverne, C. Chapman, and B.A. Baldwin. Intracranial dialysis measurement of oxytocin, monoamine and uric acid release from the olfactory bulb and substantia nigra of sheep during parturition, suckling, separation from lambs and eating. *Brain Res.* (1988a) 439:1-10.

Kendrick, K.M., E.B. Keverne, C. Chapman, and B.A. Baldwin. Microdialysis measurement of oxytocin, aspartate, gamma-aminobutyric acid and glutamate release from the olfactory bulb of the sheep during vaginocervical stimulation. *Brain Res.* (1988b) 442:171-174.

Kendrick, K.M., E.B. Keverne, M.R. Hinton, and J.A. Goode. Cerebrospinal fluid and plasma concentrations of oxytocin and vasopressin during parturition and vaginocervical stimulation in the sheep. *Brain Res. Bull.* (1991) 26:803-807.

Koob, G.F., C. Lebrun, R.M. Bluthe, R. Dantzer, and M. Le Moal. Role of neuropeptides in learning versus performance: focus on vasopressin. *Brain Res. Bull.* (1989) 23:359-364.

Le Moal, M., P. Mormede, and L. Stinus. The behavioral neuroendocrinology of arginine vasopressin, adrenocorticotropic hormone and opioids. In: Nemeroff, C.B. (ed). *Neuroendocrinology* (CRC Press, Boca Raton, FL, 1992), pg 365-396.

Levy, F., E.B. Keverne, K.M. Kendrick, V. Piketty, and P. Poindron. Intracerebral oxytocin is important for the onset of maternal behavior in inexperienced ewes delivered under peridural anesthesia. *Behav. Neurosci.* (1992) 106:427-432.

Lorenzini, C.A., C. Bucherelli, A. Giachetti, and G. Tassoni. The behavior of the homozygous and heterozygous sub-types of rats which are genetically selected for diabetes insipidus: a comparison with Long Evans and Wistar stocks. *Experientia* (1991) 47:1019-1026.

Neumann, I., J.A. Russell, B. Wolff, and R. Landgraf. Naloxone increases the release of oxytocin, but not vasopressin, within limbic brain areas of conscious parturient rats: a push-pull perfusion study. *Neuroendocrinology* (1991) 54:545-551.

Palka, Y.S., and C.H. Sawyer. The effects of hypothalamic implants of ovarian steroids on oestrous behavior in rabbits. *J. Physiol.* (1966) 185:251-269.

Parrott, R.F., I.S. Ebenezer, B.A. Baldwin, and M.L. Forsling. Central and peripheral doses of cholecystokinin that inhibit feeding in pigs also stimulate vasopressin and cortisol release. *Exp. Physiol.* (1991) 76:525-531.

Pfaff, D.W. Impact of estrogens on hypothalamic nerve cells: ultrastructural, chemical, and electrical effects. *Rec. Prog. Horm. Res.* (1983) 39:127-179.

Pfaff, D., and D. Modianos. Neural mechanisms of female reproductive biology. In: Adler, N., D. Pfaff, and R.W. Goy (eds). *Handbook of Neurobiology, Vol. 7. Reproduction* (Plenum Press, New York, 1985) pg 423-493.

Rosenblatt, J.S., A.D. Mayer, and H.I. Siegel. Maternal behavior among the nonprimate mammals. In: Adler, N., D. Pfaff, and R.W. Goy (eds). *Handbook of Behavioral Neurobiology. Vol. 7. Reproduction* (Plenum Press, New York, 1985) pg 229-298.

Ross, J., C. Claybaugh, L.G. Clemens, and R.A. Gorski. Short latency induction of estrous behavior with intracerebral gonadal hormones in ovariectomized rats. *Endocrinology* (1971) 89:32-38.

Signoret, J.P. Chemical signals in domestic ungulates. In: MacDonald, D.W. , D. Muller-Schwarze, and S.E. Natynczuk (eds). *Chemical Signals in Vertebrates 5.* (Oxford Univ. Press, Oxford, UK, 1990), pg 610-626.

Whalen, R.E., P. Yahr, and W.G. Luttge. The role of metabolism in hormonal control of sexual behavior. In: Adler, N., D. Pfaff, and R.W. Goy (eds). *Handbook of Neurobiology, Vol. 7. Reproduction* (Plenum Press, New York, 1985), pg 609-663.

Wilson, J.D., J.E. Griffin, F.W. George, and M. Leshin. The endocrine control of male phenotypic development. *Aust. J. Biol. Sci.* (1983) 36:101-128.

Witt, D.M., and T.R. Insel. A selective oxytocin antagonist attenuates progesterone facilitation of female sexual behavior. *Endocrinology* (1991) 128:3269-3276.

Van Wimersma Greidanus, T.B., B. Bohus, G.L. Kovacs, D.H.G. Versteeg, J.P.H. Burbach, and D. De Wied. Sites of behavioral and neurochemical action of ACTH-like peptides and neurohypophyseal hormones. *Neurosci. Biobehav. Rev.* (1983) 7:453-463.

Veldhuis, H.D., T.B. Van Wimersma Greidanus, and D.H.G. Versteeg. Microinjection of anti-vasopressin serum into limbic structures of the rat brain: effects on passive avoidance responding and on local catecholamine utilization. *Brain Res.* (1987) 425:167-173.

Chapter 15

NEUROENDOCRINE IMMUNOLOGY

Knowledge of the immune system has increased greatly in recent years. This knowledge includes the multiplicity of cell types and their developmental lineage as well as the secretory products of various immune system cells. It was also discovered that many cells of the immune system produce cell-regulatory polypeptides (called cytokines) that coordinate the body's response to microbiological invasion, inflammation, immunological reactions, and tissue injury. This regulatory communication among different immunological cell types has many similarities to the communication among endocrine cells. However, immune functions were formerly thought to be relatively independent of the endocrine and nervous systems, perhaps because many immunological responses could be demonstrated *in vitro* in the absence of hormones and nerves. Concurrent with the development of knowledge about cytokines and their roles in mediating and coordinating responses of immune cells, there has been an increasing appreciation that a physiological integration exists among the body's major regulatory systems (nervous, endocrine, and immune).

Introduction to the Immune System

The functions of the immune system are far too complex to be presented here in very much detail. However, certain functional elements will be discussed to provide a foundation for understanding the neuroendocrine immunology presented in this chapter. During development in mammals, a portion of embryonic stem cells in the bone marrow migrate to the thymus (T) and differentiate into ***T lymphocytes***. Another population of embryonic stem cells remain in the bone (B) marrow and differentiate into ***B lymphocytes***. The bone marrow and thymus constitute the only ***primary*** lymphoid organs in mammals (i.e., they produce lymphocytes independent of antigenic stimulation). In birds, the Bursa of Fabricius also serves as a primary lymphoid organ. ***Secondary*** lymphoid organs receive migrating cells that originate in primary lymphoid organs, and these cells are available for antigen-driven differentiation as needed for immune responses. The secondary lymphoid organs include the spleen, lymph nodes, and groups of lymphoid cells in various mucosal tissues. There may also be exchange of lymphocytes between primary lymphoid organs. Derivatives of B lymphocytes (called plasma cells) synthesize and secrete antibodies (immunoglobulins) that attack the foreign compound against which the antibody is directed. The T lymphocytes differentiate into several subtypes, most of which either attack foreign cells (cytotoxic T lymphocytes) or assist in this attack (helper T lymphocytes). There are also suppressor T lymphocytes that inhibit other immune system cells. Another type of immunological cell, the macrophage, may also internalize foreign antigens and process them for presentation to and activation of T lymphocytes. One other important lymphocyte is known as a natural killer (NK) cell. Because NK cells seem different from either B lymphocytes or T lymphocytes, they have been classified separately. However, their functions are somewhat similar to cytotoxic T lymphocytes except that they do not require antigenic sensitization.

Nomenclature for non-antibody secretory products of these various immune system cell types is complicated due to various names given to these secretory products. The generic name for the regulatory compounds produced by activated T lymphocytes and B lymphocytes is ***lymphokine***. The generic name for regulatory compounds produced by monocytes and immune cells with a macrophage lineage is ***monokine***. However, this chapter will use the generic term ***cytokine*** which encompasses both lymphokines and monokines. Moreover, the generic term interleukin, which preceded cytokine as a name for compounds that mediate interactions among leukocytes, will *not* be used in a generic sense in this chapter because specific compounds were subsequently assigned this name followed by a number (e.g., interleukin-2).

As a prelude to mounting an attack against foreign compounds, cells of the immune system must discriminate between normal endogenous compounds (i.e., self) and exogenous foreign compounds (i.e., nonself). After this discrimination, the immune response to nonself elements may involve humoral factors (e.g., secreted antibodies and cytokines) and/or cellular elements (e.g., T lymphocytes and macrophages). Figure 15-1 illustrates diagrammatically the complex network of immune cells and cytokines through which the body mobilizes its defenses to cellular, chemical and traumatic insults (Whitacre, 1990). Interaction of nonself antigens with macrophages (also termed antigen-presenting cells) initiates release of various specific cytokines including interleukin-1, interleukin-6, tumor necrosis factor (TNF), colony-stimulating factor (CSF), and transforming growth factor-β (TGF-β). These macrophage-derived cytokines promote (1) activation of B lymphocytes and T lymphocytes and (2) stimulation of hematopoesis by stem cells of the bone marrow. The activated T lymphocytes produce many of these same cytokines that amplify these processes listed above, but they also secrete the following additional cytokines: interleukin-2, interleukin-4, interleukin-5, and interferon-gamma. These cytokines derived from T lymphocytes activate the following processes: (1) additional hematopoesis, (2) activation of granulocytes, including especially eosinophils, (3) cell division and clonal expansion of T lymphocytes, (4) differentiation of activated B lymphocytes into plasma cells that secrete antibodies, and (5) activation of phagocytosis by macrophages (see Figure 15-1). NK cells produce mainly interferon-alpha that, among its many effects, further enhances activity of NK cells and also inhibits replication of viruses. Activated B lymphocytes, in addition to being the precursors of plasma cells, also produce selected cytokines that may stimulate macrophages and T lymphocytes. The network of immune cells and cytokines depicted in Figure 15-1 and described in this paragraph has been highly simplified, and the field is developing so rapidly that new concepts will undoubtedly emerge.

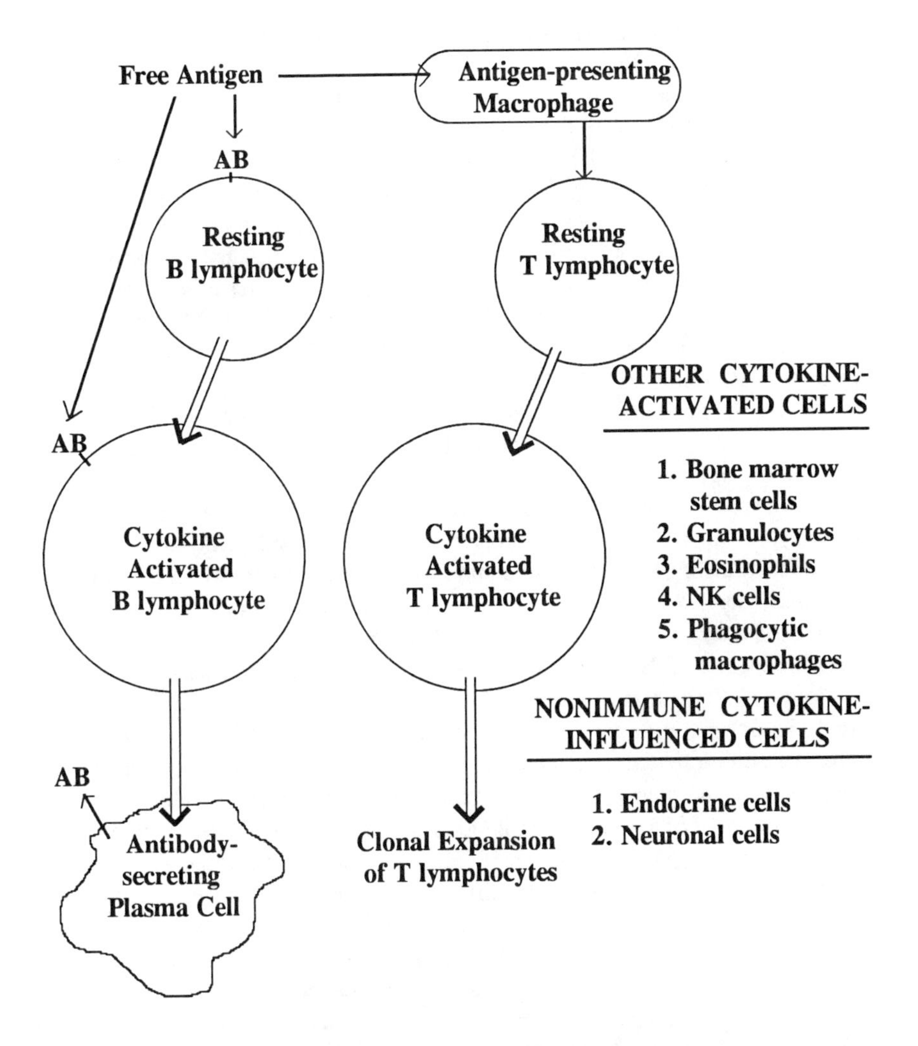

Figure 15-1. Immunological responses of the cytokine network.

Cytokine-producing cells of the immune system are diagrammatically illustrated together with lists of cells activated or influenced by the secreted cytokines. Free antigens may interact with an antibody (AB) on the surface of resting and cytokine-activated B lymphocytes (i.e., ABs function as cellular receptors) or be internalized within macrophages which process the antigen for presentation to T lymphocytes. Various interactions involving AB are denoted by $\longrightarrow$ whereas cellular activation/differentiation or division that are stimulated by cytokines are denoted in this diagram by $\Longrightarrow$. This diagram was adapted from Whitacre (1990).

Principles of Neuroendocrine-Immune Interaction

Perhaps because many immunological responses can be demonstrated *in vitro*, the possibility that the nervous system may influence immunological defenses only began to be seriously considered in the last few decades. A significant portion of the earliest scientific work in this area occurred in Russia (Korneva et al., 1978). In summary, these Russian scientists and others were able to demonstrate through clinical and experimental studies that neuropsychiatric changes and/or classical conditioning could profoundly alter immune responses to natural and/or experimental challenges (Ghanta et al., 1990). In addition, CNS ablations and exogenous hormones could influence immune responses. Because of the functional links between nervous and endocrine functions that are the subjects of this book, it should be obvious that the functional roles of neural and endocrine factors may be difficult to separate. Figure 15-2 presents a simplified diagram illustrating known ***bidirectional*** interactions among (1) the nervous system (peripheral autonomic nerves as well as neurohormones in hypophysial portal blood), (2) hormone-producing endocrine cells, and (3) cytokine-producing lymphoid cells and macrophages. As shown in the diagram, cytokines secreted into blood act on both endocrine cells and the CNS. Lymphoid cells are regulated by blood-borne hormones and by direct innervation via peripheral nerves of the sympathetic nervous system. Another interaction illustrated in Figure 15-2 consists of lymphoid cells, located within the brain or in endocrine glands, that produce cytokines for local actions on adjacent structures in a manner similar to that illustrated in the diagram for cytokine action on adjacent lymphocytes. A specialized interaction omitted from Figure 15-2 involves epithelial cells of one specific lymphoid organ, the thymus. These epithelial cells known as thymocytes are different from thymic lymphocytes because they secrete a family of hormones (e.g., thymosin and thymulin) independent of antigenic or other activation. In addition to producing thymic hormones, some thymocytes produce selected cytokines such as interleukin-1 (Hadden et al., 1989). Although the thymic hormones are not called cytokines because their secretion is not dependent on antigenic stimulation, inflammation, or tissue injury, their various actions are consistent with the types of cytokine action depicted in Figure 15-2. Upon receipt of appropriate endocrine signals, thymic hormones are secreted into blood and regulate various endocrine glands as depicted for cytokines. They may also act on the brain as do cytokines, but the evidence for CNS actions by thymic hormones is much more limited than for such CNS actions by cytokines (Millington and Buckingham, 1992).

Endocrine Effects on Immunological Functions

Administration of adrenal glucocorticoids, especially in high doses, clearly depresses immune responses and causes atrophy of the thymus (Berczi, 1986b). Various lymphoid cells also possess glucocorticoid receptors that probably

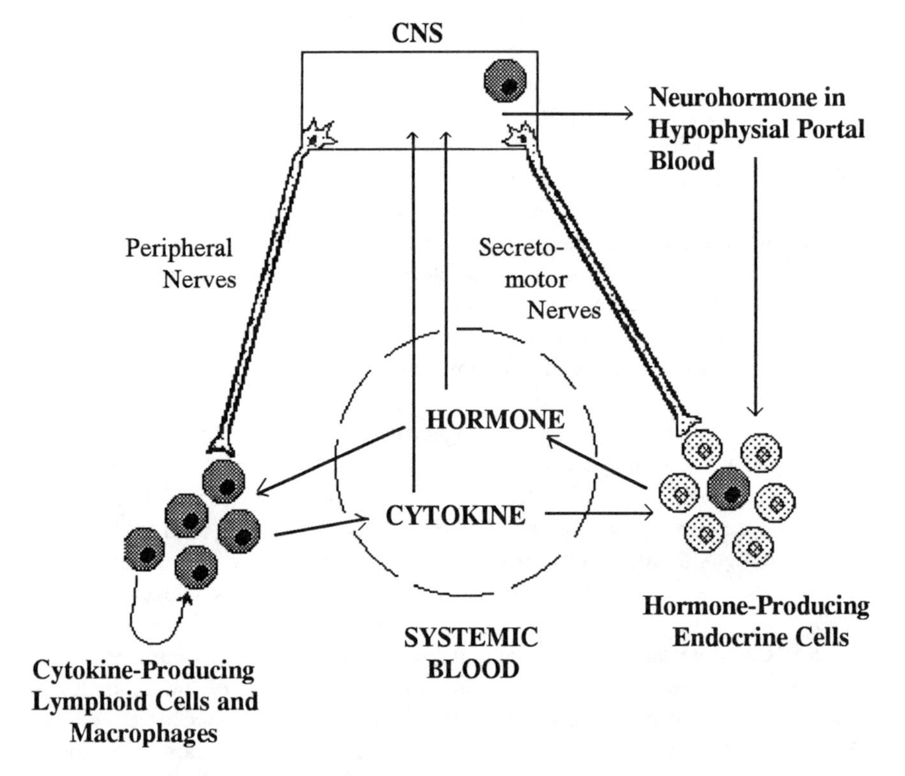

Figure 15-2. Bidirectional interactions among lymphoid cells and macrophages as well as neural and endocrine systems.

This diagram depicts various cytokine-producing cells as round ***darkened*** cells located in either (**a**) lymphoid organs, (**b**) the CNS, or (**c**) endocrine glands. Cytokines produced by activated lymphoid cells and macrophages may act locally on adjacent cells (i.e., paracrine action) or travel in the systemic blood to distant sites for action (i.e., endocrine action). This diagram also illustrates the innervation of lymphoid tissues by peripheral nerves (noradrenergic and peptidergic) as well as the action of blood-borne hormones on lymphoid cells.

mediate the immunosuppressive effects. In addition to stimulating glucocorticoid secretion, hypophysial ACTH may also exert direct effects on immune cells *in vitro*. Other products of the ACTH precursor pro-opiomelanocortin (POMC) such as α-endorphin and β-endorphin may also directly influence immune cells, but the specific type of effect appears to be cell specific (Blalock, 1989). There is also an extensive literature on the effects of stress-induced increases in endogenous ACTH and glucocorticoid on immune functions. Based on the immunosuppressive effects of exogenous hormones (POMC products and glucocorticoids), one would expect that stressors would consistently inhibit immune responses, but such is not always the case. Depending on the type and severity of stressor and on the immunological parameter measured, experimental application of stress may or may not suppress immunological reactions (Korneva et al., 1978; Cunnick et al., 1992).

Other hypophysial and neurohypophysial hormones have been shown to affect lymphocytes *in vitro* and *in vivo* (Berczi, 1986a; Blalock, 1989). Administration of GH to hormonally deficient animals can stimulate T lymphocytes and macrophages, improving their immune responses. Experimental decreases of prolactin secretion in rodents suppress selected immunological reactions that can be restored by prolactin therapy. Low dosages of vasopressin can stimulate T lymphocytes replacing a need for interleukin-2 to promote the production of interferon-gamma by these cells. In those cases in which hypophysial and neurohypophysial hormones have been shown to influence lymphocyte function, it has usually been possible to demonstrate specific receptors for that hormone on at least one or more subpopulations of lymphocytes. Notably, a transformed cell line of malignant T lymphocytes, known as Nb2, contains high levels of prolactin receptors and is mitogenically stimulated by addition of prolactin (Shiu et al., 1983). Although prolactin is not mitogenic to normal T lymphocytes, it is necessary for cytokine-driven clonal expansion of these cells (Clevenger et al., 1992).

Successful gestation usually requires intrauterine accommodation of a fetus which differs genetically from the mother in at least some of the major histocompatibility antigens. It now seems clear that antigenic contact exists between fetal and maternal tissues in the uterus. Therefore, gestational adaptations are required in order to not reject the fetal tissues as being *non-self*. As covered in Chapter 12, hypophysial, ovarian, and placental hormones play important roles in the maintenance of pregnancy. These pregnancy hormones do not suppress the systemic immunocompetence of the mother, but some of these hormones, especially progesterone, alter intrauterine mechanisms of immunological function to accomodate the fetus (Clark, 1986). Maternal tissue of the pregnant uterus undergoes decidualization, and the decidual tissue has been shown to elaborate immunosuppressive cells and compounds for local action. In summary, blood-borne gestational hormones, notably progesterone, and decidual production of specific cells and compounds combine to suppress local immunological responses within the uterine environment and to allow successful gestation.

As stated earlier in this chapter, epithelial cells of the thymus known as thymocytes secrete several thymic hormones such as thymulin and thymosin (Millington and Buckingham, 1992). Thymulin appears to enhance immunological response by promoting the cytokine-driven differentiation of T lymphocytes and by increasing cytokine release. Since release of thymulin is not activated, as are cytokines, by antigenic or tissue insults, endocrine control of release may be of major importance. In this regard, hormonal deficits of prolactin, GH, and thyroid hormones all result in reduced secretion of thymulin, and correction of these deficits restores the secretion of thymulin to normal (Fabris et al., 1989).

Abrupt increases in hypophysial secretion of ACTH, as occurs after adrenalectomy, appear to increase thymulin release, and glucocorticoids exert only a minor, often synergistic effect with ACTH.

Cytokine Effects on Neuroendocrine Function

Activation of the cytokine network (Figure 15-1), as occurs when the immune system detects non-self antigens or there is tissue injury, initiates a cascade of cellular and hormonal events, some of which involve the neuroendocrine system. It is not always possible to determine experimentally which specific cytokine is responsible for each neuroendocrine change because administration of one exogenous cytokine compound may trigger the release of a different endogenous cytokine that actually causes the neuroendocrine change.

The neuroendocrine axis that is most affected by blood-borne cytokines is the hypothalamus-hypophysis-adrenocortical axis. Figure 15-3 depicts this axis and the multiple sites where cytokines have been shown to stimulate it. Blood-borne cytokines such as interferon-alpha, interleukin-1, and interleukin-2 can probably increase glucocorticoid release by independent actions at the adrenal cortex, the corticotroph cells of the hypophysis, and the CRH neurons of the hypothalamus. Cytokine stimulation of CRH release in the hypothalamus and of ACTH release in the adenohypophysis is also modulated by inhibitory feedback of glucocorticoid (Cambronero et al., 1992). In addition to activation of the adrenocortical axis by blood-borne cytokines, there is strong evidence that interleukin-1β and perhaps other cytokines are synthesized within the brain in response to local application of other cytokines (Higgins and Olschowka, 1991). In the absence of stimulation, intracerebral stores of interleukin-1β and interleukin-2 are found mostly in microglial cells, but some interleukin-1β can be stained within axonal projections of the rat forebrain (Lechan et al., 1990; Merrill, 1990). Such local production of cytokines within the hypothalamus could directly or indirectly (via stimulation of interneurons or glial functions) stimulate or inhibit hypophysiotrophic neurons (Figure 15-3). Local release of cytokines within the brain may also be important for neuronal growth/repair and especially for wound healing (Rothwell, 1991). Receptors for interleukin-1 and interleukin-2 are readily demonstrated on glial cells and may also exist on neurons.

Suppression of the reproductive axis is another major consequence of immunological activation. As illustrated in Figure 15-3, blood-borne cytokines such as interleukin-1 inhibit LHRH neurons. Because interleukin-1 is also produced within the brain, paracrine/neurocrine actions of this molecule may also be involved in LHRH suppression. In order to distinguish between the actions of blood-borne and CNS-derived endogenous interleukin-1, Ebisui et al. (1992) attempted to antagonize the LH-suppression caused by endogenous interleukin-1 using both specific antibodies and a receptor antagonist. Through

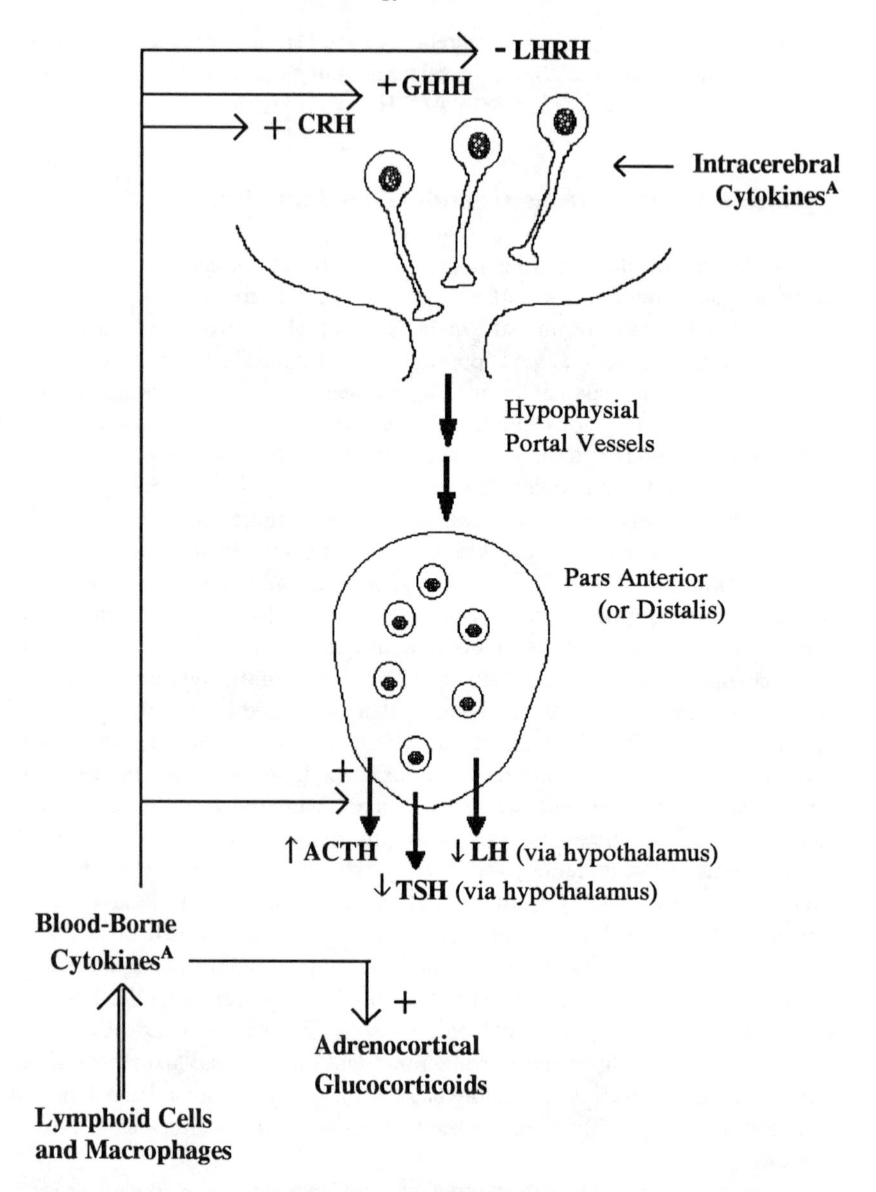

Figure 15-3. Cytokine modulation of hypothalamic hypophysial function.

Those blood-borne and intracerebral cytokines[A] shown to have one or more of these effects on the pituitary-adrenal axis include: interferon-alpha, interleukin-1, and interleukin-2. Supression of LHRH/ LH has also been demonstrated for interleukin-1, and cytokine-stimulated increase of GHIH to suppress TSH has also been demonstrated for interleukin-1, interleukin-6, and TGF-β (Scarborough, 1990).

a comparison of systemic and intracerebral administration of these anti-interleukin-1 compounds, these authors concluded that at least part of the LH-suppression caused by endogenous interleukin-1 was mediated locally within the CNS.

Another effect of infectious or inflammatory responses is the suppression of the hypothalamus-hypophysis-thyroid axis. The resulting decrease in secretion of thyroid hormones may minimize the catabolic side-effects of infectious diseases. Although several cytokines may be involved, the primary effect appears to be stimulation of GHIH neurons, as shown in Figure 15-3. The resulting increase in GHIH release decreases adenohypophysial secretion of TSH and also GH (Scarborough, 1990).

Innervation of Lymphoid Organs

Not all of the immunomodulation caused by factors thought to be strictly neural can be explained by changes in blood-borne neurohormones and hormones. As depicted in Figure 15-2, lymphoid cells and tissues receive some peripheral innervation. Therefore, direct innervation of lymphoid organs may mediate some CNS influences on immune responses. Most of the direct innervation involves postganglionic sympathetic axons containing norepinephrine as a neurotransmitter. Notably, many lymphoid cells have been shown to possess receptors for such adrenergic compounds and also to be modulated by norepinephrine (Bellinger et al., 1990). As stated in Chapters 4 and 5, corticotropin-releasing hormone (CRH) acts within the CNS to activate the sympathetic nervous system. Stressful stimuli decreased cytotoxic NK cells, and CNS immunoneutralization of CRH blocked this response without decreasing the stress-induced activation of ACTH and glucocorticoid secretion (Irwin et al., 1990). These results suggest that stress-induced release of CRH in the brain selectively suppressed cytotoxic NK cells, perhaps via the sympathetic innervation of lymphoid organs.

In addition to noradrenergic innervation, lymphoid cells also receive inputs from several types of peptidergic neurons; these peptides include vasoactive intestinal peptide (VIP), substance P, and others (Bellinger et al., 1990). Receptors for VIP, neurotensin, and substance P have also been demonstrated in lymphoid tissue. The effects of VIP are generally inhibitory while those of neurotensin and substance P are usually stimulatory to immunological function (Bar-Shavit and Goldman, 1990). Local release of substance P is thought to play a role in modulation of inflammatory responses (McGillis et al., 1990).

Production of Hormones by Lymphoid Cells

As stated earlier, neurons and glial cells can produce cytokines formerly thought to only be produced by lymphoid cells of the immune system. It has also been discovered that lymphoid cells can produce certain polypeptide hormones and neurohormones that formerly were thought to be restricted to the neuroendocrine system. The most thoroughly studied of these hormones are those derived from POMC, the precursor of ACTH discussed in Chapters 5 and 6. The gene for POMC appears to be transcribed in both lymphocytes and macrophages, although the peptide products of translation in some lymphocytes may be processed more extensively (Blalock, 1989; Harbour and Smith, 1990). Therefore, these POMC-derived peptides would be available wherever the lymphoid cells occurred to locally stimulate glucocorticoid secretion or to interact with opioid receptors whenever the adjacent cells contained the appropriate receptors and mechanisms. Other adenohypophysial hormones shown to be synthesized in at least one type of lymphoid cell includes TSH, GH, and prolactin. In addition, neurohormones shown to be present in specific lymphoid cells include VIP, GHIH, vasopressin, and oxytocin (Blalock, 1989). Luteinizing hormone-releasing hormone (LHRH) coexists in lymphoid cells with its receptor (Marchetti et al., 1989). Moreover, exposure of thymic and splenic cells to LHRH or its agonist stimulated their proliferation and expression of receptors interleukin-2 (Batticane et al., 1991). In summary, a variety of hormones and neurohormones may be produced in lymphoid tissue for local paracrine/autocrine actions. The functional importance of these various lymphoid-derived compounds is unclear at present, but the strongest evidence for physiological relevance exists for LHRH and for POMC-derived peptides.

In summary, the interaction between the immune and neuroendocrine systems is termed ***bidirectional*** for valid reasons. Neural cells can produce cytokines. Lymphoid cells can produce hormones and neurohormones, and these cells may also be innervated. Endocrine and neuroendocrine cells are influenced by both blood-borne and locally produced cytokines (derived from lymphoid cells and macrophages). In addition, both basal (especially those of the thymus) and activated lymphoid cells are influenced by blood-borne hormones.

REFERENCES

Bar-Shavit, Z., and R. Goldman. Modulation of phagocyte activity by substance P and neurotensin. In: Freier, S. (ed). *The Neuroendocrine-Immune Network.* (CRC Press, Boca Raton, FL, 1990) pg 177-186.

Batticane, N., M.C. Morale, F. Gallo, Z. Farinella, and B. Marchetti. Luteinizing hormone-releasing hormone signaling at the lymphocyte involves stimulation of interleukin-2 receptor expression. *Endocrinology* (1991) 129:277-286.

Bellinger, D.L., D. Lorton, T.D. Romano, J.A. Olschowka, S.Y. Felton, and D.L. Felton. Neuropeptide innervation of lymphoid organs. In: O'Dorisio, M.S., and A. Panerai (eds). *Neuropeptides and Immunopeptides: Messengers in a Neuroendocrine Axis* (*Ann. N.Y. Acad. Sci.* Vol. 594, 1990) pg 17-33.

Berczi, I. *Pituitary Function and Immunity.* (CRC Press, Boca Raton, FL, 1986a), pp 347.

Berczi, I. The influence of pituitary-adrenal axis on the immune system. In: Berczi, I (ed). *Pituitary Function and Immunity.* (CRC Press, Boca Raton, FL, 1986b) pg 49-132.

Blalock, J.E. A molecular basis for bidirectional communication between the immune and neuroendocrine systems. *Physiol. Rev.* (1989) 69:1-32.

Cambronero, J.C., F.J. Rivas, J. Borrell, and C. Guaza. Interleukin-1-Beta induces pituitary adrenocorticotropin secretion: evidence for glucocorticoid modulation. *Neuroendocrinology* (1992) 55:648-654.

Clark, D.A. Role of hormonal immunoregulation in reproduction. In: Berczi, I. (ed). *Pituitary Function and Immunity.* (CRC Press, Boca Raton, FL, 1986) pg 261-272.

Clevenger, C.V., A.L. Sillman, J. Hanley-Hyde, and M.B. Prystowsky. Requirement for prolactin during cell cycle regulated gene expression in cloned T-lymphocytes. *Endocrinology* (1992) 130:3216-3222.

Cunnick, J.E., D.T. Lysle, B.J. Kucinski, and B.S. Rabin. Stress-induced alteration of immune function. Diversity of effects and mechanisms. In: Fabris, N., B.D. Jankovic, B.M. Markovic, and N.H. Spector (eds). *Ontogenetic and phylogenetic mechanisms of neuroimmunomodulation.* (*Ann. N.Y. Acad. Sci.* Vol. 650, 1992) pg 283-287.

Ebisui, O., J. Fukata, T. Tominaga, N. Murakami, H. Kobayashi, H. Segawa, S. Muro, Y. Naito, Y. Nakai, Y. Masui, T. Nishida, and H. Imura. Roles of interleukin-1α and 1β in endotoxin-induced suppression of plasma gonadotropin levels in rats. *Endocrinology* (1992) 130:3307-3313.

Fabris, N., E. Mocchegiani, M. Muzzioli, and M. Provinciali. Neuroendocrine-thymus interactions. In: Hadden, J.W., K. Masek, and G. Nistico (eds). *Interactions among CNS, Neuroendocrine and Immune Systems.* (Pythagora Press, Rome, 1989) pg 177-189.

Ghanta, V.K., H.B. Solvason, and R.N. Hiramoto. Augmentation of natural immunity by conditioning and possible mechanisms of enhancement. In: Freier, S. (ed). *The Neuroendocrine-Immune Network.* (CRC Press, Boca Raton, FL, 1990) pg 103-113.

Hadden, J.W., A. Galy, H. Chen, Y. Wang, and E. Hadden. The hormonal regulation of thymus and T lymphocyte development and function. In: Hadden, J.W., K. Masek, and G. Nistico (eds). *Interactions among CNS, Neuroendocrine and Immune Systems.* (Pythagora Press, Rome, 1989) pg 147-164.

Harbour, D.V., and E.M. Smith. Immunoregulatory activity of endogenous opioids. In: Freier, S. (ed). *The Neuroendocrine-Immune Network.* (CRC Press, Boca Raton, FL, 1990) pg 141-162.

Higgins, G.A., and J.A. Olschowka. Induction of interleukin-1β mRNA in adult rat brain. *Molec. Brain Res.* (1991) 9:143-148.

Irwin, M., W. Vale, and C. Rivier. Central corticotropin-releasing factor mediates the suppressive effect of stress on natural killer cytotoxicity. *Endocrinology* (1990) 126:2837-2844.

Korneva, E.A., V.M. Klimenko, and E.K. Shkhinek. *Neurohumoral Maintenance of Immune Homeostasis* (translated and edited by Corson, S.A. and E.O. Corson. The University of Chicago Press, Chicago, 1978 (original), 1985 (translation)), pp 253.

Lechan, R.M., R. Toni, B.D. Clark, J.G. Cannon, A.R. Shaw, C.A. Dinarello, and S. Reichlin. Immunoreactive interleukin-1β localization in the rat forebrain. *Brain Res.* (1990) 514:135-140.

Marchetti, B., V. Guarcello, M.C. Morale, G. Bartoloni, Z. Farinella, S. Cordaro, and U. Scapagnini. Luteinizing hormone-releasing hormone binding sites in the rat thymus: characteristics and biological function. *Endocrinology* (1989) 125:1025-1036.

McGillis, J.P., M. Mitsuhashi, and D.G. Payan. Immunomodulation by tachykinin neuropeptides. In: O'Dorsio, M.S., and A. Panerai (eds). *Neuropeptides and Immunopeptides: Messengers in a Neuroimmune Axis* (*Ann. N.Y. Acad. Sci.* Vol. 594, 1990) pg 85-94.

Merrill, J.E. Interleukin-2 effects in the central nervous system. In: O'Dorisio, M.S., and A. Panerai (eds). *Neuropeptides and Immunopeptides: Messengers in a Neuroimmune Axis* (*Ann. N.Y. Acad. Sci.* Vol. 594, 1990) pg 188-199.

Millington, G., and J.C. Buckingham. Thymic peptides and neuroendocrine-immune communication. *J. Endocrinol.* (1992) 133:163-168.

Rothwell, N.J. Functions and mechanisms of interleukin 1 in the brain. *Trends Pharmacol. Sci.* (1991) 12:430-436.

Scarborough, D.E. Cytokine modulation of pituitary hormone secretion. In: O'Dorisio, M.S., and A. Panerai (eds). *Neuropeptides and Immunopeptides: Messengers in a Neuroimmune Axis* (*Ann. N.Y. Acad. Sci.* Vol. 594, 1990) pg 169-187.

Shiu, R.P.C., H.P. Elsholtz, T. Tanaka, H.G. Friesen, P.W. Gout, C.T. Beer, and R.L. Noble. Receptor-mediated mitogenic action of prolactin in rat lymphoma cell line. *Endocrinology* (1983) 113:159-165.

Whitacre, C.C. Immunology: a state of the art lecture. In: O'Dorisio, M.S., and A. Panerai (eds). *Neuropeptides and Immunopeptides: Messengers in a Neuroimmune Axis* (*Ann. N.Y. Acad. Sci.* Vol. 594, 1990) pg 1-16.

INDEX

D

E

F

G

H

I

K

L

M

N

O

P

R

S

T

U

V